Human Geography

A Short Introduction

Third Edition

John Rennie Short and Lisa Benton-Short

*University of Maryland, Baltimore County,
and George Washington University*

OXFORD
UNIVERSITY PRESS

Oxford University Press is a department of the University of Oxford.
It furthers the University's objective of excellence in research, scholarship,
and education by publishing worldwide. Oxford is a registered trade mark
of Oxford University Press in the UK and in certain other countries.

Published in the United States of America by Oxford University Press
198 Madison Avenue, New York, NY 10016, United States of America.

For titles covered by Section 112 of the US Higher Education Opportunity
Act, please visit www.oup.com/us/he for the latest information about
pricing and alternate formats.

Library of Congress Cataloging-in-Publication Data
Names: Short, John R., author. | Benton-Short, Lisa, author.
Title: Human geography : a short introduction / John Rennie Short and Lisa
  Benton-Short, University of Maryland, Baltimore County, and George
  Washington University.
Identifiers: LCCN 2023044787 (print) | LCCN 2023044788 (ebook) | ISBN
  9780197662809 (paperback) | ISBN 9780197662847 (ebook)
Subjects: LCSH: Human geography.
Classification: LCC GF41 .S488 2024 (print) | LCC GF41 (ebook) | DDC
  304.2—dc23/eng/20231019
LC record available at https://lccn.loc.gov/2023044787
LC ebook record available at https://lccn.loc.gov/2023044788

Printed by Integrated Books International, United States of America

This book is dedicated to the memory of two of John's grandparents.

His paternal grandmother, Janet Adamson Craig Short (1895–1966), had a profound respect for learning and social justice, an endless devotion to his education, and an almost infinite patience with his spelling lessons. He thinks of her often, especially when his thoughts turn to the connections between love and teaching, devotion and education, wonder and learning.

His maternal grandfather, John Rennie (1903–1963), was a coal miner who spent all his working years in the darkness and the danger. Perhaps because he worked in such cramped and tight spaces, in one memorable holiday, just the two of them, he gave John a sense of limitless freedom.

Brief Contents

Contents

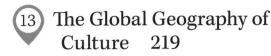

Preface

The aim of this book is to introduce students to a wide range of important and exciting work in human geography. The primary audience is students in colleges and universities. We decided to write this book because many of the standard texts are too big and increasingly too expensive to provide the accessible and affordable base most of us need for our human geography courses. The overly large and expensive books available now have grown into, to use Henry James's description of many nineteenth-century novels, "loose and baggy monsters." There is room for a more interesting and subtle book than the standard texts. This briefer and more accessible alternative is written in a more familiar style that can be augmented by other resources.

We can use as metaphor the attempts on the big Himalayan peaks. In the 1970s, the attempts were increasingly organized as large teams with many climbers and elaborate systems of camps and base camps. Then, in the late 1970s, a number of climbers dispensed with the large teams and sought to climb alone or with one other climber. Less burdened by organizational weight, they were much more successful in reaching the summits in quick, direct assaults. This book adapts a similar ethic of "light and fast" that affords more flexibility to instructors than a traditional textbook: not an exact metaphor, to be sure, but close enough to give you a sense of the book's character and mission.

The title, *Human Geography: A Short Introduction*, employs the word "short" in two ways. First, it indicates a relatively brief introduction rather than a wide survey—although only in the word-rich world of college textbooks can a 100,000-word text be described as short; "concise" may be a more appropriate term. Second, the play on our own name is to signal that it is a book with a distinctive authorial voice. Textbooks are written at specific times in specific places by specific people, and these three basic facts color and shape the material covered and the nature of the coverage. We have drawn heavily on much of our own work conducted over the past fifty years. The main title announces the subject matter, while the subtitle lets the reader know that it is the world of human geography as seen by just two people. This is less an act of egotism than a reminder to the reader that the text is not revealed truth but the singular vision of just two scholars.

The aim is to be both engaging and comprehensive. The text is intended to be both student- and instructor-friendly. The structure, while providing a coherent whole, also allows sectional choices to meet the different needs, time constraints, and interests of individual instructors. Each chapter has a list of further readings and websites that instructors can employ in teaching and develop as resources in ways suitable for the size and constitution of their particular classes.

This is an ambitious book that gives readers a sense of the complex human geography of the contemporary world. It brings together a global perspective with an understanding of national concerns and the growth of select urban regions. Broad arguments are enlivened with detailed case studies. The writing style is accessible to the general reader and the scholarship is comprehensive, so that different interpretations are presented. Six large themes dominate:

- the relationships between people, environment, and resources
- the geography of population
- the economic organization of space
- the cultural organization of space
- the political organization of space
- the urban organization of space.

Running through a discussion of these broad themes are case studies that include examples of specific places as well as examples of the geographical imaginations—models, ideas, and theories—that inform and shape the relationship between people and their environments.

The first chapter discusses major themes in the development of the discipline of human geography. It asks the basic question, What is human geography? Not all human geography courses include this topic, but we feel to understand contemporary human geography, it is necessary to have a basic grounding in the physical geography of the planet and the intellectual history of the discipline. This section is an elective for those with the time to set the course in its broader intellectual history. Some may elect to move straight to Part 1, which looks at the environmental context to make the connections between population, environment, and resources. Part 2 discusses different facets of the geography of population. Part 3 examines the economic organization of space. Part 4 considers the cultural organization of space and focuses on the important geographies of religion, language, and culture. Part 5 examines the political organization of space. Part 6 discusses the urban organization of space, and specific chapters examine trends of urbanization, urban networks, and the internal structure of the city.

The text is constructed so instructors can tailor the readings to suit class needs. The entire book may be used, but some may want to exclude the introductory chapter, while others may want to focus on just four of the five major topic areas. Those with a more economic interest, for example, will certainly want to include Part 3, while those with more cultural emphasis will definitely use Part 4. Those with a focus on political geography will use Part 5 and those with an interest in cities will use Part 6. Each of the sections is self-contained, so an instructor may elect to choose any four or five and leave more time for other class activities.

New to the Third Edition

The text has been substantially changed and improved in successive editions. The second edition, for example, added an entirely new section entitled "The Cultural Organization of Space" that included new chapters on the geographies of religion, language, and culture. For the third edition we reorganized some of the chapters into the different parts so that the material in each part is more pedagogically coherent. We made a full revision of the entire text, making it an even more up-to-date source of the latest geographical work. We added new boxes and contemporary material on the key issues of climate change, social difference and inequality, and sustainability.

Teaching and Learning Package

This book is supported by a carefully crafted ancillary package designed to support both professors' and students' efforts in the course. Instructors should contact their Oxford University Press representative for more information about the supplements package.

Acknowledgments

We would like to thank the many thoughtful scholars who, during the writing process, dedicated valuable time to reviewing and offering comments on the manuscript. Their comments improved the book significantly.

Jeff Baldwin, *Sonoma State University*
Randy Bertolas, *Wayne State College*
Brian Blouet, *The College of William and Mary*
Gina M. Bryson-Prieto, *Hialeah–Miami Lakes Senior High School, Miami, FL*
Karl Byrand, *University of Wisconsin Colleges*
Dorothy Cassetta, *Carroll High School, Southlake, TX*
Thomas Chapman, *Old Dominion University*
Patrick Clancy, *Strath Haven High School, Wallingford, PA*
Timothy W. Collins, *University of Texas at El Paso*
Daniel J. Dempsey, *College of the Redwoods*
Mike DeVivo, *Grand Rapids Community College*
Elizabeth Dunn, *University of Colorado, Boulder*
Owen Dwyer, *Indiana University–Purdue University Indianapolis*

Kyle T. Evered, *Michigan State University*
Ann Fletchall, *Western Carolina University*
Adrienne Goldsberry, *Michigan State University*
Richard Grant, *University of Miami*
Steven M. Graves, *California State University, Northridge*
Joshua Hagen, *Marshall University*
Benjamin Harris, *Bishop Kelly High School, Boise, ID*
Stephen Healy, *Worcester State University*
Peter R. Hoffman, *Loyola Marymount University*
Valentine Udoh James, *Clarion University of Pennsylvania*
Richard E. Katz, *Roosevelt High School, Seattle, WA*
Heidi Lannon, *Santa Fe College*
Jonathan Leib, *Old Dominion University*
Donald Lyons, *University of North Texas*
Heather J. McAfee, *Clark College*
Richard Medina, *George Mason University*
Douglas C. Munski, *University of North Dakota*
Nancy J. Obermeyer, *Indiana State University*
Linda F. Pittman, *Richard Bland College*
Nathan J. Probasco, *Briar Cliff University*
Darren Purcell, *University of Oklahoma*
William Rowe, *Louisiana State University*
James C. Saku, *Frostburg State University*
Arun Saldanha, *University of Minnesota*
Richard H. Schein, *University of Kentucky*
Colleen Schmidt, *Carnegie Vanguard High School, Houston, TX*
Roger M. Selya, *University of Cincinnati*
Jeremy Slack, *The University of Texas at El Paso*
Jill Stackhouse, *Bemidji State University*
Thomas Sullivan, *University of Montana*
Selima Sultana, *University of North Carolina at Greensboro*
Kate Swanson, *San Diego State University*
Rajiv Thakur, *University of Tennessee*
Scott Therkalsen, *Grossmont Community College*
Paul M. Torrens, *University of Maryland, College Park*
Amy Trauger, *University of Georgia*
Erika Trigoso, *University of Denver*
Nicholas Vaughn, *Indiana University*
Jamie Winders, *Syracuse University*
David J. Wishart, *University of Nebraska–Lincoln*
Ryan Weichelt, *University of Wisconsin–Eau Claire*
Max D. Woodworth, *The Ohio State University*

The book in your hands is the product of a collaborative effort. We are very fortunate to work with a team of great professionals at Oxford University Press. Over the years the team has included Joan Kalkut, Tracey MacDonald, Liezl Roux, McKenna Lay, Kathaleen McCormick, Dan Kaveney, Christine Mahon, Megan Carlson, Sarah Goggin, David Jurman, Marianne Paul, and Michele Laseau. We also want to tip our hat to the tireless band of salespeople for promoting the book on campuses all across the country.

Welcome to human geography: an endlessly fascinating and always rewarding subject!

Human Geography

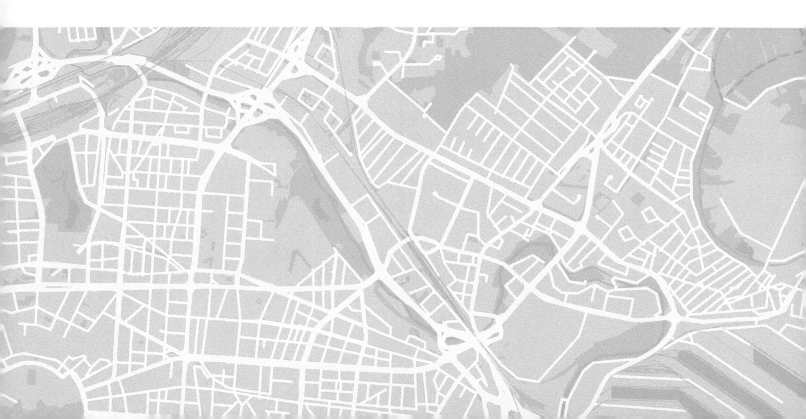

Introduction

We would like to introduce you to human geography. The term "geography" derives from the Greek for "earth description." The term "human" implies a concern with human activity. Chapter 1 looks at the evolution of earth description from its earliest roots to its current concerns.

1

What Is Human Geography?

LEARNING OBJECTIVES

1.1 Identify the key historical developments that have shaped the nature of geography overall and human geography specifically.

1.2 Explain the contributions made by ancient world scholars and early explorers in expanding geographical knowledge.

1.3 Describe geography's human applications as well as the expressions of the discipline in early modernity.

1.4 Describe the variety of approaches in contemporary human geography and the major critiques of the discipline.

An editorial opinion piece in the influential British newspaper *The Guardian* effectively summarizes the importance of human geography today:

> *Geography is a subject for our times. It is inherently multidisciplinary in a world that increasingly values people who have the skills needed to work across the physical and social sciences. Geographers get to learn data analysis. . . . They learn geographic information systems. They can turn maps from a two-dimensional representation of a country's physical contours into a tool that illustrates social attributes or attitudes: not just where people live, but how, what they think and how they vote. They learn about the physics of climate change, or the interaction of weather events and flood risk, or the way people's behaviour is influenced by the space around them. All these are not just intrinsically interesting and valuable. They also encourage ways of seeing and thinking that make geographers eminently employable.*
>
> (Editorial, 2015)

In this chapter we will place the contemporary study of human geography in a deeper historical context. Human geography of today is shaped by the debates of the past. We identify five themes that resonate down through the years: coordinating absolute space, the emergence of geography from **cosmography**, the discussion of relative space, environment and society, and the links between geography and society. A discussion of these themes is a revealing entry point into the history of human geography as they continue to inform a geographic perspective. We will also explore the contemporary concerns of human geography.

1.1 Mapping Absolute Space

A significant and recurring theme in geography is the attempt to accurately represent the world we live in through maps. It is an important element of the geographical imagination. Absolute space is the mapping of the surface of the earth through points, lines, areas, and configurations whose relationships can be fixed with mathematical systems of coordinates, such as latitude and longitude.

One of the earliest geographers who sought to measure and coordinate the world was Claudius **Ptolemy**. He was a Greek Egyptian who worked in the great library of Alexandria. He lived sometime between 90 CE and 168 CE and spent his adult life in Alexandria, a Greek city, now located in Egypt, founded in 331 BCE and named after Alexander the Great. The city was one of the wonders of the classical world, and its population grew to almost half a million. It featured palaces and large public buildings, and at the heart of its intellectual life was the library, a major center of scholarship with over 700,000 volumes. Euclid and Archimedes worked at the library, and Eratosthenes was, for a time, the chief librarian.

In one of his major books, *Geography*, Ptolemy defined **latitude** and **longitude**, discussed methods of mapmaking, and posed the problem of how to represent the round world on a flat surface. He proposed two map projections, a **conic projection** and a partial conic projection, on which to represent the habitable world. Conic projections map the Earth on a cone and then show it as flattened, with lines of latitude as concentric circles (Photo 1.1). He compiled lists of the latitude and longitude of places in the world known to him, basically a world centered on the Mediterranean.

In the European Dark Ages, Ptolemy's influence quickly disappeared. His work, however, was kept alive in the Arab world. Ptolemy's work was translated by the Arab cosmologists of the Abbasid court in Baghdad. Al-Battani (ca. 880) restated the *Almagest*, Ptolemy's work on astronomy, with improved measurements, and *Geography* was revised by al-Khwarizmi around 820 in his *Face of the Earth*, which included a map of the world. Al-Idrisi (ca. 1100–1166) was a practicing geographer who traveled through the Arab world. He constructed a celestial sphere and a map of the world. He was part of a larger geographical school of scholarship in the Arab world, involving book knowledge as well as geographical fieldwork and travel, that included the *Book of Countries* by al-Yaqubi (ca. 891), al-Balkhis's *Figures of the Climates* (921), al-Muqaddasi's *The Best Description for an Understanding of All Provinces* (ca. 985), and al-Bakri's *Book of Roads and Kingdoms* (ca. 1050).

In the early 1400s, Byzantine scholars brought their Greek manuscripts, including manuscripts of Ptolemy's writings, to Italy. The first printed copy of *Geography*

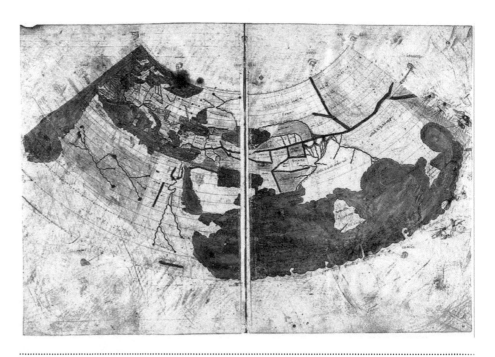

PHOTO 1.1 Ptolemy's map of the world in a conic projection, ca. 1300.

The measuring of the world is a significant part of the early history of geography. By the fifth century BCE, the Greeks knew that the world was a sphere. Eratosthenes (ca. 276 BCE–ca. 194 BCE), who, like Ptolemy, worked at the great library of Alexandria, calculated the circumference of the Earth. Assuming a spherical Earth, he calculated the circumference with only a 1–2 percent difference from the correct figure of 24,901 miles.

Ptolemy's measurement of the circumference was much less accurate, around 18,000 miles. The wide diffusion of Ptolemy's writings meant that this figure was taken as fact during the early modern period. Columbus probably had this much smaller distance in mind when he was planning to sail to the East Indies; Ptolemy's shorter distance made the voyage that much less intimidating.

In 1669–1670, the Frenchman Jean Picard tried to accurately measure the Earth by extrapolating from the distance of one degree of latitude measured in the French countryside. However, similar surveys of a degree of latitude undertaken in Peru and Scandinavia at the same time had different values. This was not the result of an instrumentation error, but because the Earth is oblate, flattened at the poles. We live on a less than perfect sphere in all its misshapen complexity.

appeared in Vicenza in 1475. Other editions appeared in Rome (1477), Florence (1480), and Ulm (1482). The Ulm edition, with its richly colored woodblock maps featuring deep-blue seas and bright-yellow borders, is arguably one of the most beautiful. The printing of Ptolemy's *Geography* stimulated new knowledge and innovative cartographic techniques. After 1508, editions of Ptolemy's *Geography* included maps that incorporated the discoveries of the New World. The first printed map of the New World appeared in an edition of *Geography*. Ptolemy's continuing legacy includes orientating maps with north at the top, the grid of latitude and longitude, using gazetteers in maps and atlases, and employing a variety of map projections. The map shown in Photo 1.1 employs a conic projection. Ptolemy introduced us to the idea and practice of a coordinated world.

It is important to remember, however, that Ptolemy's view of the world was centered on the West. It was not the only view. Photo 1.2 is one of the older extant world maps from East Asia, created around 1389, roughly the same time as the world's map shown in Photo 1.1. This map was initially produced in China. Notice how China sits in the center of the map, the pivot of the world's landmass, with a hazily outlined Europe located on the far western periphery.

PHOTO 1.2 Map of the world, ca. 1389.

1.2 The Shift from Cosmography to Geography

Geography was part of a much wider cosmographical understanding of an interconnected cosmos that also included astronomy, astrology, and magic. The history of geography in the past 500 years is the dismemberment of this cosmological enterprise.

It begins in the Renaissance, where there is evidence of both continuances and ruptures. There was an esoteric side to the Renaissance that included alchemy, astrology, and magic. These concerns were not marginal, but rather central to the life and work of influential Renaissance scholars. John Dee (1527–1608) was a mathematician, mapmaker, and astrologist. His writings combine an interest in exploration and travel, magic and alchemy, mathematics and geography. Dee, like Ptolemy, had a concern with a cosmography that linked the Earth and the heavens, people and their wider environment.

At roughly the same time, the geographer Xu Xiake (1587–1687) was traveling through China. His travel writings combine both detailed observations and poetic reflections. He devised the idea of the watershed, conducted detailed measurements of caves, and expressed his emotional connection to the landscape. Like Dee, his idea of geography encompassed a wide spectrum of approaches and concerns.

We sometimes assume that the Scientific Revolution marked a major change: before was superstition; after was rational, enlightened thought. Recent scholarship has questioned the traditional assumption of such a radical break.

The shift from magus to scientist and from cosmography to geography was neither sudden nor abrupt. The new scientific order bears the mark of the old cosmologies. These connections are clear when we look at some of the early cosmographer/geographers. The great mapmaker Gerardus Mercator (1512–1594) had a religious dimension to his work. Born Gerard Kremer into a modest household in the small town of Rupelmonde near Antwerp, he went to Louvain in 1530 and Latinized his surname to Mercator. He was always interested in theology; he wrote religious tracts, and his first map was a map of the Holy Land. In 1544 he was accused of heresy, one of forty-three accused in Louvain. Two of the accused were burned at the stake, one was beheaded, and a woman was buried alive. Supported by friends and colleagues at the university, he was released. Mercator was a friend of John Dee; they corresponded, and Dee visited him in 1547. Mercator is best known today for his maps. In 1585 he published volumes covering France, Belgium, and Germany. In 1589 he published twenty-two maps, including those of Italy, Slovakia, and Greece. He died in 1594, but his son Rumold collected all the maps in one volume, entitled it *Atlas*, and published it in 1595. We continue to use the term **atlas** to describe a collection of printed maps.

Perhaps the last cosmographer and first modern geographer was Baron Friedrich Wilhelm Heinrich Alexander von Humboldt (1769–1859; see Photo 1.3). Born in Berlin, he was privately tutored as a young boy and then studied at universities in Frankfurt, Göttingen, Hamburg, and Freiburg. He traveled widely. He arrived in Paris shortly before the storming of the Bastille; he wrote later that it

PHOTO 1.3 Portrait of Baron von Humboldt, 1843.

"stirred his soul." He had a lifelong social concern with social reform and improvement. A recurring theme of his intellectual curiosity was to identify the "life force." His belief in universal harmony echoed concerns of John Dee. He was also deeply empirical; measurement and numbers were important to him. His work was a great influence on Charles Darwin, who also traveled widely, carefully collecting and measuring things. Humboldt showed Darwin and later scientists the importance of geographical fieldwork to a deeper understanding of the world.

After 1796, when Humboldt inherited money after his mother's death, he was able to pursue his dream of traveling, exploring, and writing about the world. He traveled to France and Spain—a "measuring expedition," he called it—so he could practice surveying techniques. He traveled to South America in 1799 and stayed until 1804. He passed through Cuba, Colombia, Ecuador, Mexico, and Venezuela, recording, measuring, and explaining. He gathered 60,000 plant specimens, made maps, and amassed a range of data from climate to linguistics. His careful mapping of plants in different places demonstrated the effect of altitude and latitude on plant communities.

Humboldt was an influential figure. His fame and reputation stimulated the adoption of measurement and

1.2 Thinking About Maps

Maps and mapmaking are an integral part of the history of geography. The history of cartography was long dominated by a narrative that stressed mapping as a steady rise in greater knowledge and increased understanding of the world. It was a tale of the increasing scientific rationality of mapmaking from a dim and distant premodern past to an increasingly enlightened present. Maps were milestones along this journey. This discourse was undermined by the postmodern turn, which emphasized maps as social constructions, stories with a purpose, full of erasures and silences as well as inscriptions and disclosures. Maps have since been deconstructed according to their ideological basis, political undertones, and social contexts. They are no longer seen as uncomplicated pictures of the world. They reflect power relations and embody the knowledge and ignorance, articulations and silences, of the wider social world. The accuracy or provenance of maps is no longer the only consideration in the new history of cartography. It is important to uncover their narrative context—their truths as well as their lies. Mapping is not innocent science—it is a political act.

REFERENCES

Edney, M. H. 2019. *Cartography: The Ideal and Its History*. Chicago: University of Chicago Press.

Harley, B. 2001. *The New Nature of Maps*. Baltimore: Johns Hopkins University Press.

Short, J. R. 2003. *The World through Maps*. Toronto: Firefly.

Wood, D. 2010. *Rethinking the Power of Maps*. New York: Guilford Press.

observation in various expeditions and surveys throughout the world. He stimulated geographical measurement and observation. Photo 1.4 is an 1823 **isothermal map** of the world, as the text notes, "drawn from the account of Humboldt and others." Humboldt traveled throughout his life, in later years venturing east to Siberia and covering almost 9,000 miles at a time when transport was rudimentary. His travel journal consisted of thirty-four volumes.

His best-known work, *Cosmos*, was published in four volumes from 1845 to 1862. It deals with responses to the physical environment and explores the world of perception and feelings. There is extensive treatment of poetic descriptions of nature, from the Greeks to modern travelers, and more standard scientific works that discuss astronomy, geology, and physical geography. Humboldt thought of calling his work *Cosmography*, but stayed with *Cosmos* because it implied order. On the pivot of intellectual change, he shared some of the concerns of the older cosmographical tradition while inaugurating modern geography.

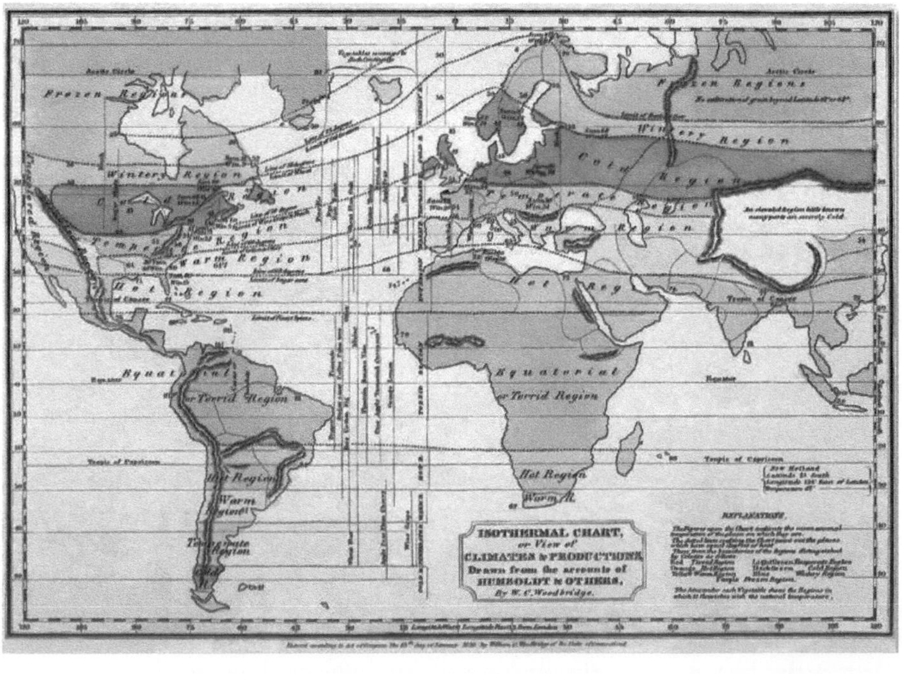

PHOTO 1.4 Isothermal chart, 1823.

1.3 Mapping Relative Space

Absolute space is the grid of latitude and longitude. It is now easy to note that as we type these lines, we are located at 38.87 degrees north, 77.00 degrees west. We have come very far in giving an accurate spatial fix. In absolute terms, the Earth is accurately mapped, gridded, and coordinated. However, this locational fix says little about the neighborhood where we live. What kind of people live here? What are the housing conditions and neighborhood characteristics? Absolute location tells us nothing about the character of places. Much of contemporary human geography is concerned with the **relative space** of social connections, political arrangements, and economic conditions. We can consider the development of relative space by sampling some varied endeavors: the history of the mappings of crime, health, and social deprivation.

Crime

The term **moral statistics** first appears in an essay by André-Michel Guerry in 1833. It was used to refer to crime, pauperism, and a wide range of social phenomena. Such statistics were an important part of nineteenth-century thematic mapping. Maps of crime in France first appeared in 1829, when Guerry used data from 1825 to 1827 to plot, for each of the *départements* in the country, the incidence of crimes against persons, crimes against property, and educational instruction. Mapping in the early nineteenth century predated the development of statistical techniques. Maps and mapping were thus an important way to identify and suggest causal connections. Belgian statistician Adolphe Quetelet also used maps to suggest connections. His carefully gradated shading maps of France, the Low Countries, and parts of the Rhineland, published in 1831, show statistics on crimes against property and people. Quetelet was an influential figure in Europe, and his work and maps were translated into other languages. In the United States, Chicago was an early laboratory of social analysis and spatial mapping.

Today, the mapping of crime is an important element in policing. Keith Harries summarizes the now vast material on this practice. Maps of crime allow a spatial visualization vital to patrol officers, investigators, police managers, policy makers, and community organizations. Improvements in geographic information systems (GIS) now allow data to be mapped, presented, and correlated. Maps are used to identify areas of high crime activity (so-called hot spots), and mapping the timing and spacing of crimes allows a more efficient use of police resources. **Crime maps** of local areas are regularly posted by police departments.

Mapping is also used to solve crimes. **Geographic profiling**, like psychological profiling, has become an important part of law enforcement. Geographic profiling assumes that criminal activity is place-specific. Criminals do not travel very far from their anchor points to commit crime. The dates and places of committed crimes can thus be mapped to create a probability surface of clues to the location of criminals.

The background to the mapping of these "moral" statistics is the concern not only to know where social deviance occurs but also to look for the covariance of other factors so that the deviance can be understood, controlled, and negated. Mapping moral statistics is not innocent of wider political considerations. "Deviance," for example, is a politically freighted term whose definition tells us about those with power, their concerns, and perceived threats. Contemporary human geographers no longer concern themselves just with mapping crime. A large body of work is now concerned with exploring the socio-spatial nature of crime and the policing of space. Phil Hubbard, for example, looks at the geographies of prostitution, while Katherine Beckett and Steve Herbert study how the spatial strategies of policing in Seattle banish certain types of people, especially the poor and indigent, and enforce zones of exclusion in the city.

Public Health and Disease

The first maps of the incidence of disease, at least in the United States, were by Valentine Seaman (1770–1817), who identified individual cases of yellow fever along the New York City waterfront. Mapping the spread of disease became more common in the nineteenth century. Public anxiety over the health of industrial cities in Britain led to the creation of the Poor Law Commission, which prompted the Ordnance Survey, the government mapping agency, to produce detailed maps of towns and cities.

One of the most famous sets of medical maps was drawn by John Snow, a doctor working in mid-nineteenth-century London. He was convinced that cholera was communicated by contaminated water. The 1855 second edition of his work *On the Mode of Communication of Cholera* contained two maps. The first showed the areas of London served by two water companies. These companies used different sources for their water. While the area served by the company using fresh water had death rates due to cholera of 5 per 1,000, the area that drew polluted water from the river Thames had rates of 71 per 1,000. Snow's second map plotted the distribution of cholera cases at a more finely grained level and showed that they clustered around particular pumps. People using the water pump in Broad Street, poisoned by a leaking cesspool, were more likely to contract cholera than those using water from unaffected pumps. The maps highlighted the fact that cholera was caused by contaminated water and not, as the prevailing view at the time believed, contact with polluted air.

Social geographers have long mapped diseases and epidemics, and the interest in mapping disease continues. The human geographer, Peter Haggett and colleagues, has produced a large body of work plotting the spread of diseases. They have produced studies of disease diffusion through islands as well as case studies of individual diseases such as AIDS, influenza, and measles. Medical geographer Andrew Lawson, for example, works on **disease mapping** and analyzes how diseases cluster. He showed how minority and low-income children in South Carolina were more likely to live in areas with lead-contaminated soils. The work confirmed the existence of **environmental racism** and reaffirmed the connection between lead exposure in children and increased incidence of intellectual disabilities.

The COVID-19 epidemic that spread around the world in 2020 became an important topic for researchers. There were numerous studies of its global diffusion, its regional patterns, its impact on social differences, and the efficacy of polices such as travel restrictions. David Tosi and Alessandro Campi, for example, used a predictive model to forecast the behavior of COVID-19 in the Lombardy region of Italy. John Rennie Short has a broadcast lecture, on Vimeo, which links the emergence and diffusion of the virus to broader geographical themes.

There is more to the contemporary **social geography** of health and disease than mapping. More recent work examines the geographies of health inequalities, healthcare provision, health, and well-being.

Deprivation and Inequality

London in the 1880s was a place of turmoil: an economic depression in the middle of the decade increased unemployment and heightened social unrest. Charles Booth, a wealthy ship owner, established a social survey, one of the first. His multivolume *Life and Labour of the People in London* (1892–1903) quantified social problems in the metropolis. Booth's use of **social statistics** to measure and map urban inequality inaugurated an important strand of urban research in which empirical methods were tied to welfare reform objectives. This strand of urban investigation became a common feature of urban studies in the early twentieth century.

Jane Addams was a social activist (Photo 1.5). In 1889 she founded Hull House on Chicago's Near West Side. She worked there until her death in 1935. She and the other residents of the settlement provided services for the neighborhood, such as kindergarten and day care facilities for children of working mothers, an employment bureau, an art gallery, libraries, and music and art classes. Hull House surveys of the local areas, similar to Booth's survey, led to the construction of maps of household income levels and ethnicity (Photo 1.6). One interesting feature of Hull House was the important role of women. Eight of ten of the contributors to the 1895 volume *Hull-House Maps and*

PHOTO 1.5 Jane Addams, 1914.

Papers were women. The Hull House residents and their supporters forged a powerful reform movement that launched the Immigrants' Protective League, the Juvenile Protective Association, the first juvenile court in the nation, and a Juvenile Psychopathic Clinic. They lobbied the Illinois legislature to enact protective legislation for women and children and to pass, in 1903, a strong child labor law and an

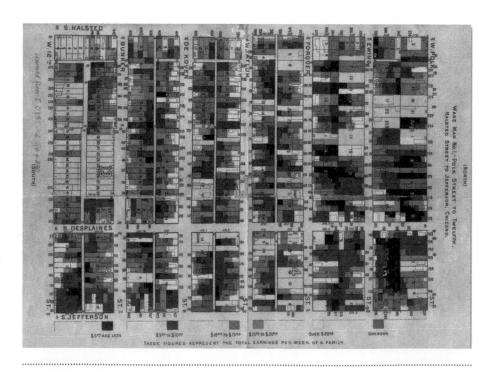

PHOTO 1.6 Wage map published in *Hull-House Maps and Papers* in 1895. Courtesy of the Newberry Library, Chicago, Illinois.

1.3 Depicting Countries in Relative Space

You are perhaps familiar with the standard atlas that shows countries of the world. Most depict nation-states in absolute space. All map projections are distortions, so, depending on the map projection, some areas are exaggerated, while others are minimized. In the **Mercator projection**, for example, the size of countries in more northerly and southerly locations is exaggerated, while the size of countries in the tropical zone is minimized.

It is also useful to compare the relative size and comparative size of countries in relative space. On one website (http://www.worldmapper.org/), the size of a country is shown not in relation to land area but in relation to the data under consideration. These maps provide revealing pictures of world geography.

The maps are more accurately termed cartograms. In the world cartogram of total population, China and India "expand" in size because of their large populations. In the map of gross domestic product, by contrast, the United States, European countries, and Japan enlarge dramatically. In a world map of poverty (Map 1.1), the size of a country is a function of the proportion of the world's poor living in that country. Notice the small sizes of North America, Europe, and Japan compared to the inflated sizes of Africa and countries in the Indian subcontinent.

This cartogram is both accessible and strange. We can immediately make out the shapes of South America and Africa, so we know it depicts our world. But it is a weird world: North America is thin and spindly, western Europe has shrunk, and Australia has all but disappeared from view, while other regions have expanded. National territory is depicted as a proportion of the world's poor living in that territory. Poverty is measured for most countries with a human poverty index based on life expectancy, literacy, access to clean water, and hunger. Because these indicators make sense in the poor world but not in the rich, thirty of the richest countries use measures of life expectancy, literacy, income, and unemployment. Our understanding of the geography of the world depends on the measures we use. In this depiction, India, much of South and Southeast Asia, and tropical Africa balloon into major significance as the diagram allows us to identify quickly the geography of global poverty. Effective antipoverty measures in these parts of the world will have a global impact.

REFERENCES

Dorling, D., M. Newman, and A. Barford. 2010. *The Atlas of the Real World.* Rev. ed. London: Thames & Hudson. http://www.worldmapper.org/.

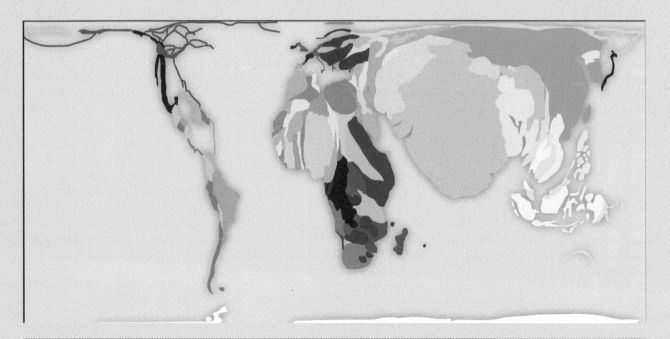

MAP 1.1 The world's poor.

accompanying compulsory education law. The Keating–Owen Child Labor Act of 1916 was the legislative result of their efforts.

Jane Addams was tireless in her dedication. She wrote many popular books and numerous articles, maintained speaking engagements around the world, and played an important role in many local and national organizations, such as the Consumers League, the National Conference of Charities and Correction (later the National Conference of Social Work), Campfire Girls, the National Playground Association, the National Child Labor Committee, the National Association for the Advancement of Colored People, and the American Civil Liberties Union. She was awarded the Nobel Peace Prize in 1931.

The concern with urban social difference continues to be an important part of urban social geography. While the maps and techniques have become more sophisticated, there remains at the core of human geography a consistent interest in documenting, mapping, and explaining the socio-spatial patterns of inequality. Inequality is a major theme within human geography research and scholarship.

1.4 Environment and Society

One goal of early-twentieth-century geography was to study the causal effects of environment on society. Ellsworth Huntington (1876–1947), a professor of geography at Yale University, argued that certain climatic conditions are especially favorable to human progress. His **environmental determinism** was a form of racism. In his book *Civilization and Climate*, he argued, "We know that the denizens of the torrid zone are slow and backward, and we almost universally agree that this is connected with the damp, steady heat." (Huntington 1915, 2). Climate, for Huntington, created culture.

Geography, like all intellectual disciplines, reflects and embodies current beliefs and ideologies. In the late nineteenth and early twentieth centuries, geography was often used to justify imperial adventures as well as to back up belief systems that ranked humans and societies. At the top were the white Protestants and societies of northwest Europe and North America. At the bottom were the Black and brown peoples of the world. Racism and imperialism were an integral part of geography.

Writing at approximately the same time as Huntington, Ellen Churchill Semple (1863–1932), who taught geography at the University of Chicago and Clark University, developed a more probabilistic view of the role of environment in human progress. She was concerned with examining the relationship less deterministically

and with less emphasis on racial differences and cultural hierarchies. Her writing needs to be placed in context. At a time when technological triumphalism was dominant, she was suggesting a counternarrative that humans are connected to the land. In her classic work *Influences of Geographic Environment* she writes, "Man has been so noisy about the way he has 'conquered Nature,' and Nature has been so silent in her persistent influence over man, that the geographic factor in the equation of human development has been overlooked" (Semple 1911, 2). Semple's book, a product of its time in its use of such terms as "savage," is still worth reading for both its eloquent style and its sensitivity to the relationship between people and environment.

Modern geography has long moved past issues of environmental determinism. However, a socially sensitive environmentalism still has some value. At the macro level, for example, climate change plays an important role in the rise and fall of civilizations. Mesopotamian civilization collapsed around 3,400 years ago, in part because of a severe 200-year drought. The Mayan civilization of Central America collapsed almost 1,100 years ago, again the result of a combination of social and environmental factors. The environmental context is an important one for looking at the big picture of societies' rise and fall. Echoing the basic theme of Huntington, but with a very different emphasis, Christian Parenti considers in his book *Tropic of Chaos* the area between the Tropics of Cancer and Capricorn. The **tropical zone** is one where issues of climate change and especially water availability are partial causes of social conflict and political ruptures. Parenti draws a connection between climate change and violence through preexisting conditions, such as the legacy of Cold War militarism and the destructive consequences of neoliberal economics. He describes a catastrophic convergence of societies littered with cheap weapons and fractured by the lack of social connectivities now worsened by the wrenching effects of climate change. In an even broader perspective, Jared Diamond identifies the links behind social collapse and environmental degradation. This is not the simple determinism of earlier geographers but a more nuanced account that shows how social change is embedded and embodied in environmental relations.

1.5 Geography and Society

The history of an academic discipline is tightly bound up in broader social developments. Three themes are evident in the evolving relationship between geography and society: the connection with wider intellectual debates, the varied relationships to political power, and contemporary concerns.

Intellectual Debates

One historian of geography, David Livingstone, describes the discipline as a contested tradition. He argues for a history of geography that assumes not an unchanging metaphysical core, but rather a series of situated geographies, ones that shape and are shaped in turn by the time and place of their making. To write a global geography in London in 1910 or Washington, DC, in 1945 is a very different enterprise from writing a global geography in twenty-first-century Beijing. Practices and procedures vary, and this variation is not incidental to the evolution of geographic writing, but rather fundamental. What interests are advanced, what assumptions are made, and what underlying model of how the world hangs together all depend on the time and place. Historians of the discipline, for example, are often appalled at the **casual racism** prevalent in the work of geographers writing in decades past. In other words, there is a historical geography to the evolution of **human geography**.

The situated and compromised nature of knowledge production is best summarized in the title and subtitle of Steven Shapin's 2010 book on the history of science, *Never Pure: Historical Studies of Science as If It Was Produced by People with Bodies, Situated in Time, Space, Culture, and Society, and Struggling for Credibility and Authority*. The history of geographic thought is intimately bound up in the historical geography of ideas.

Geography and Politics

The development of human geography is intimately connected to the interests of powerful groups, whether they are private and corporate or public and governmental. John Dee was an important intellectual figure in Renaissance England. He cast horoscopes for Queen Elizabeth I, promoted the expansion of English mercantile interest, and "proved" English claims in the New World. The production of geographical knowledge is intimately connected to national economic interest. Fast-forward over 300 years: geography, according to Sir Halford Mackinder (1861–1947), was an aid to **statecraft**. What he meant by that was that geography could help frame geopolitical strategies for the nation-state, in his particular case, Britain and the British Empire. He developed a **heartland theory** to argue that whoever ruled the central part of the Eurasian landmass, the heartland, ruled the world. The idea was first formulated in 1904 and appeared in book form in 1919. At the time he was a fervent anti-Bolshevik, and so his argument can be seen in one light as a rationale for providing a cordon around Bolshevik Russia. His ideas influenced Nazi strategic thinking and US Cold War strategy.

Isaiah Bowman (1878–1950), a contemporary of Mackinder, was director of the American Geographical Society for over twenty years. He was also the chief territorial advisor to President Woodrow Wilson and to the US State Department during the First World War. He had a strong belief in US expansionism based on penetration of foreign markets by US capitalism. He provided a **geographical imagination** to ideas of US global superiority and dominance. His book *The New World*, written in 1921, is still a very well-written and accessible account of geopolitical issues and problems facing the world in the aftermath of the First World War and in the wake of the fragmentation of prewar empires.

Dee, Mackinder, and Bowman are at one end of the applied geography scale, their work and ideas directly shaping and influencing national interests. At the other end are more antiestablishment figures. Peter Kropotkin (1842–1921) provides an interesting contrast. He was born into the privileged life of pre-Revolutionary Russian aristocratic society; his father owned vast tracts of land and over 1,200 serfs. In 1864 he led a geographical expedition to Siberia, and on his return to St. Petersburg he became secretary to the Imperial Russian Geographical Society. His interests widened to include a more radical vision of society: an **anarcho-communist** society based on mutual aid and voluntary associations. In 1875 he was imprisoned, but because of his social status he was allowed to finish a geographical report on the last Ice Age, based on the fieldwork he had conducted earlier in Finland and Sweden. Students who are late with their work might like to think of Kropotkin, who finished his geography report even while in jail. He managed to leave Russia and lived in Switzerland and England, but returned in 1917 after the February Revolution overthrew the czar. He spoke out against the authoritarian socialism inaugurated with the October Revolution and the Bolshevik accession to power. Since 1957 a Moscow subway station has borne his name.

A more intellectual radicalism shapes the work of some contemporary human geographers. David Harvey (b. 1935) is a human geographer whose PhD work was on hops production in the English county of Kent. His first book, *Explanation in Geography* (1969), was an argument for adopting the deductive approach in geography. The context for his next book, *Social Justice and the City* (1973), was the wider world of the Vietnam War, the persistence of poverty in the richest countries, intractable racial divisions, and the decline of the long postwar economic boom, exposing inequalities and social discontent. There was a growing dissatisfaction in the academy with traditional methods of scholarship. In *Social Justice*, Harvey reflects on and informs the debates surrounding these societal and academic issues. *Social Justice*, then, is really two books. The first part, "Liberal Formulations," focuses geographical inquiry on socially relevant topics. The second, "Social Formulations," marks a distinct **epistemological break**. The concept of the city is radically altered between the first and second parts. Rather than being an independent object of inquiry, it is an important element that mediates and expresses social processes. In subsequent books, Harvey continues to provide radical critiques of capitalism. *The Enigma of Capital* outlines the tendency of crisis in capitalism, while *Rebel* argues for the importance of the city and the neighborhood as sites of political struggle. The title of his 2014 book, *Seventeen Contradictions and the End of Capitalism*, summarizes his focus on exposing the internal working of the capitalist system and highlighting its inherent long-term instability.

1.6 Contemporary Debates

Human geography has undergone profound change since the 1960s. A number of "turns" can be identified. The turn of **quantitative geography**, associated with such scholars as Peter Haggett and Brian Berry in the 1960s, was concerned with measurement, calibration, the creation of spatial models, and the testing of hypotheses. It was not just a fascination with numbers. Waldo Tobler, for example, sought to represent flows in maps and to add a much-needed temporal dimension to cartographic representation and the geographical imagination. He is also credited with devising **Tobler's law**, which states, "Everything is related to everything else, but near things are more related to each other." The quantifiers also excavated previously neglected scholars such as the 1930s German geographer Walter Christaller, whose work we will consider in Chapter 17. This excavation of previously ignored work is a consistent theme of all the different approaches and leads to a constant recreation of the history of human geography.

But the quantitative turn, which built on a long tradition of quantitative geography, was not a revolution that came to dominate the entire discipline. It is more appropriate to think of it as a seismic shudder with multiple epicenters: Cambridge and Bristol in Britain and the University of Washington in the United States. In many places the seismic shift was scarcely recorded, and if it was, it was dismissed as being of minor, ephemeral significance. And here we come to another trait of contemporary human geography: no turn has involved all of the discipline, because it is just too big and too varied.

The quantifiers had an uphill task against the entrenched forces of reaction. Old men (and most of them were men) with established reputations rarely take much heed of young scholars in a hurry who make very negative statements about the work of their elders. It was in one sense a generational struggle for power and status. But scarcely had the quantifiers established a beachhead in the discipline, through creating mandatory courses on

statistical techniques and establishing their own publishing outlets, before countercritiques and new turns emerged from the early 1970s. We can identify five from the many.

First, the notion of dispassionate, objective observers was criticized for its lack of social relevance and its connections with the corporate world. In effect, the "scientific" endeavor was neither socially neutral nor politically innocent. The emphasis shifted from identifying the spatial structure of society—a world of surfaces, edges, nodes, hierarchies, and flows—to the analysis of the social organization of space. This involved a more radicalized human geography.

Second, there was a criticism of the quantifiers from a humanist perspective. The **humanist critique** pointed to the lack of concern with personal geographies, the fact that statistical averaging ignored the whole realm of feeling, perception, and intersubjectivity. The 1984 work of Peter Jackson and Susan Smith showed what a radically informed **cultural geography** could look like. In the United States, cultural geography was revived through a concern with landscape as a socially contested, materially produced cultural artifact. Geographers such as Jim Duncan reinvigorated the study of landscape by integrating the new interest in social theory.

Third, in the 1980s feminists appeared as a significant voice in the struggle for the hearts and minds of human geography. The **feminist critique** drew attention to the bias in the academic practice of geography and geographical writings. It was a discipline dominated by men and a discourse that rarely connected with the reality of gendered spaces and gendered lives. There is now a considerable body of work on the gendered separation of productive and reproductive spheres, issues of domestic labor, work–home separation, and the overall living and working experience of women. The feminist perspective also shows how gender interacts with race, class, and sexuality to produce socio-spatial patterns of inequality and marginalization. Geographers such as Doreen Massey, Linda McDowell, and Gill Valentine, to name just a few feminist scholars, have widened the angle of vision.

Fourth, the adoption of **postmodernism** in human geography was a significant development. Postmodern geographers such as Michael Dear, Gillian Rose, and Ed Soja criticized the notion of objective universal constructs. Knowledge was not revealed to socially objective viewers but socially constructed by the perspective of the viewer. The postmodern turn shifted the debates by introducing notions of hybridity, alternative rather than dominant narratives, and a multiplicity of ways of knowing.

Fifth, there is now a greater concern with the marginal and the marginalized and the range of inequalities among people and places. For much of the history of geography in the past 200 years, the subject was written about by a relatively small sample of people—generally men in positions of power and authority in the richer countries of the

The "spatial turn" is the name given to a greater emphasis on space and place in the social sciences and humanities. It was promoted by two trends. The first is an increasing recognition that space and place play important roles in the issues and concerns important to social scientists and humanity scholars. Of course, it is at the heart of the geographical enterprise, so we have been practicing the spatial turn long before it became fashionable in other disciplines. A spatial turn is noted in anthropology, architecture, history, psychology, religion, literature, sociology, and history. Topics such as maps and mapping, territoriality, landscape, spatial change, and the transformations of place are the heart of new ventures in other disciplines that have found a shared interest in describing and explaining the culturally inscribed meanings of place and power relations of space.

The second trend is the growing use of GIS in the social sciences and humanities—for example, mapping in the digital humanities. One research project generated a 1.5-million-word corpus from eighty texts written about the Lake District in England between 1622 and 1900. Places that were described with terms such as "beautiful," "majestic," and "sublime" were mapped. The analysis showed the influence of the Romantic sensibilities, how "sublime" is used to refer to wide sweeps of landscapes, while "majestic" is more specific. The mappings reveal the different ways used at different times to describe and evaluate the landscape of the Lake District.

REFERENCES

Gregory, I., and C. Donaldson. 2016. "Geographical Text Analysis: Digital Cartographies of Lake District Literature." In *Literary Mapping in the Digital Age. Digital Research in the Arts and Humanities,* eds. Cooper, D., Donaldson, C., and Murrieta-Flores, P. 67–78. London: Routledge.

Warf, B., and S. Arias, eds. 2008. *The Spatial Turn: Interdisciplinary Perspectives*. London: Routledge.

world. Geography was linked to the existing global and national power structure. In recent years, however, there has been a growing body of work that we can call **subaltern geographies**. More geography is being written from outside the circle of usual suspects. More female scholars and scholars from developing countries are beginning to shift the angle of vision. There are also more geographical studies of marginalized groups: women, children, migrant workers, peasants, slum dwellers, sexual minorities, and so forth. Geography is enriched by these developments and is adding these voices to our understanding of the contemporary world.

Today, human geography is a rich area of intellectual inquiry. The discipline continues to be contested in a lively fashion.

1.7 The Concerns of Human Geography

Geography has some enduring concerns. Geographers are interested in the reasons behind and the consequences of the spatial distribution of human activities and human environmental interactions. Geographers look at space and place as containers, stages, sites, and outcomes of human activities and environmental interactions. They are always asking two major and related questions: where and why?

Fundamental Concepts

Space and place are fundamental concepts in human geography. Space is a more generalized place such as the state, the region, the city, or the neighborhood, while places include the United States, southeast England, Shanghai, and Bukchon, South Korea.

Understanding how space becomes place and how places are incorporated into space is at the heart of the human geography enterprise. Culture turns space into places. Place is space made more intimate by culture. But places are in turn incorporated into space, as when the United States is linked to the global economy by flows of people, goods, and ideas. Space and place are more than mere background; they are stages for performances, given meaning and significance by social actions, political conflict, and economic processes. Space and place are active moments, not static categories.

Geographers are concerned with the spatial organization of society and the social organization of space: how spaces and places are produced, used, consumed, destroyed, represented, managed, and controlled. They are concerned with spaces of hope and spaces of exclusion, the reconstructions as well as the destructions of place.

Emerging Themes

Human geography is a richly diverse discipline that employs a variety of methods to look at a diversity of topics. In recent years, however, several major themes have appeared that cut across topical and methodological differences. These include environment and **climate change**, **globalization** and its discontents, inequality, widening the discourse through listening to other voices (gender, race, class, etc.), and **sustainability**. We will explore these in more detail in the boxes in the individual chapters that follow.

Spatial Categories

There is a range of spatial categories that hold special meaning for geographers: they include the region, landscape, city, neighborhood, and home. Other spatial categories include the frontier, boundary, territory, and state. Geographers look at the physical and discursive construction of these categories and their changing usage and meaning. The categories are employed not as reified spatial abstractions, but as social constructions full of cultural meaning and political significance. They are not unchanging verities but the outcome and site of negotiation and conflict.

There is spatial imagination at work in the study of human geography, and so a wide range of spatial categories are employed, including location, which is a point or place on the Earth's surface. Location can be used in absolute terms, such as latitude and longitude, as well as in relative terms, such as describing a country or region as rich or poor.

Distributional categories employed include density, concentration, and pattern. Typical questions include: What is the density of population in a certain place and why? Why are certain economic activities concentrated in certain locations? Is there a pattern to events such as the distribution of towns and cities?

Movement across space and between places is another important topic. Geographers are interested in how the friction of distance influences things. The time taken to travel is improved by transport improvements, leading to what is known as space-time convergence, the collapse of the time taken to travel the same distance.

The term "diffusion" is used to describe the spread of things, ideas, people, or practices across space. Diffusion occurs in different ways. It can occur through relocation, as when people often take their language and religion with them when they move, diffusing it more widely; through contagion, as when diseases spread through contact; or through hierarchical diffusion, as when ideas flow from large to small cities.

Geographers study different types of regions, which we can define for the moment as demarcated areas of the Earth's surface. Regions are defined by different criteria and so we refer to cultural regions and economic regions. Hollywood, the Rust Belt, and the Asian Tigers are all examples of some of the regional definitions used in economic geography. Human geography employs different types of regions at different spatial scales.

The last few paragraphs are very dense. However, think of them as mere pointers that introduce you to spatial categories that are more carefully defined and strategically employed throughout the rest of this book.

Methods

Geographers work at a variety of scales, from the global and national to the more intimate spaces of our personal interaction. Human geography has a large bandwidth and so employs a variety of techniques. There is an enduring concern with spatial representation and mapping. Map 1.2 maps the Human Development Index, a composite index that combines measures of a long and healthy life, being knowledgeable, and having a decent standard of living. From Map 1.2, what conclusion can you make about the global distribution of human development? In which countries would you choose to live, and why? How does this map underscore inequality?

In recent years, mapping has been dramatically altered by the rise of **Geographic Information Systems (GIS)**, systems that store geographic data and allow their manipulation and representation. GIS is used by governments and companies to store, access, and display data. Geographers also use remote sensing, which involves the collection of data taken from aircraft, satellites, and, more recently, drones. For example, a recent study by geographer Do-Hyung Kim and colleagues used satellite imagery to examine tropical forests in thirty-four counties. They discovered that the rate of loss was higher than that suggested by the United Nations, which had been using self-reported data. While the UN data showed deceleration in forest loss, Kim and colleagues detailed study showed a 62 percent acceleration in deforestation in the humid tropics from the 1990s to the early 2000s.

Remote sensing allows us to utilize data that would be difficult to map, such as ice sheet shrinkage or the emergence of slums at the edge of cities in the developing world. It also gives unique perspectives. Photo 1.7 shows a space-based perspective of the Kansas City metro area taken in September 2014 by astronauts on the International Space Station.

Geographic profiling is now an often-used technique. Initially used in crime studies to identify patterns of crime and locate criminals, it is now used to identify disease outbreaks and control invasive species. It was used to solve an art mystery: Who was the insurgent artist Banksy? In 2011 it was also used to help narrow down the location of Osama bin Laden while he was in hiding. Assumptions included a distance-decay function: he would not move far from his last sighting in Afghanistan, he would be in Pakistan because it was a similar culture, and he would be in a town or city in a dwelling large enough to house his retinue with high walls to provide privacy and security. His precise location in Abbottabad was confirmed

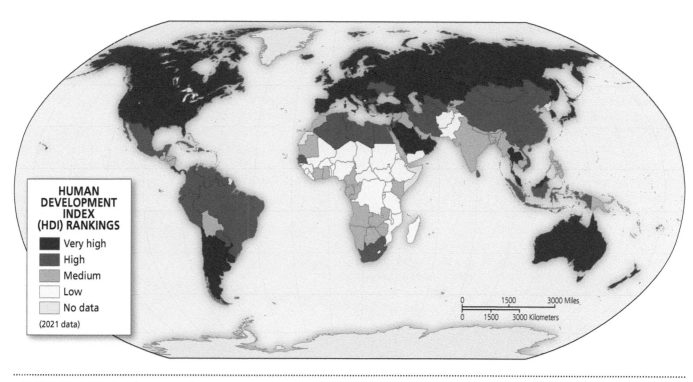

MAP 1.2 Human Development Index.

PHOTO 1.7 Remotely sensed image of Kansas City metro area.

1.5 Geography and Inequality: GIS, Remote Sensing, and the Democratization of Mapping

While it is often used in the worlds of national security and as part of corporate research, GIS can also be used to map social inequalities and to identify areas of deprivation.

There also has been a democratization of mapping. OpenStreetMap is a collaborative effort with 2 million registered users who generate a free map of the world. The community plays an important role in **crisis mapping**. During the 2010 Haiti earthquake, volunteers used satellite data to provide a real-time digital atlas of the road system, vital information in a country desperate to get aid supplies to affected communities. The project was so successful that now the OpenStreetMap community regularly teams up with aid organizations in the wake of disasters to provide urgent and much-needed spatial data.

Open-source software and global volunteer networks are now an important part of crisis mapping. GIS and social media are being used to provide real-time georeferenced data in places experiencing violence or disruption. Patrick Meier provided examples from Haiti, Japan, Kenya, Russia, Somalia, and Syria.

There are also more focused examples. Geography students at the University of Montana used satellite images to construct a detailed topographic map of a tiger reserve in Nepal, which allowed people on the ground to reconcile their position on the map and track down poachers.

GIS and remote sensing can also be used in more open and democratic ways, especially for societies that are closed and undemocratic. Google Earth has been used to create maps of North Korea that highlighted missile-storage facilities, mass graves, labor camps, and the palaces of the elite. Crowdsourced atlases of secretive states can open up even the world's most isolated and brutal countries to greater public scrutiny.

REFERENCES

Andrews, M. R., et al. 2020. "Geospatial Analysis of Neighborhood Deprivation Index (NDI) for the United States by County." *Journal of Maps* 16:101–112.

Chaney, R. 2015. "Nepali Tiger Poachers Fear University of Montana Mapmakers." *Missoulian*, December 16, 2015. http://missoulian.com/lifestyles/recreation/nepali-tiger-poachers-fear-university-of-montanamapmakers/article_40c44f52-bbe2-5c87-80f0-ab5668c58ad0.html.

Meier, P. 2012. "Crisis Mapping in Action: How Open Source Software and Global Volunteer Networks Are Changing the World, One Map at a Time." *Journal of Map and Geography Libraries: Advances in Geospatial Information, Collections and Archives* 8:89–100.

PHOTO 1.8 Two young people studying a globe in a park in Vienna, Austria.

by following a courier. DNA evidence from people in the house provided **ground truthing**.

A range of techniques is employed in contemporary human geography. Quantitative analysis relies on data and statistical analysis, while qualitative techniques include storytelling and the use of focus groups and participant observation. The combination of qualitative and quantitative provides sympathetic and grounded accounts. In John's own case study of Aboriginal art production and racial tensions in Alice Springs, Australia, he mapped peripheral settlements and studied government data and city records, but also visited the city over many years to interview people, cultivate connections, and immerse himself in the world of art production and sale (Short 2012).

Human geographers look at space and society from a variety of scales, encompassing approaches that range from GIS, remote sensing, and statistical analysis techniques to qualitative ethnographic studies. They intersect with other researchers to consider such important topics as the flexible identities of gender, ethnicity, and nationality; the measurement, explanation, and understanding of globalization; the connections between space and place; the recurring question of how we represent the world around us; and the pressing concerns of sustainability, environmental transformation, and climate change. At its very best, the discipline exhibits an environmental sensitivity with a social awareness. And like the two young people in Vienna shown in Photo 1.8, it looks at the world as a constant source of fascination and interest.

At its very best, the study of geography exercises the spatial imagination and cultivates a profound sense of place. Welcome to the most exciting and engaging of subjects.

Cited References

Beckett, K., and S. Herbert. 2009. *Banished: The New Social Control in Urban America*. New York: Oxford University Press.

Booth, C. 1903. *Life and Labour of the People in London*. London: Macmillan.

Bowman, I. 1921. *The New World: Problems in Political Geography*. Yonkers, NY: World Book Company.

Diamond, J. 2005. *Collapse: How Societies Choose to Fail or Succeed*. New York: Penguin.

Duncan, J. S. 1990. *The City as Text*. Cambridge: Cambridge University Press.

Editorial. 2015. "The Guardian View on Geography; It's the Must Have A-Level." *The Guardian*, August 13, 2015. http://www.theguardian.com/commentisfree/2015/aug/13/the-guardian-view-on-geography-its-the-must-have-a-level?CMP=share_btn_tw.

Harries, K. 1999. *Mapping Crime: Principle and Practice*. NCJ 178919. Washington, DC: Department of Justice.

Harvey, D. 1969. *Explanation in Geography*. London: Arnold.

Harvey, D. 1973. *Social Justice and the City*. London: Arnold.

Harvey, D. 2010. *The Enigma of Capital*. London: Profile.

Harvey, D. 2012. *Rebel Cities: From the Right to the City to the Urban Revolution*. London: Verso.

Harvey, D. 2014. *Seventeen Contradictions and the End of Capitalism*. Oxford: Oxford University Press.

Hubbard, P. 1999. *Sex and the City: Geographies of Prostitution in the Urban West*. Aldershot, UK: Ashgate.

Humboldt, A. 1897. *Cosmos: A Sketch of a Physical Description of the Universe*. New York: Harper.

Huntington, E. 1915. *Civilization and Climate*. New Haven, CT: Yale University Press.

Jackson, P., and Smith, S. J. 1984. *Exploring Social Geography*. London: Allen and Unwin.

Kim, D. H., J. O. Sexton, and J. R. Townshend. 2015. "Accelerated Deforestation in the Humid Tropics from the 1990s to the 2000s." *Geophysical Research Letters* 42:3495–3501.

Livingstone, D. N. 1993. *The Geographical Tradition: Episodes in the History of a Contested Enterprise*. Oxford: Blackwell.

Parenti, C. 2011. *Tropic of Chaos: Climate Change and the New Geography of Violence*. New York: Nation.

Semple, E. C. 1911. *Influences of Geographic Environment on the Basis of Ratzel's System of Anthropogeography*. New York: Holt. https://archive.org/details/influencesofgeog00semp.

Shapin, S. 2010. *Never Pure: Historical Studies of Science as If It Was Produced by People with Bodies, Situated in Time, Space, Culture, and Society, and Struggling for Credibility and Authority*. Baltimore: Johns Hopkins University Press.

Short, J. R. 2012. Representing Country in the Creative Postcolonial city. *Annals of Association of American Geographers* 102:129-150.

Short, J. R. 2020. "COVID-19 in the Context of Human and World Regional Geography." Oxford University Press. Video. https://vimeo.com/416056090.

Snow, J. 1855. *On the Mode of Communication of Cholera*. London: John Churchill. http://www.ph.ucla.edu/epi/snow/snowbook.html.

Tosi, D., and A. Campi. 2020. "How Data Analytics and Big Data Can Help Scientists in Managing COVID-19 Diffusion: Modeling Study to Predict the COVID-19 Diffusion in Italy and the Lombardy Region." *Journal of Medical Internet Research* 22(10): e21081.

Select Guide to Further Reading

ON MAPS AND MAPPING:

Crampton, J. W., and J. Krygier. 2005. "An Introduction to Critical Cartography." *ACME International Journal of Critical Geography* 4:11–33.

Edney, M. H. 2019. *Cartography: The Ideal and Its History*. Chicago: University of Chicago Press.

Kraak, M. J., and F. Ormeling. 2021. *Cartography: Visualization of Geospatial Data*. 4th ed. Boca Raton, FL: CRC Press.

Krygier, J., and D. Wood. 2011. *Making Maps.* 2nd ed. New York: Guilford.

Short. J. R. 2003. *The World through Maps.* Toronto: Firefly.

Tyner, J. 2015. *The World of Maps.* New York: Guilford.

FOR CONTEMPORARY DISCUSSION OF THE GEOGRAPHIES OF HEALTH:

Anthamatten, P., and H. Hazen. 2011. *An Introduction to the Geography of Health.* New York: Routledge.

Brown, T., S. McLafferty, and G. Moon. 2010. *A Companion to Health and Medical Geography.* Malden, MA: Wiley–Blackwell.

Cliff, A. D., P. Haggett, and M. S. Raynor. 2004. *World Atlas of Epidemic Diseases.* London: Hodder Arnold.

Crooks, V. A., G. J. Andrews, J. Pearce, and M. Snyder, eds. 2020. *Introducing the Routledge Handbook of Health Geography.* New York: Routledge.

Koch, T. 2011. *Disease Maps: Epidemics on the Ground.* Chicago: University of Chicago Press.

Prior, L., D. Manley, and C. E. Sabel. 2019. "Biosocial Health Geography: New 'Exposomic' Geographies of Health and Place." *Progress in Human Geography* 43:531–552.

Short, J. R. 2020. "COVID-19 in the Context of Human and World Regional Geography." Oxford University Press. Video. https://vimeo.com/416056090.

A SAMPLE OF PAPERS ON CRIME:

Andresen, M. A., and P. L. Brantingham. 2012. "Visualizing the Directional Bias in Property Crime Incidents for Five Canadian Municipalities." *The Canadian Geographer/Le Geographe Canadien* 57:31–42.

Melo, S. N. D., M. A. Andresen, and L. F. Matias. 2017. "Geography of Crime in a Brazilian Context: An Application of Social Disorganization Theory." *Urban Geography* 38:1550–1572.

Mordwa, S. 2016. "The Geography of Crime in Poland and Its Interrelationship with Other Fields of Study." *Geographia Polonica* 89:187–202.

Rossmo, D. K., and K. Harries. 2011. "The Geospatial Structure of Terrorist Cells." *Justice Quarterly* 28:221–248.

A USEFUL ENTRY POINT INTO THE HISTORY OF AN INTELLECTUAL DISCIPLINE IS TO READ THE BIOGRAPHIES AND AUTOBIOGRAPHIES OF ITS PRACTITIONERS. HERE IS A VERY SMALL SAMPLE OF A RANGE OF HUMAN GEOGRAPHERS:

French, P. J. 1987. *John Dee: The World of an Elizabethan Magus.* London: Routledge.

Gould, P. 1999. *Becoming a Geographer.* Syracuse, NY: Syracuse University Press.

Gould, P., and F. Pitts, eds. 2002. *Geographical Voices.* Syracuse, NY: Syracuse University Press.

Haggett, P. 1990. *The Geographer's Art.* Oxford: Blackwell.

Hubbard, P., and R. Kitchin, eds. 2011. *Key Thinkers on Space and Place.* London: Sage.

Hughes, R. G., and J. Heley. 2015. "Between Man and Nature: The Enduring Wisdom of Sir Halford J. Mackinder." *Journal of Strategic Studies* 38:898–935.

Kearns. G. 2009. *Geopolitics and Empire: The Legacy of Halford Mackinder.* Oxford: Oxford University Press.

Kropotkin, P. (1899) 1927. *Memoirs of a Revolutionist.* London: Smith Elder. http://theanarchistlibrary.org/library/petr-kropotkin-memoirs-of-a-revolutionist.

Moss, P., ed. 2000. *Placing Autobiography in Geography.* Syracuse, NY: Syracuse University Press.

Smith, N. 2004. *American Empire: Roosevelt's Geographer and the Prelude to Globalization.* Berkeley: University of California Press.

Wulf, A. 2015. *The Invention of Nature: Alexander von Humboldt's New World.* New York: Knopf.

ON HUMAN GEOGRAPHY, ANCIENT AND MODERN:

Agnew, J. A., and J. S. Duncan, eds. 2016. *The Wiley–Blackwell Companion to Human Geography.* Oxford: Wiley–Blackwell.

Agnew, J. A., and D. N. Livingstone, eds. 2011. *The Sage Handbook of Geographical Knowledge.* Thousand Oaks, CA: Sage.

Association of American Geographers. 2017. *International Encyclopedia of Geography.* Chichester, UK: Wiley–Blackwell.

Cox, K. 2014. *Making Human Geography.* New York: Guilford.

Cresswell, T. 2013. *Geographic Thought: A Critical Introduction.* Chichester, UK: Wiley–Blackwell.

Gregory, D., R. Johnston, G. Pratt, M. Watts, and S. Whatmore, eds. 2011. *The Dictionary of Human Geography.* Chichester, UK: Wiley–Blackwell.

Martin, G. J. 2005. *All Possible Worlds: A History of Geographical Ideas.* 4th ed. Oxford: Oxford University Press.

Thomson, J. O. 2013. *History of Ancient Geography.* Cambridge: Cambridge University Press.

ON GIS:

Bolstadt, P. 2016. *GIS Fundamentals.* Ann Arbor, MI: XanEdu.

Lü, G., M. Batty, J. Strobl, H. Lin, A. X. Zhu, and M. Chen. 2019. "Reflections and Speculations on the Progress in Geographic Information Systems (GIS): A Geographic Perspective." *International Journal of Geographical Information Science* 33:346–367.

Nyerges, T. L., H. Couclelis, and R. McMaster, eds. 2011. *The Sage Handbook of GIS and Society.* London: Sage.

Pavlovskaya, M. 2018. "Critical GIS as a Tool for Social Transformation." *The Canadian Geographer/Le Géographe Canadien* 62:40–54.

GLOBALIZATION:

Baylis, J. 2020. *The Globalization of World Politics: An Introduction to International Relations.* Oxford: Oxford University Press.

Beck, U. 2018. *What Is Globalization?* London: Wiley.

Campbell, J. L. 2021. *Institutional Change and Globalization.* Princeton, NJ: Princeton University Press.

Scholte, J. A. 2017. *Globalization: A Critical Introduction.* London: Bloomsbury.

Singer, P. 2016. *One World Now: The Ethics of Globalization.* New Haven, CT: Yale University Press.

Steger, M. B. 2017. *Globalization: A Very Short Introduction.* Oxford: Oxford University Press.

CLIMATE CHANGE:

Dessler, A. E., and E. A. Parson. 2019. *The Science and Politics of Global Climate Change: A Guide to the Debate*. Cambridge: Cambridge University Press.

Dow, K., and T. E. Downing. 2016. *The Atlas of Climate Change*. Berkeley: University of California Press.

Hughes, S., E. K. Chu, and S. G. Mason. 2020. *Climate Change and Cities*. Oxford: Oxford University Press.

SUSTAINABILITY:

Clayton, T., and N. Radcliffe. 2018. *Sustainability: A Systems Approach*. London: Routledge.

Matson, P., W. C. Clark, and K. Andersson. 2016. *Pursuing Sustainability: A Guide to the Science and Practice*. Princeton, NJ: Princeton University Press.

Mulligan, M. 2017. *Introduction to Sustainability*. London: Taylor & Francis.

Portney, K. E. 2015. *Sustainability*. Cambridge, MA: MIT Press.

PART 1

The Environmental Context

This section sets the environmental stage. Chapter 2 gives a brief account of the planet we call home. Attention is paid to the evolution of the Earth, its emerging physical geography, and its humanization. Chapter 3 examines the complex cultural relationships between people and the environment. Chapter 4 focuses more on the economic aspects of the environment as an economic resource.

OUTLINE

2

The Home Planet

We inhabit a tiny blue dot in a vast inky darkness. It has been in existence for 4.6 billion years, but our occupancy is more recent. Our species emerged around 200,000 years ago. Later, our immediate ancestors migrated out of Africa to populate the world. Today we live in a humanized world deeply impacted by our presence. The health of the planet is our primary responsibility.

2.1 The Big Picture

According to the aptly named **Big Bang theory**, it all started with a very big explosion that produced enough energy to expand a single point outward to infinity. From a singular point in space-time of intense heat and pressure, the universe began to expand, cooling as it spread outward. The universe is still moving outward from this specific moment and particular location. Measuring the speed of the expansion allows us to calculate the approximate moment of the "birth" of our universe, around 13.7 billion years ago. A billion years after the Big Bang (a term coined by cosmologist Fred Hoyle in 1949), galaxies first came into being from differential gravitational pull in the young universe.

The Big Bang expansion exacerbated minutely small differences in density into sites of star clusters and galaxies. For the first 7 to 9 billion years, the attraction of this matter slowed down the rate of expansion of the universe in a cosmic pull of competing forces, the explosive energy of the Big Bang dampened by the gravitational tug of the matter that it created in its wake. Then, the expansion of the universe began to accelerate as a mysterious energy source, with the foreboding name of **dark energy**, overcame gravity. Almost three-quarters of the universe is made up of this unknown force; of the rest, a little under a quarter is made up of dark matter that neither reflects nor emits light, and only around 4 percent is the matter in the universe that we can see and, as yet, understand. We live in a dark universe. And to add to the pervasive strangeness of it all, there is also the intriguing proposition that there was more than one Big Bang. We are living in the aftermath of a Big Bang, but perhaps there were more, possibly an infinite number as the universe expands and contracts, each Big Bang leading to a giant implosion followed in turn by another Big Bang in an endless cycle.

The universe consists of around 200 billion galaxies and 30 billion trillion stars. Our home planet is situated in the Milky Way, a galaxy composed of 100 to 400 billion stars bound together by gravity and stretching across 100,000 light years. One **light year** is the equivalent of 5.8 trillion miles; it is calculated from the speed of light in a vacuum, 186,000 miles per second, or 700 million miles per hour. One of the stars, located 24,000 light years from the center of this galaxy, is the Sun, the center of our planetary system and rightfully deserving its capitalization. Eight planets revolve around this star. From near to far they are Mercury, Venus, Earth, Mars, Jupiter, Saturn, Uranus, and Neptune. The four nearest to the Sun are relatively small and composed of rock and metal; the four farthest, the gas planets, are larger, with planetary rings of particles and cosmic dust. Jupiter and Saturn are composed of hydrogen and helium, and the two farthest planets, Uranus and Neptune, traveling the darker, colder edges of the solar system, are ice giants made up of water, ammonia, and methane.

In 2016, scientists at the California Institute of Technology presented evidence that there may be a ninth planet with a mass of five to ten times that of the Earth. This planet, only known as Planet Nine, has not been observed directly but only inferred from the orbits of surrounding objects. The same type of inductions occurred in the nineteenth century when observations of Uranus indicated that there may be another planet out there. There was—Neptune. The 2021 launch of NASA's James Webb Space Telescope may allow us to delve deeper into the darkness of the universe; perhaps one day Planet Nine may be sighted.

We are situated close enough to the Sun to get more heat and light than the farthest planets but not so close as to be burned up by the intense heat experienced by the two planets closer to the Sun. The third planet out from the Sun has enough oxygen and water to support life.

We live on a planet in motion. It revolves around the Sun once every 365.25 days. A year marks how long it takes the Earth to complete one full movement around the Sun. The Earth is tilted approximately twenty-three degrees from the perpendicular in its orbit. This creates the seasons, especially marked farther away from the equator, where the distance from the Sun varies more substantially when the tilt is angled away or toward the Sun. The tilt of the Earth divides this yearly cycle into seasons; closer to the poles, the seasonal effects are exaggerated, as intense cold turns into a distinct warming when the long, dark days of winter become the light-filled days of summer. At the equator the seasons are less pronounced and the daily division into light and dark is more even throughout the year. At the poles, the annual cycle moves from a Sun that never sets to a Sun that is barely visible over the dark horizon. At the equator, the Sun is a more constant and reliable presence.

Our planet rotates on its own axis on a roughly twenty-four-hour cycle. The daily cycle is created as the Earth turns toward and then away from the Sun. Although we continue to use the terms "rising" and "setting" Sun, it is the Earth that moves. Our planet revolves around the Sun and rotates on its own axis.

The Earth does not revolve in a perfect circle. It wobbles, the angle of tilt moves, and its orbit around the Sun varies in its eccentricity (departure from circularity). These small differences in distance from the Sun may account for very long-term climate changes on the Earth, especially the rise and fall of ice ages.

Soon after the formation of the solar system—after only 30 to 50 million years—a giant **asteroid**, almost half the size of the Earth, hit our planet. The impact created the Moon, which now revolves around the Earth every 27.3 days, its pitted surface a silent witness to the destructive forces still at work in the universe. The gravitational effect of the Moon on the Earth is responsible for the **tides** that move the oceans

2.1 Pluto's Controversial Journey

From 1930 to 2006, there were nine planets; during this period, Pluto was discovered, identified as a planet, and then dropped from the list of major planets. The American astronomer Clyde W. Tombaugh discovered the planet in 1930 from the Lowell Observatory in Arizona. Before this date it was too small—only 18 percent of the Earth's diameter—and too far away to be visible. It was designated as the ninth and farthest planet from the Sun, but in 2006 it was reclassified as one of the dwarf planets, now classified as *plutoids*, and dropped from the list of major planets. Public reception to this decision was mixed, and the controversy has continued. While denied membership in the club of major planets, poor Pluto still makes its eccentric orbit at the edge of the solar system. In 2020, the "I Heart Pluto Festival" in Flagstaff, Arizona, commemorated the ninetieth anniversary of the discovery of Pluto and featured NASA scientists who argued it should be reinstated as a planet. In a touching moment of solidarity with the marginalized and the shunned, the supporters of the Turkish soccer club Beşiktaş carried banners that proclaimed, "We Are All Pluto."

and seas in ceaseless and regular vertical motion. The **lunar cycle** is the source of our division of time into months.

The yearly, monthly, and daily cycles that are such important rhythms of our lives are caused by the movement of our planet and the Moon. Even in a more electronic age we are affected by the beats of the cosmos. Two researchers, Scott Golder and Michael Macy, examined millions of tweets over a two-year period. Using selected words in messages to connote moods, they found that there was distinct periodicity, with more positive words in the early morning. Each new dawn offers the promise of a new beginning. Our moods also vary over the seasons. **Seasonal affective disorder** is the tendency for more negative moods among normally healthy people during distinct seasons. Winter blues are more common in northern latitudes because of the rapid decrease in sunlight. Seasonal affective disorder varies with latitude, with only 1.4 percent of people experiencing it in Florida but 9.7 percent in colder, darker New Hampshire.

2.2 Shaky Ground: Plate Tectonics

Look at any map of the world. The continents sitting in the blue seas and vast oceans look solid, firm, permanent. From a very long perspective the image is deceptive. The present distribution of landmasses is just the most recent frame in a dynamic, complex picture of continents forming, reforming, splitting up, and moving across the Earth's surface.

Earth came into being around 4.55 billion years ago. Soon afterward, around 4.51 billion years ago, the Earth was hit by a giant asteroid, turning it into a fiery ball of intense heat. Over the years, the surface cooled more than the interior. The Earth's surface, the "solid ground" of so many metaphors, is in fact a thin, brittle crust, no more than four to sixty-five miles thick, that floats precariously on a viscous mass of molten metal. The crust that we occupy is the cold top level that formed on the Earth's surface, just as a skin forms when boiling milk cools. Below is the hot mantle and, even deeper, the extremely hot core. The solidified crust fractured across the large round object of the Earth, breaking up into distinct **tectonic plates**. There are nine large plates and numerous small ones (Map 2.1). They sit atop powerful currents; below them, the mantle of molten metal heaves as hotter liquid moves up in convection currents from the boiling mass at the Earth's core, while the cooler liquid closer to the surface sinks to the bottom. This continuous subterranean motion moves the plates on the surface, like bumper cars in a fairground ride that move in a restricted space. Driven by the upward convection currents deep in the Earth's mantle, they bump against, slide past, and move away from each other. Mountains are formed, trenches are created, **earthquakes** occur, and boiling magma spews out in volcanic eruptions to reveal the fiery material that shapes the

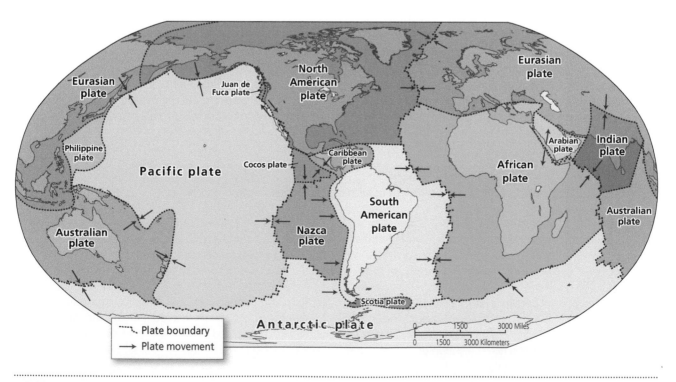

MAP 2.1 | Plate boundaries.

Earth's geological formations. The surface that we live on is a fragile membrane across deep and powerful subterranean forces. Volcanic and earthquake activity are particularly severe at the edges where plates meet (Map 2.2 and Photo 2.1).

The plates are always on the move. In the geological equivalent of political empires, giant landmasses rise and fall. Over 1.4 billion years ago, small masses of land collided to form the supercontinent of Rodinia. This landmass was rent asunder into different blocs including Avalonia; Baltica; Laurentia, eventually to become North America; and Gondwana, the source continent for South America, Africa, India, Antarctica, Australia, and New Zealand. When Baltica and Laurentia joined up with Gondwana around 300 million years ago, a new supercontinent, Pangaea, was created. It took up almost one entire side of the globe—the other side was mostly water—and was centered on the South Pole. This began to break up, but the large landmass of Gondwana remained. Beginning around 200 to 160 million years ago, it too began to break up, as South America broke off and drifted west, and a landmass of what was to become India began a 150-million-year journey northward to Asia, forming the Himalayas when it crashed into the Eurasian plate. Some plate boundaries are particularly active such as the **Ring of Fire** around the Pacific and the plate boundary that bisects Iceland (Photo 2.2).

Around 60 million years ago, another landmass, what we now call Australia, detached from what is now Antarctica and began a 45-million-year, 500-mile journey through various climate zones toward the warmth of a more northerly location. Australia was on the move. Slowly. It is still moving northward away from Antarctica at a rate of around 2.7 inches per year. There are many traces of the long journey. Australia shares with South America similar species of marsupials, turtles, and lungfish that are over 400 million years old, a biotic reminder of when the two continents were joined in Gondwana. And the distribution of an extinct fern, *Glossopteris*, in Antarctica, Africa, Australia, India, and South America is a biological reminder of the supercontinent.

The Indo-Australian plate eventually crunched up against the Eurasian plate around 15 million years ago. The border area between the two plates is still one of the most active volcanic zones in the world. Let us consider three events. Around 70,000 years ago, a giant volcano exploded at Mount Toba in Sumatra (now part of Indonesia). Almost a billion tons of ash and dust were thrown into the atmosphere, darkening the skies and reducing global temperatures by around 5°C for up to five years. One anthropologist, Stanley Ambrose, writes of a long volcanic winter, an instant ice age that severely reduced early human populations. The total human population in the world was probably reduced to around 10,000, so that local variations

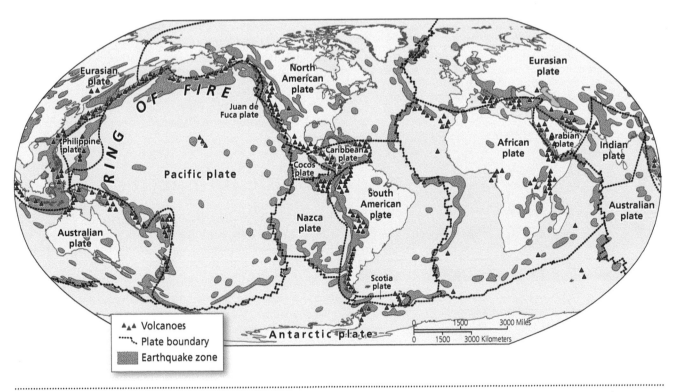

MAP 2.2 Location of seismic activity (earthquakes and volcanoes). They cluster along plate boundaries. Notice the Ring of Fire around the Pacific.

PHOTO 2.1 Along the line of mountains in Central America is a zone of active volcanic and earthquake events. This is an active volcano just outside the city of Antigua in Guatemala.

PHOTO 2.2 There is both a **volcanic hotspot** and a tectonic plate boundary in Iceland, giving the island nation a haunting and fascinating landscape.

became more pronounced in human **evolution**, creating the contemporary racial differentiation. The huge, violent explosion caused **global climate change** that further

impacted human populations. The extent to which the consequent **environmental stressors** may have promoted social cooperation and thus facilitated the human dispersal from Africa is a debatable but intriguing proposition.

The region saw a second event, another massive volcanic explosion, when the island of Krakatoa exploded on August 27, 1883. Not as violent as the Mount Toba explosion, it was still devastating; the resultant tsunami killed approximately 36,500 people in the region as tidal waves engulfed towns and villages along the coast. The air pollution caused by all the dust and ash blotted out the Sun's heat and light and reduced the world's temperature by about 0.5°C. Simon Winchester argues that Krakatoa also influenced political developments. He suggests that the 1888 Banten Peasants' Revolt, which pitted local people against Dutch colonial rule, was partly influenced by the cataclysm that befell the Indonesians and fed into an anticolonial and Islamic fundamentalist narrative that "explained" the explosion and mass deaths as the work of a wrathful Allah signaling displeasure at colonial control and lax religious practices. Indonesian scholars have shown that the environmental catastrophe took place as peasant resistance was increasing against the increasingly exploitative colonial economic system of the Dutch.

On December 26, 2004, movement along the plate caused the third event—another major disaster. The Sumatra–Andaman earthquake occurred nineteen miles below sea level off the coast of Sumatra. Along a 1,000-mile zone, plates shifted almost 50 feet and raised the surface of the seabed, displacing millions of gallons of water that caused the **tsunami**, the postearthquake waves that suddenly overwhelmed coastal communities in fourteen countries fringing the Indian Ocean. As waves cresting 100 feet crashed into unsuspecting communities, more than 230,000 people died, 125,000 were injured, and almost 1.7 million were displaced from their homes.

2.3 Life on Earth

For half of its entire existence the Earth did not sustain life. It took a long time for the planet to cool down and its landmasses to become more stable. The beginnings of life around 2.2 billion years ago were in the modest form of blue–green algae. Evolution was slow until the development of sexual reproduction. An explosion of life took place around 545 million years ago in the Cambrian period of the early Paleozoic era as living forms with skeletons first appeared. The growth of the amount and diversity of life was not a simple upward trajectory. The same era also marked the first global mass extinction, because many species of shallow water fauna were killed off by rising sea levels.

Around 418 million years ago, the first land animals appeared. Insects and amphibians also emerged. In the Mesozoic era, from 248 to 65 million years ago, the first dinosaur appeared, as did the first mammal, a tiny, shrew-like animal dwarfed by the giant reptiles around it. In the late Mesozoic there was another mass extinction of animal life. Dinosaurs became extinct, and birds lost almost 80 percent of their species, as did the marsupials. Some scientists suggest that colossal volcanic eruptions in a mountain chain in eastern India spewed huge amounts of carbon dioxide and **sulfur dioxide** into the atmosphere, poisoning the air. Other scientists point to an asteroid hitting the Earth around the same time, 65 million years ago. A crater measuring 100 miles wide and 6 miles deep was discovered in the Yucatan peninsula near the town of Chicxulub. The impact of the asteroid created global fires that sucked oxygen out of the atmosphere and caused huge shock waves that triggered volcanoes and earthquakes. The volcanic eruptions in India and the asteroid strike in the Yucatan created devastating global environmental conditions. Dust covered the Earth, leading to temperature decrease. **Acid rain** fell. Dark and coldness settled on the Earth, plants stopped growing, and animals died. More than 90 percent of all marine life died as the seas became acid baths. The dinosaurs became extinct, because they could not survive in the wasteland of permanent winter. They starved to death.

The Tertiary period, from 65 million to 1.8 million years ago, is the age of mammals. Whales appeared, as did elephants, cats, and dogs. Around 50 million years ago, the first primate appeared, a little lemur-like creature barely more than two pounds in weight. The primates diversified and spread. Four million years ago, the first **hominid** to stand upright, *Australopithecus*, walked on the Earth. Between 2 and 3 million years ago, at least eight different hominid species emerged from this one species, correlating with a time of increased climate change. The rapidly changing conditions rewarded species that could adapt quickly, especially those that had the brain capacity to adjust to constant **environmental change**. Adaptability to diverse environmental conditions is one of the defining hallmarks of our human species.

Mass Extinctions

The fragility of life on Earth is revealed most dramatically during episodes of **mass extinctions**. Five such events are recorded. Figure 2.1 depicts five spikes in extinction events. Table 2.1 provides more detail about the causes and consequences of these events. In the first one, climate change in the form of an ice age disrupted the oceans' chemistry during the Ordovician–Silurian extinction event, killing a majority of sea life species. Impact events, either an asteroid hitting the Earth and/or giant volcanic eruptions around 65 million years ago, were the principal cause behind

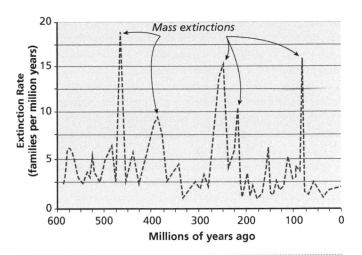

FIGURE 2.1 Mass extinctions.

TABLE 2.1 Mass Extinction Events

TIME PERIOD (MYA)	CAUSES	CONSEQUENCES
Ordovician–Silurian (443)	Climate change	85 percent of all sea life was wiped out
Late Devonian (359)	Impact events	75 percent of species wiped out
Permian (248)	Multiple	96 percent of species wiped out
Triassic–Jurassic (200)	Multiple	50 percent of animals wiped out
Cretaceous–Tertiary (65)	Multiple	75 percent of species wiped out

Note: mya, million years ago.

the more recent extinction event during the Cretaceous–Tertiary, when the dinosaurs were killed off. The catastrophic events caused a massive increase in greenhouse gases and rapid warming that made it unlivable for many terrestrial species, even the fearsome *Tyrannosaurus rex*.

These mass extinctions show that life is fragile, but also resilient, because new life forms emerge in their aftermath.

2.4 A Humanized World

Tracing human origins was long the preserve of archaeologists and physical anthropologists. Digging in old sites and dating the old bones was the preferred method. Dates were always provisional, and the ambiguity of the information always provided lots of room for debate. Things began to change when scientists could use live humans to chart the past. Our bodies, it appears, are the equivalent of an archaeological site, containing memories and traces of the far-distant past. Plotting the distribution of the information in our bodies allows us to map possible sites of human origin and paths of human dispersal.

In the 1950s, differences in blood group protein were identified that allowed a picture of human ancestry. A more sophisticated analysis was made possible in 1987, when scientists first used the DNA in mitochondria, the rod-shaped "power plants" of cells, to construct a genetic tree. Using this method allowed scientists to examine genetic materials passed from mother to child and led them to the startling conclusion that we are all descended from one woman who lived in Africa around 200,000 years ago.

Out of Africa

After decades of competing ideas and dueling theories, there is an emerging consensus that all humans are descended from a group of people who left Africa. Human origins lie in Africa. The first humans stepped out of Africa to populate and humanize the world (Map 2.3). Evidence of **genetic variation** sustains this **Out of Africa hypothesis**. Peoples in central and southern Africa have more genetic variation than elsewhere in the world. The highest levels of genetic diversity in the world are among the Namibian and Khomani Bushmen of southern Africa, the Biaka Pygmies of Central Africa, and the Sandawe of East Africa. There is a decrease of genetic diversity with increasing distance from Africa, suggesting a founder's group in Africa as the

basis for all subsequent human population. As smaller subgroups broke off to settle in different places, the amount of genetic diversity decreased. The amount of genetic variation declines with distance from Africa—a classic case of the **distance-decay effect**—suggesting that humans came out of Africa to populate the Earth. It was the first wave of globalization.

The human diffusion from Africa, around 120,000 to 70,000 years ago, initially involved a tiny population of probably between 3,000 and 5,000 people. Notice the implied leeway in the range. We have gotten a general idea, not a sure fix. The movement out of Africa took different routes at different times. The oldest was a group that moved out through the Horn of Africa into the Middle East. One group then moved along a coastal route skirting the Indian Ocean. They took around 20,000 years to migrate around the Indian Ocean, taking the beachcomber route into Southeast Asia and Oceania. The migration took the form of **budding**, as small subgroups broke away to settle new places. Traveling along the coast and up rivers, they avoided the difficulties of passing through deserts and across mountains. Consider, as just one example, the settling of Australia. Around 50,000 years ago, when sea levels were lower, Australia was linked to Papua New Guinea, and there was only a relatively short distance, probably no more than sixty-five miles, between the edge of Asia and the coast of Australia. Bamboo rafts probably were constructed to make the sea crossing. Current archaeological evidence dates the human presence in Australia to between 55,000 and 42,000 years ago. Most of the early Australian sites have deep charcoal deposits that

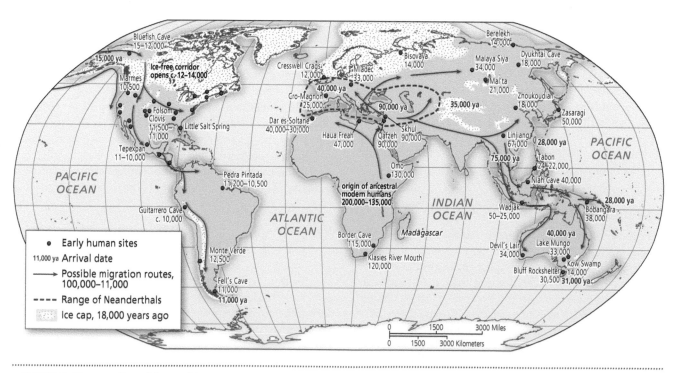

MAP 2.3 The peopling of the world.

suggest both early settlement and a long record of occupation. The migrants traveled quickly up the rivers and along the coast. Colonization was rapid, and there are sites scattered throughout the vast continent in the north, northwest, southwest, and southeast and as far south as Tasmania, all dated to around 32,000 to 40,000 years ago. Central Australia was settled last, because there were no easy coastal or riverine routes to follow. People had to venture into a dry, forbidding interior. But venture they did. We have a site, the Puritjarra Cave Rock Shelter, with rock art suggesting occupancy as far back as 32,000 years ago.

Another later wave of human migration around 50,000 years ago moved into Eurasia. Giant walls of ice restricted movement north; it was still the Ice Age, also known as the Pleistocene era. However, sea levels were much lower in the Pleistocene, exposing land bridges that are now covered with seawater. The English Channel was dry land, allowing people and animals to move from Europe into what is now Britain and Ireland. On the other side of the world, the **land bridge** between the northeastern-most edge of Eurasia and the northwest edge of America allowed people and animals to make the journey across the continents around 22,000 to 25,000 years ago.

Human groups were restless, continually moving to and settling new areas. People even sailed across the vast expanses of the Pacific Ocean to settle on such distant specks of islands as Hawaii. The early humans explored the world, settled in different habitats, and impacted their environment.

These colonizers' impact was immediate. Despite early interpretations of them as docile elements in the landscape, they were very much active agents in modifying and changing the environment. Early humans first shared the Eurasian landmass with other hominids such as the Neanderthals and the Denisovans. These were cousins to humans on the evolutionary tree, emerging as separate subspecies around 500,000 years ago. The Neanderthals were more intelligent than common characterizations assume. They walked upright and effectively communicated with each other. They and the Denisovans emerged in and lived throughout Eurasia, but with the entry of *Homo sapiens*, these two hominid species died out around 30,000 years ago. As the human population spread across the Eurasian landmass, we effectively annihilated our closest competition for food and environmental resources. One competitive advantage was the division of labor. Among human populations, women and young children were responsible for skill-intensive crafts such as making clothing and shelter, as well as gathering food. Men were primarily responsible for hunting. The division of labor allowed access to a wider range of food sources compared to the other hominids, who, because they did not have the gender and age division of labor that allowed economic specialization, were almost entirely dependent on hunting game. When the game disappeared, they starved and died out. The gender and age division of labor in early human populations created more specialized economic roles that allowed a more efficient use of labor and gave access to a wider food resource base.

The other hominid groups died out, but they live on in our genetic makeup. They must have been close enough to humans that interbreeding occurred. We have a genetic record of the process. All non-Africans have between 1 and 4 percent of Neanderthal DNA. The people of New Guinea have around 5 percent of Denisovan DNA. Most of the world's non-African population carries this genetic reminder of the width of human sexual appetite and willingness to "interact" with their evolutionary cousins. The distinction between human and nonhuman hominids was more of a liminal area than a sharp and fixed boundary.

Pleistocene Overkill

The early human colonization of the Earth was also marked by ecological changes. The most pronounced is the **Pleistocene overkill**, which refers to the extinction of **megafauna** such as the mammoth and the mastodon that occurred around 11,000 years ago. This extinction occurred on every continent and was experienced by every community of large terrestrial vertebrates at a time of humans' increased technological sophistication in hunting. While there has been some debate about the relative role that climate change played in the mass extinction, human hunting definitively contributed. On some continents, both hunting and climate change were important, while on others hunting was the primary cause. The extinction was particularly marked in North America around 20,000 years ago.

America is a test case of the **overkill**, since human entry into the region was relatively late, and there are no records of other hominids. John Alroy created a grid of 754 spatial cells of one degree of latitude and one degree of longitude. His computer simulation model for North America clearly shows that human population growth and hunting led to mass extinction 1,200 years after humans first appeared.

The larger and slower megafauna were very vulnerable to skilled hunters. The megafauna, weighing more than 220 pounds and thus slower than other animals, became especially vulnerable. In Australia, for example, many megafauna became extinct around 46,000 years ago, soon after the entry of humans into Australia. We will never again see the ten-foot-tall giant kangaroo, the rhino-sized wombat, or the marsupial lion. We do have a visual record from rock paintings in Arnhem Land in northern Australia, which may depict a *Palorchestes*, a large browsing animal; a *Genyornis*, similar to an emu but three times as tall; and a marsupial lion. Farther west in Kimberley, rock paintings depict giant kangaroos and huge echidnas. The paintings are a reminder of the time when megafauna and humans shared the same space and an early example of the immense role that humans played and continue to play in shaping the global ecosystem.

2.2 Wallace's Line: A Biogeographical Boundary

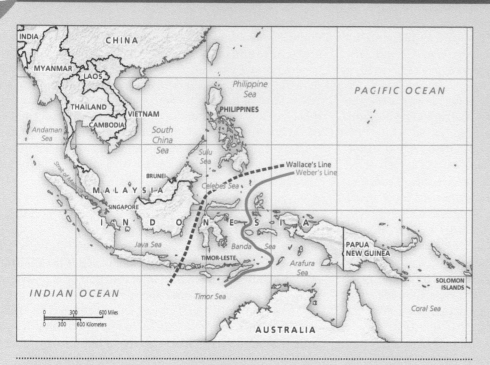

MAP 2.4 Wallace's line.

Max Carl Wilhelm Weber drew his own line based on surveys of mammalian fauna. His line is farther to the east than Wallace's line. The difference between the two lines is that flora found it easier to move across the "line," while the seas constituted a larger barrier to animal movement. The area between the Wallace and Weber lines can be considered a zone of transition.

The biogeographical boundaries identified by Wallace and Weber are a result of **plate tectonics**. The northward drift of the Australian plate made it crash against the Eurasian plate around 15 million years ago. Previously separate biological realms were brought into contact. The Australian plate was like a raft filled with distinctive flora and marsupial animals now abutting the Asian flora and fauna.

In the mid-nineteenth century, Alfred Russell Wallace (1828–1913) spent eight years travelling the Malay Archipelago. He traveled over 14,000 miles and collected over 100,000 samples. He was less well known than Darwin, and his letters back to members of England's scientific community indicating a theory of natural selection for the origin of species prompted some to counsel Darwin to publish his own work in 1859. In 1863, Wallace published a paper in the *Journal of the Royal Geographical Society* that identified a boundary line between two realms. This line, now known as Wallace's line, divides the Australian and Asian biotas (Map 2.4).

Wallace's work was part of a surge of mapping in the nineteenth-century natural sciences. Maps were, and still are, used as a conceptual framework and as a tool for presenting vast amounts of data in simple yet suggestive ways. Wallace's line evolved from papers he wrote in 1855 and 1860, as well as the later 1863 paper. Later, the zoologist and biogeographer

na. Without a theory of plate tectonics (Alfred Wegener did not moot it until 1912, or publish it in 1915), Wallace could not discern the geological basis for the division. However, his careful mapping highlighted the benefits of fieldwork and the representation of these data in maps. Wallace's line is the biological equivalent of plate boundaries, even though Wallace was not aware of the plates and the theory of continental drift.

REFERENCES

Camerini, J. R. 1993. "Evolution, Biogeography and Maps: An Early History of Wallace's Line." *Isis* 84:700–727.

Vetter, J. 2006. "Wallace's Other Line: Human Biogeography and Field Practice in the Eastern Colonial Tropics." *Journal of the History of Biology* 39:89–123.

Wallace, A. J. 1863. "On the Physical Geography of the Malay Archipelago." *Journal of the Royal Geographical Society* 33:217–234.

2.5 Climate Change

Over the long history of the Earth, its climate has changed. Over the past 100,000 years the Earth has warmed, ice sheets have melted, and sea level has risen. In the past 1,000 years the climate warmed during the Medieval Warm Period from 1100 to 1300 and then cooled from 1300 to 1700, in what is known as the Little Ice Age. In the past 100 years our planet's average temperature has increased by 1.5°F and is projected to rise even more in the next 100 years because of human impact, the most important being **greenhouse gas** emissions (especially **carbon dioxide**), which warm the atmosphere.

The full implications of climate change have yet to be unraveled, but a number of major trends are now identified.

2.3   A New Climate Regime

The sixth and most recent report of the Intergovernmental Panel on Climate Change in 2021 noted that climate change is widespread, rapid, and intensifying. The report noted, "It is unequivocal that human influence has warmed the atmosphere, ocean and land. Widespread and rapid changes in the atmosphere, ocean, cryosphere and biosphere have occurred" Climate change has gone from an inconvenient truth to an undeniable reality.

Warming is twice as high in polar regions, leading to liquefying of permafrost, loss of seasonal snow cover, and the melting of glaciers and ice sheets. The impacts on the ground are varied. Global warming is leading to increasing heat waves, longer warm seasons, and shorter cold seasons. There is likely to be more intense rainfall and associated flooding in some regions of the world, as well as more drought in others, leading to desertification. The impacts are also complex. In higher latitudes, warming will mean shorter snow seasons, but because warmer air holds more moisture, the warmer winter air can result in more intense snowfalls. Therefore, while the winter season may be shorter, snowstorms may be more intense. Rainfall is likely to decrease over large parts of subtropical areas. Coastal areas are threatened by sea level rise. The maritime regions of the world will be impacted by marine heatwaves, ocean acidification, reduced oxygen levels, and more intense hurricanes and typhoons. The overall warming trend will have multiple expressions in different parts of the world, from melting permafrost, to increased snowfall, to more drought, greater incidence of flooding, and, in general, more extreme weather events. It will also herald new diseases and their global transmission.

We are entering a new climate regime.

REFERENCES

Dessler, A. E., and E. A. Parson. 2019. *The Science and Politics of Global Climate Change: A Guide to the Debate*. Cambridge: Cambridge University Press.

International Panel on Climate Change. 2021. *Synthesis Report of the Sixth Assessment Report*. https://www.ipcc.ch/ar6-syr/.

At the higher latitudes and altitudes, glaciers are shrinking and **permafrost** is warming and thawing. Sea level rise is impacting coastal areas. Animal life has responded in a variety of ways, including shifts in geographic range (with heat-loving species able to live farther north in the Northern Hemisphere and farther south in the Southern Hemisphere), migration patterns, and species interaction. There is mounting evidence that climate change is increasing the risks of extinctions and the potential impact on human health as changing ecologies provide new opportunities and vectors for viruses and the diseases they cause.

A number of respected scientists now argue that we are in a sixth mass extinction event, caused by climate change and human activity (see Figure 2.2). There is a background rate of around two mammal extinctions per 10,000 species per 100 years, but in the past 100 years the rate is now 110 times higher than this level. Over the past 500 years there has been a 10 percent reduction in land-based wild animals and a 13 percent reduction in the number of bird species. The rapid loss of biodiversity means that a sixth extinction event may be underway.

More extreme weather events including drought, floods, and major storms are also more likely because of climate change. The consequences vary with location. In North America there is greater risk of drought in the west and southwest, a higher chance of more powerful tornadoes in the central region, and more punishing storms along the east and southeast coasts. This increased likelihood is not a straight-line relationship such that each year

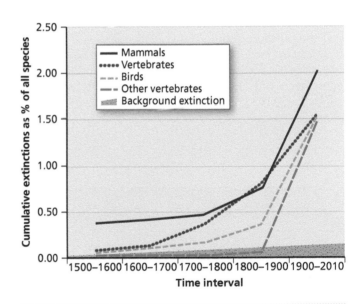

FIGURE 2.2 The sixth extinction.

sees more frequent extreme weather events; rather, the probability of their happening is greater, so that "100-year storms" occur once every 10 years.

Some ecosystems are especially threatened. Coral reefs survive in only a very narrow temperature range and are thus highly vulnerable to warming oceans. Low-lying coral atolls such as Atafu Atoll in the South Pacific are especially vulnerable (see Photo 2.3). Climate change is particularly

PHOTO 2.3 Atafu Atoll, South Pacific.

marked in the high latitudes because of what is known as the Arctic amplitude, which causes greater near-surface warming than at middle and low latitudes.

Humans have responded directly to climate change in two main ways: adaption and mitigation. Adaption involves engineered and technological solutions, such as building storm barriers, or changing behaviors, such as the shift in garden preferences from lawns to desertscapes in the southwestern United States. Mitigation involves trying to reduce our **carbon footprint** by reducing the emission of greenhouse gases. There is also the more indirect response of climate-induced migration. In many parts of the world, the negative impacts of climate change such as desertification, floods, drought, and warming temperatures are generating climate refugees seeking to escape the deteriorating environmental conditions and find a better life.

2.4 Geography and Climate Change

One of the impacts of climate change is sea level rise. The small, low-lying island nations of the Pacific are very vulnerable to sea level rise, as well as the increased cyclonic activity associated with climate change.

The highest point in the Marshall Islands is only 33 feet above sea level, Kiribati has an average elevation of only 7 feet, and the Maldives averages 4 feet in elevation. On many islands, the majority of people live close to the shore and most of the infrastructures on the island chains are within a quarter mile of the coastlines. Some climate models predict as much as a 13-inch rise in sea level by 2050 and a 3.5-foot rise by 2100. Waves will become higher and more of the land will be inundated with salt water, even during normal tides and especially during storms, killing off vegetation, plants, and crops and polluting fresh water sources. In addition, there are no hills to escape to higher ground. For many of these island nations in the South Pacific, climate change could force the evacuation of the entire island population.

A major challenge for the small island countries of the South Pacific is that they are small in population, resources, and wider political influence. They are so small and so distant from centers of political power that they have little voice in climate change discussion and negotiations. With the lowest carbon footprints on the planet, they are paying the heaviest price: the threat of territorial extinction.

Cited References

Alroy, J. 2001. "A Multispecies Overkill Simulation of the End-Pleistocene Megafaunal Extinction." *Science* 292:1893–1896.

Ambrose, S. 1998. "Late Pleistocene Human Population Bottlenecks, Volcanic Winter, and Differentiation of Modern Humans." *Journal of Human Evolution* 34:623–651.

Golder, S. A., and M. W. Macy. 2011. "Diurnal and Seasonal Moods Vary with Work, Sleep, and Daylength across Diverse Cultures." *Science* 333:1878–1881.

Winchester, S. 2003. *Krakatoa: The Day the World Exploded: August 27, 1883*. New York: Viking.

Select Guide to Further Reading

Cawood, P. A. 2020. "Earth Matters: A Tempo to Our Planet's Evolution." *Geology* 48:525–526.

Condie, K. C. 2022. *Earth as an Evolving Planetary System.* 4th ed. London: Academic Press.

Hazen, R. M. 2012. *The Story of Earth: The First 4.5 Billion Years, from Stardust to Living Planet.* New York: Viking Penguin.

Henson, R. 2019. *The Thinking Person's Guide to Climate Change.* 2nd ed. Boston: American Meteorological Society.

Kolbert, E. 2014. *The Sixth Extinction: An Unnatural History.* New York: Holt.

Marean, C. W. 2015. "An Evolutionary Anthropological Perspective on Modern Human Origins." *Annual Review of Anthropology* 44:533–556.

Stern, R. J. 2018. "The Evolution of Plate Tectonics." *Philosophical Transactions of the Royal Society A: Mathematical, Physical and Engineering Sciences* 376 (2132): 20170406.

Stix, G. 2008. "Traces of a Distant Past." *Scientific American* 299 (1): 56–63.

Tattersall, I. 2009. "Human Origins: Out of Africa." *Proceedings of National Academy of Sciences* 106:16018–16021.

3

People and the Environment

The environment is a source of economic opportunity, a repository of meaning, a sometime hazardous context, and always the background to our individual and collective lives. Examining the relationship between people and environment, which we will define as the world around us, is at the heart of human geography. In this chapter we will look at three broad themes: the environment as a source of cultural meaning, the environment's **affect** on human feelings and activity, and the impacts of human activity on the environment.

3.1 Environment and Cultural Meaning

We can make a distinction between three basic forms of social–spatial–economic relations: hunting-gathering, traditional agriculture, and urban–industrial systems. Each has overlapping but also distinct conceptions of environmental space. Respectively, they tend to inflate the importance of sacred space, productive space, and managed space.

Sacred Space

In hunting-gathering societies, people derive their individual and collective sustenance by hunting animals and gathering food. There was and often still is a gendered and age division of labor, with men responsible for hunting and women and younger children responsible for gathering other food. In hunting-gathering societies people need to be in tune with the seasonal rhythms of flora and fauna. The close connection with the environment provides both the material basis of life and spiritual sustenance. The environment is not an inert container of resources, but a living organism filled with beings, gods, and spirits that provide comfort and support as well as irritation and trials. Nature is the ground of being, powerful and present, and human–nature relations are not simply functional and economic but also filled with spiritual meaning and cosmological significance.

Cultural Encounters

Hunting-gathering societies came and continue to come into contact with people and societies with different environmental views and cosmologies. Here are just two examples.

Canadian explorer and geographer David Thompson (1770–1857) was traveling in western North America in 1807–1811 when he recorded the following encounter.

> Both Canadians [French Native Americans] and Indians often inquired of me why I observed the sun, and sometimes the moon, in the day-time, and passed whole nights with my instruments looking at the moon and stars. I told them it was to determine the distance and direction from the place I observed to other places; neither the Canadians nor the Indians believed me; for both argued that if what I said was the truth, I ought to look to the ground.
>
> (Tyrrell, 1916, 141)

This encounter encapsulates the difference between the search for the coordinates of the grid and a reliance on a more intuitive sense of place. We now look with more sympathy on the "Canadians and Indians" than the Enlightenment figure of David Thompson struggling to measure and map the world.

Today, we have far more accurate measurements of the Earth. Latitude and longitude can be easily and precisely calibrated. Handheld global positioning systems allow observers to get a very accurate and almost immediate reckoning of their coordinates in the grid. But perhaps because this avenue has been so fully explored and developed, there is a renewed interest in the areas outside the arc of traditional geography. We also need to develop the more empathetic understanding that made the Canadians and Indians tell Thompson to "look to the ground."

Maps have been made throughout human history. Across the globe, a rich variety of maps were made. Sometimes they were employed in colonial appropriations. Map-making Europeans "explored" and mapped large parts of Africa, America, and Asia. Despite the traditional view that Europeans created such maps on their own, local and Indigenous peoples helped considerably. It is more accurate to consider the notion of cartographic encounters involving Europeans and local peoples, rather than a simple cartographic appropriation by Europeans. The mapping of the continents was underpinned by native knowledge. There is a hidden stratum of Indigenous geographical knowledge that is only now being uncovered.

Take the case of North America. The Europeans landed in a populated place. To find their way and move around the unfamiliar landscape, they relied on the Indigenous people to provide information, advice, and guidance. The maps made by the explorers drew heavily on an Indigenous cartographic contribution. In return, the Indigenous people obtained trade goods, spatial information, and the possibility of new alliances against long-standing enemies. This exchange was invaluable to the Europeans over both the short and the long term. For the Native Americans, the exchange over the short and medium term could prove useful, but over the long term it sealed their fate. The cartographic encounters resulted in Native Americans losing control over their territory. No longer needed, they were moved and marginalized, and often killed. In the case of North America, the mapping of the land embodied a *symbiotically destructive* relationship. It allowed Native Americans to parlay their deeper and wider knowledge of the land into a strong bargaining position, but in this very exchange, where they had some leverage, lay the roots of their loss of the land. They used what resources they had to gain a short-term advantage, but over the long term the cartographic encounter gave the newcomers more power.

Productive Space

Joseph Campbell made a distinction between the environmental cosmology of the hunter-gatherers, concerned more with the "way of animal powers," and that of agricultural peoples, concerned more with the "way of the seeded earth." It is a shift from the notion of sacred space to that of productive space (see Photo 3.1). The distinction is perhaps too clear-cut. In reality, many traditional agricultural societies grew out of hunting-gathering groups and still retained elements of the earlier environmental beliefs. Along the Atlantic seaboard of what is now the United States, people hunted the mammoth, bison, and other megafauna, but as the climate warmed at the ending of the Pleistocene Ice Age, farming became more intensive. Hunting-gathering cosmologies, concerned with the relations with animals, were carried over into agricultural cosmologies concerned with the fertility of the land. Maize, beans, squash, tobacco, potatoes, and corn were grown throughout the central Atlantic region. The Iroquois referred to corn, beans, and squash as the "Three Sisters" or "Our Supporters." Growing the three crops in the same field was an agricultural innovation that increased productivity and crop yields among Indigenous peoples. The corn provided structure and support for the other plants, the beans were a valuable source of nitrogen for the soil, and the squash provided protective mulch.

An annual cycle of festivals and rituals, especially at planting and harvesting times, respectively, invoked continued agricultural plenty and gave thanks for it. Animal spirits were still respected; for instance, the panther was an animal of great spiritual significance for the traditional Iroquois.

The productive space of rice paddies in Luzon, Philippines.

The agriculturalists tended more to stress the sacrifice, symbolic and real, to promote and sustain fertility. Ensuring the fertility of the land is vital to agricultural societies and central to their belief systems. In some societies, such as the Aztec, the sacrifice took on a literal meaning as humans were offered up to the gods on altars in public displays. From its agricultural roots the practice grew to become an instrument of political coercion and control. More than 4,000 human sacrifices were made at a reconsecration of the Great Pyramid of Tenochtitlan in 1487. The idea of sacrifice continues into the twenty-first century in contemporary religious practices. The Christian Eucharist, which signifies the body and blood of Christ, is an updated and particularized version of the older and more widespread sacrifice narrative of agricultural societies.

As we move toward more **commodified agriculture** where we grow things for profit rather than simply to eat, we shift from the idea of land as a primary source of meaning and identity to land as a commodity, a vehicle to accumulate wealth. In the early twenty-first century, modern industrial agriculture relies on chemicals and fertilizers rather than ritual sacrifice.

Managed Space

The rise of cities and the development of industry create new environmental belief systems. On the one hand, there is a strengthening of the belief in technological mastery of the world. The success of urban industrialization in transforming our experience of the world suggested that nature could be managed and controlled rather than revered and respected. A technological triumphalism saw few limits and no constraints. Nature was less a living thing and more a resource to be exploited, a backdrop for human desires and social needs. As land was commodified, the particularities of place as spiritual significance were replaced by the generality of market space. Particular places had economic value but not cosmological significance. The environment was merely a calculus of economic costs and benefits, and nature was now a commodity to be bought, sold, and traded.

Yet, on the other hand, the managed spaces also included religious structures. And the environmental degradation caused by the exploitative attitude did generate counterclaims and alternative visions. **Romanticism** was a reaction to the scientific rationalism of nature and the commodification of land. It saw nature as a source of profound aesthetic experiences and a perfect vehicle for a personal connection to the infinite. As factories were being built and cities replaced fields, William Wordsworth was singing the praises of daffodils and Henry David Thoreau was espousing the virtues of the woods and lonely places. All kinds of public health reforms and environmental movements emerged, contesting the dominant narrative of nature simply as resource. In the United States, for example, because industrial growth was accelerating, John Muir was extolling the virtues of protecting wilderness. Rapid industrialization and urbanization generated resistances and new ideologies that stressed preservation and conservation. Scientific discourses such as **ecology** developed out of the concern to consider the links that bind all living things. Public policies such as wilderness preservation and land conservation were also developed. The promotion of sustained economic growth now competed with a belief in environmental protection.

In the early twenty-first century we have competing claims. Powerful forces promote continued economic growth and tend to see nature primarily as a commodity to be developed and exploited. Arguments about jobs or economic growth are often used to trump "merely" environmental concerns. Yet we also have an increasingly powerful awareness of the environmental consequences of unrestrained growth. We live in a world where economic growth and global environmental impacts are unprecedented, but also in a time when a new environmental ethic, one shaped both by past belief systems and by current practices, is taking shape. Its main ideas include a greater awareness of our environmental impact, a less triumphalist optimism in technology, and a greater concern with long-term sustainability than short-term growth.

The link between economic growth and environmental quality is sometimes represented as a bell-shaped curve termed the **environmental Kuznets curve** (Figure 3.1). (The adjective is placed in front of the proper noun because there is another Kuznets curve that depicts a similar relationship between per capita income and inequality.) At the early stages of economic growth, when per capita income is still relatively low, the environment worsens as aggregate growth is pursued despite the environmental effects. Then, as incomes increase, more people place priority on

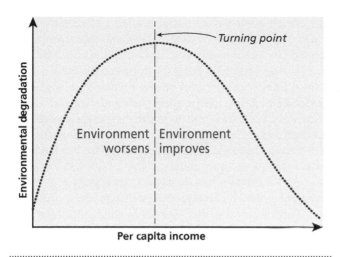

FIGURE 3.1 Environmental Kuznets curve.

environmental quality. Evidence can be found that both contests and confirms this dynamic between income and environment. In terms of counterargument, for example, levels of carbon emission continue to increase with rising per capita incomes. Moreover, richer regions and countries may export the negative environmental consequences of their high growth through shipping pollution waste or offshoring the more environmentally damaging economic practices. Poorer countries and regions also may be able to affect environmental quality much earlier than the curve suggests, as social movements and governments are made acutely aware of the damaging consequences and the false nature of the jobs-versus-environment slogan. Eco-communities in regions across China have promoted environmental improvement despite low per capita income.

In terms of environmental awareness and appropriate policy responses, there is now a wide variety of opinion. At one extreme is **deep ecology**. The term was first used by Norwegian philosopher Arne Næss to refer to a deep reverence for Nature that implies the radical restructuring of our current lifestyles to be more biocentric. At the other end of the continuum is an environmental sensitivity that suggests less radical change but a more pronounced commitment to alternative technologies, reducing pollution, and sustainable development. The deep ecologist questions the morality of growth, while the environmentalist/conservationist wants sustainable growth.

There is now a vigorous environmental debate with an increasing number of strands of social ecology that link environmental issues more directly to political issues, and indeed see the environment as a deeply political issue. For example, **ecofeminism** highlights the gendered nature of many environmental ideologies and practices and tries to engage feminist and ecological issues in a shared dialogue. What they all share is a questioning of the belief in unbridled economic growth and a commitment to a new relationship between humans and nature.

Landscape

One line of research in the society–nature theme looks more specifically at landscapes. Landscapes reflect the basic forms of social–spatial–economic relations. In his classic 1955 study *The Making of the English Landscape*, W. G. Hoskins explored the English landscape much as an archaeologist explores an excavation. The different levels in the landscape, beginning with the pre-Roman and continuing all the way through the industrial, were brought to the surface for inspection, cleaned of dust, and examined. In direct reference to Hoskins's work, the geographer Michael Conzen's 2010 book is entitled *The Making of the American Landscape*.

Landscapes can be viewed as archive, nature, habitat, artifact, ideology, history, place, and aesthetic. Landscape is a complicated text, telling us who has power and how and why that power is wielded. Landscape making has connection with deeper economic and political relations. We can make a distinction between dominant and alternative cultures. Landscapes reflect the dominant culture, but have traces of alternative cultures. Don Mitchell connects the making of the California landscape to agricultural workers and their labor processes. This labor history of landscape highlights the conflicts between workers and their employers.

Landscapes can be interrogated for their connection with capitalism, national identity, social justice, race, and class. Landscape is important in the creation of personal, regional, and national identity and historical memory.

3.2 Environmental Impacts on Society

Recent geographical scholarship is developing the idea of "affect," or an emotional effect, to refer to the force that makes us feel, think, or act. At the level of individuals, we also know that certain very local environments have an affect. People recovering in hospital rooms that look out onto attractive scenes of nature get better more quickly than those whose views are only dismal walls. The COVID-19 pandemic reinforced the value of parks and open spaces to counter depression or frustration. The affect of the microenvironment in influencing moods and behaviors is exploited by restaurants that play discordant music to stop people from lingering too long and by stores that play soothing music to make consumers shop longer. Placing mirrors in public spaces so that people can see themselves reduces vandalism. We act differently in public than when we are in more private spaces. Our mood and behavior are both affected by our environment. There is a very positive relationship between the amount of green space in the local environment and people's self-reported indicators of physical and mental health. The relationship is strongest for reports of anxiety. People tend to feel less anxious if there is more green space in their local neighborhood. Local environments do play a role in people's sense of well-being.

There is also the **bystander effect**. It is a well-known finding that individuals often do not help people in distress if, for example, they scream for help in a public place. The effect is more noticeable in dense urban areas than in rural areas. The difference is not because rural dwellers are more caring; rather, it is the result of more stimuli in urban environments, so that people may not hear or see the distress, and because there tend to be more people, so there is a wider diffusion of responsibility.

Different environments not only exaggerate and encourage certain forms of behavior over others; they also have specific forms of behavior encoded. Our dress and demeanor at a job interview, for instance, are different than when we are in a club with friends.

The context of place is important. People who meet for the first time in stressful situations have greater attraction to each other than if they meet first in less stressful situations: same people, different context, and so different outcomes. The title of Sam Summers's book, *Situations Matter*, summarizes a very interesting body of work. Place matters in group creativity. Colleagues who work closer together tend to do better work than when they are situated farther apart. Small, flexible spaces are places where creativity between people can flourish, a point not lost on designers of buildings for creative industries, who work to ensure close contact and random encounters in everyday work life.

There is no simple answer to the question of what role the local environment plays in guiding human feelings and behaviors. There is a complex social construction of the environment as people and groups give meanings to space and place, which change over time.

Consider the city of Tunis. While it was under French occupation, a major avenue was constructed to link the city to the port area (Photo 3.2). It was built to represent the French presence, a central boulevard surrounded by roads and bordered by wide pavement that specifically echoed the Champs-Élysées, so that Tunis could be seen as a smaller version of Paris, both linked by their expression of French culture and power. French and local elites paraded along the street to show their embrace and display of modernity. It was built to make visible the connection between France and Tunisia, to show the rationality and technological superiority of France compared to medieval religiosity. After independence, the street was renamed after a revolutionary hero and was known as Avenue Habib Bourguiba, which came to represent a modern, independent Tunisia and was the place for celebrations of national identity and solidarity. In late 2010 it became the scene of demonstrations against the regime of Zine El Abidine Ben Ali. As part of the Arab Spring, young people flooded into the street, gaining strength from each other's presence, hardening their resolve of the protestors, and highlighting the lack of support for the status quo. The street has now become a place for young people to show their defiance of the regime and display their resistance. This is one street with multiple meanings and very different affects.

3.3 Human Impacts and Environmental Change

The relationship between society and nature can also be considered from the opposite direction by looking at the impacts of humans on the environment. There are two distinct approaches. The first is an accounting of human impacts on the environment. This has a long history. George Perkins Marsh (1801–1882) was one of the first to argue that deforestation could lead to desertification. His book *Man and Nature*, first published in 1864, indicates some of the connections between human actions and environmental impacts. A classic and foundational work by William Leroy Thomas, *Man's Role in Changing the Face of the Earth*, deals with the rising dominance of nature by humans and looks at such issues as agricultural intensification, deforestation, urbanization, and the depletion of resources (Photos 3.3 and 3.4). This work has influenced generations of geographers. An important aim of this ongoing body of work is to examine and measure the major impacts of human action on the environment, including:

1. Destruction of habitats and loss of biodiversity;
2. Pollution of water and soil;
3. Land use changes (e.g., desertification, deforestation, loss of productive farmland urbanization); and
4. Global climate change.

PHOTO 3.2 Avenue Habib Bourguiba in Tunis, Tunisia.

PHOTO 3.4 Land use change 2: the urbanization of the rural landscape, Maryland.

ples: a study of land use changes in Colombia since 1500. The researchers identified seven periods of landscape change, listed in Table 3.1. From 1500 to 2000, there was marked long-term decline in forest cover (see Figure 3.2). There was also a drastic reduction in the nomadic hunting-gathering in the savannas and a large increase in the amount of grazing on cleared land and natural grasslands. The periodization used in Table 3.1 references political and economic conditions such as conquest, colonialism, the development of the state, and the growth of international markets. These changes have an impact on land use. Land use change, as this example clearly shows, is linked to political and economic changes.

Ground Truthing

The impact of human land use on ecosystems is measured by using traditional methods, such as map comparison, as well as more modern means like **remote sensing**. If we

PHOTO 3.3 Land use change 1: creating an intensive agricultural landscape in the Netherlands.

The best geographical research is not simply an accounting and mapping of impacts but also a deeper consideration of the social and political effects and the drivers of these changes. We will consider just one of the very many exam-

TABLE 3.1 Land Use Changes in Colombia, 1500–2000		
ERA OF LANDSCAPE CHANGE	DATE	MAIN LAND USE CHANGES
Pre-Spanish	Before 1500	80 percent nomadic hunting-gathering in savanna
Conquest	1500–1600	Reversion to forest as a result of demographic collapse
Colonial period	1600–1800	Forest clearance
Independence and state formation (1800–1850)	1800–1850	Clearance of land for cattle grazing
International markets	1850–1920	Marked forest clearance for grazing
Early urbanization	1920–1970	Deforestation
Industrialization	1970–2000	Deforestation

Source: After Etter, McAlpine, and Possingham (2008).

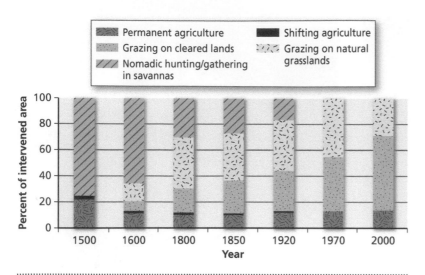

FIGURE 3.2 Land use change in Colombia.

use remotely sensed data, it is important to identify finer-grained processes to see the effects and impacts on households and their vulnerabilities and coping strategies. This is sometimes referred to as **ground truthing**. It is an evocative phrase that encapsulates the very nature of human geography. Geographers seek to identify the truth on the ground by utilizing different scales of analysis, from the telescope to the microscope, and by employing aggregate analysis as well as detailed interview techniques.

There is bias in much of ground truthing. Laura Martin and colleagues looked at the location of 2,573 terrestrial study sites that were the basis of studies published from 2004 to 2009 in the top ten major ecological journals. They found a distinct locational bias toward protected areas and temperate deciduous woodlands in wealthy countries. Even when studies considered urbanized and settled areas, they focused on fragments of "natural" areas. Large parts of the global south and the densely populated areas of the world—the places where most of us live and work—are underrepresented in ecological studies. Not quite truthing of all the ground.

Political Ecology

There is now a large body of work that looks specifically at the relationships between socioeconomic factors and environmental changes. One line of work is termed **political ecology**, since it highlights political as much as ecological issues. Political ecology scholarship examines the processes, policies, and practices that reproduce conditions of poverty, inequality, and oppression around the world and how these link to the environment. There is a focus on the poor and less powerful, who are most adversely affected by exclusionary power relations and practices. It views capitalism as the problem rather than the solution. Theirs is a more radical discourse.

The term "political ecology" was first used in the 1930s but came to higher prominence in the 1980s. Piers Blaikie looked at soil erosion in developing countries. In his work on Nepal, he showed that rather than being the result of overpopulation or mismanagement, soil erosion resulted from marginalization and poverty. Soil erosion was not a function of peasant farmers' mismanagement but a reflection of their precarious position in commodity markets. In his 1983 book *Silent Violence: Food, Famine, and Peasantry in Northern Nigeria*, Michael Watts looked at the political context of household vulnerability to hunger. He built on previous work on the social construction of hunger and famine that we will discuss in Chapter 7 and examined the role of colonialism and agrarian capitalism in subjecting the Hausa-speaking peasantry to food insecurity. Hunger was not the result of drought or resource scarcity but of political relations.

Political ecology looks at the struggle over resources, who controls them, and how the costs and benefits are apportioned. Political ecology situates land use changes in their socioeconomic context and links strategies of corporate profitability to how households are made more or less vulnerable to environmental hazards such as deforestation, social erosion, or irregular food supply. Rather than a neutral backdrop, the environment is the very field in which power is revealed, contested, and enforced.

Other fields of study have also emerged as political ecology has deepened its concerns and areas of research. While many of the early political ecology studies focused on developing societies and rural land use changes, more recent work widens the approach to look at cities. Urban political ecology is now a vibrant part of geographical research. Another example of a subfield within political ecology is the work of Jody Emel and Jennifer Wolch, who have studied how animal rights can be understood within a political ecology framework. A third example is the emergence of feminist political ecology whose work focuses attention on the intersection of capitalism, patriarchy, and the environment. Vandana Shiva examined how rural Indian women experience and perceive ecological destruction and its causes and how they have conceived and initiated processes to arrest the destruction of nature and begin its restoration. Her work makes direct links between ecological crises, colonialism, and the oppression of women.

Climate Change

One of the most significant impacts of human activities on the environment is global climate change. There is now a large body of work that measures and seeks to explain climate change. Climate change is caused by many factors, but one important factor is the heavy use of fossil fuels that increase atmospheric carbon dioxide, which traps sunlight that warms the atmosphere and the Earth's surface. In broad outline, the effect includes an increase in global temperatures by about 1.5°F since 1880, a decrease in the amount of sea and land ice, and rising sea levels on an average of 3.19 millimeters per annum. Much of the basic data and some striking images are available at the NASA website (http://climate.nasa.gov/). Climate change produces a variety of effects in different places. The official federal site that collates data on the United States has a regional breakdown. In the American Southwest, key issues include decreasing water supplies, increasing forest wildfires, and flooding. In the Southeast, major elements include significant sea level rise, higher seawater temperatures, and ocean acidification. In the Northeast, changes include increased warming and poorer air quality, severe flooding as a result of sea level rise, and reductions in fish stocks, especially cod. Some of the complicated suite of impacts on the United States is illustrated in Map 3.1.

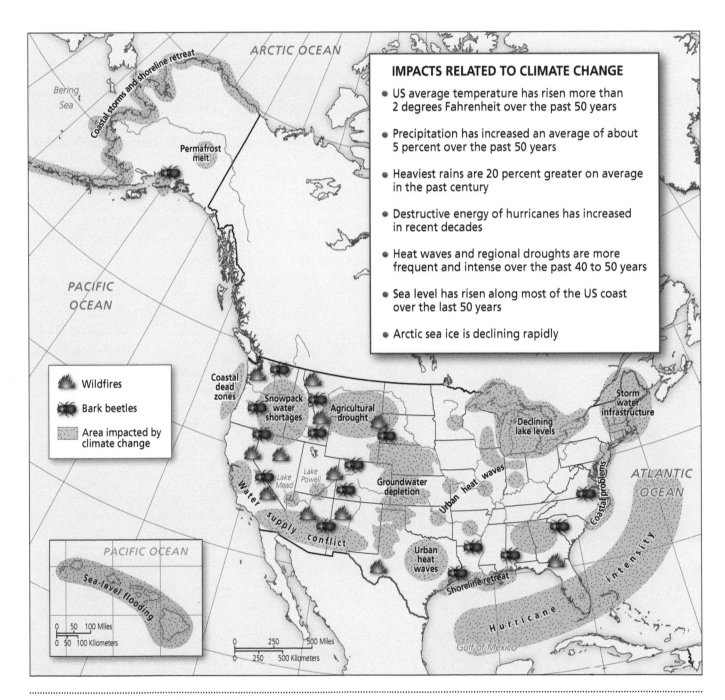

MAP 3.1 Impact of climate changes in the United States.

A volume on climate change, edited by Richard Aspinall, gives a sense of the range of global dynamics. It contains articles examining glacier melt in the Andes and the effects on household vulnerability; increasing drought in Jamaica and the impact on local farmers; the **double exposure** of climate change and economic restructuring in California's Central Valley; and the renewed geopolitical rivalries in the Arctic because warming is leading to new shipping routes, easier access to natural resources, and the ability to project a military presence. The general conclusion to this exciting body of work is that climate change is enmeshed in social structures and economic processes at all levels, from the households to the firm and the state. Climate change generates new household vulnerabilities and stimulates new coping strategies as well as changes in geopolitical relations.

Hazards and Disasters

One of the most visible and dramatic examples of people–nature relations is the experience of hazards and disasters. The tsunami wave that sweeps away the coastal settlements, the storm surges that flood cities, and the flash flood that washes away entire neighborhoods are vivid and searing images that remind us of our environmental vulnerability. We live in a hazardous world of earthquakes, floods, volcanic eruptions, and violent storms. We have a better sense of the underlying dynamics and hence predictability of storms, floods, and eruptions. Yet our knowledge is not absolute. The earthquakes that struck Haiti in 2010 and Italy in 2016 or the tsunamis that overwhelmed towns across the Indian Ocean coastline in 2004 and the Fukushima power plant in Japan in 2011 were unforeseen events. We do know the location of fault zones, but the precise timing of plate shifts cannot be predicted within actionable certainty.

We need to be very careful in using the term "natural disasters," because they tend to be very social events in terms of who is affected most. Vulnerability to hazards magnifies as the population increases, settlements become more dense, and more people live in marginal places. We use "marginal" in the double sense implied by "on the edge of things." First, places like the Caribbean and southern Atlantic seaboard of the United States are zones of hurricane activity (Map 3.2). If we build houses and cities in hurricane zones, it should come as no surprise that someday storms will wash away beachfront property. Hazards become disasters when we build on fault lines and develop along hurricane coasts—especially if we build cheap and quickly without due deference to possible hazards. Marginality also occurs when people are forced to inhabit places such as the squatter settlements on steep slopes or slums on floodable plains and thus are more vulnerable to environmental hazards. The poor are made more vulnerable because they are

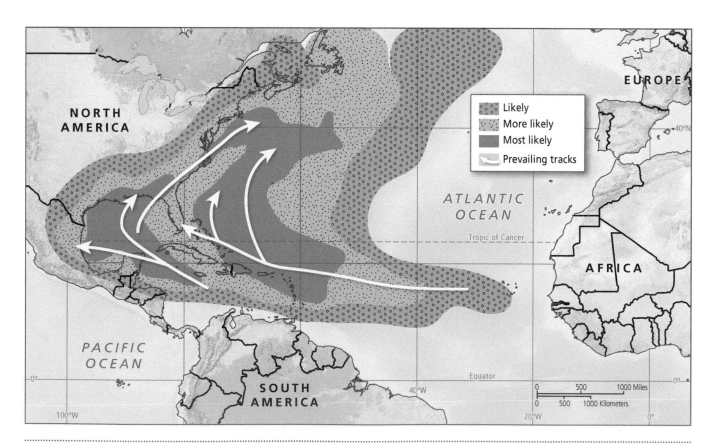

MAP 3.2 | Hurricanes in the Caribbean and eastern United States.

3.1   Inequality and Environmental Injustice

Those most vulnerable to hazards are the poor and people of color, raising issues of environmental injustice. Environmental inequalities are pervasive. At the global scale, we can apply the concept to scenarios such as wealthy countries exporting their waste to lower-income countries, where many of those living near waste sites are exposed to hazards and low wages. Another example of environmental injustice occurs around mining; the extraction of iron in Guinea or gold in Burkina Faso generates highly toxic chemicals that affect land and water resources. It also poses harm to miners. Around the world, the poor and the marginal more often than not live in the areas of the worst environmental quality and the highest risk of hazards.

The existence of environmental inequalities, whether in terms of siting of waste plants or the absence of green spaces, raises the issue of environmental justice. In 1994, the US EPA defined the term as the "fair treatment and meaningful involvement of all people regardless of race, color, national origin, or income with respect to the development, implementation, and enforcement of environmental laws, regulations, and policies."

"Fair treatment" implies that no single group should bear a disproportionate share of negative environmental consequences or a disproportionate lack of positive environment consequences. "Meaningful involvement" implies that people should have the opportunity to participate in environment decisions and effect regulatory bodies. The international community has largely adopted the US EPA's definition of environmental justice.

Environmental justice issues arise from the obvious fact that there is a correlation between the siting of hazardous facilities and low-income communities and/or minority communities. However, legal redress is difficult if procedures were correctly followed or permits were legally obtained. In many cases, environmental injustices are created and maintained through the everyday operation of the market.

An interactive database called the Environmental Justice Atlas, also known as EJAtlas, provides information on major global issues, including nuclear, mineral ores and building extractives, waste management, biomass and land conflicts, fossil fuels and climate justice, water management, infrastructure and built environment, tourism recreation, biodiversity conservation conflicts, and industrial and utilities conflicts. EJAtlas offers an inventory of data as well as information about communities struggling for environmental justice around the world.

REFERENCE

EJ Atlas: Mapping Environmental Justice https://ejatlas.org.

---

often forced to live in the more environmentally hazardous areas. "Natural" disasters are less acts of nature and more the result of human actions. It is important to consider the social context of vulnerability to hazards.

How hazards impact households or become disasters is mediated through the prism of social and economic power. Disasters do not affect everyone equally. They often have a noticeable racial and class bias. Hurricane Katrina in 2005, for example, hit poor Black neighborhoods more than the rich white areas of New Orleans in the United States. Impacts also vary by gender as well as race and class. Based on a sample of 141 countries, Neumayer and Plümper show that disasters killed more women than men and that the higher the women's socioeconomic status, the smaller the gender gap in mortality. The most impacted were low-income women. Natural disasters, on closer inspection, are very social in their outcomes and effects.

The fault lines of power, class, and race are revealed in these three brief examples:

1.   In the Grand Cayman Islands, situated firmly in a zone of hurricane activity (Photo 3.5), there are two responses to the threat of hurricane. The very wealthy, with the warning provided by accurate forecasting, hop on a plane and simply fly away from the island to safety. The poorer inhabitants of the island, unable to flee, have to wait out the storm.

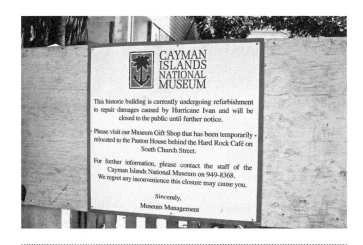

PHOTO 3.5 Hurricane effects in Grand Cayman.

2. Rising temperatures brought about by **global warming** have different effects. Increasingly hot days have less impact on mortality in the United States because it is a relatively rich country with access to air conditioning, either individually or collectively. In rural India, in contrast, an increase from 70 to 90 degrees increases mortality by more than 1 percentage point above the baseline rate. The hazard of global warming is the same, but with very different outcomes, depending on who you are and where you are.

3. The drowning of New Orleans in the wake of Hurricane Katrina in 2005 was anything but a natural disaster. The levees were overwhelmed because they were inadequately built and poorly maintained. The poor environmental management that allowed marshland to erode and the construction of a navigation canal heightened the damaging impact of the storm surge. The disaster was not a natural disaster caused by unforeseen, unpredictable events. It was the result of building a city in an area of environmental hazard, pursuing a destructive environmental management policy that enhanced risk, failing to build and maintain the necessary infrastructure to protect the city, and doing little to help the most vulnerable citizens of that city in the event of flooding. There was a disproportional impact on the poor, on Blacks, and on Vietnamese Americans.

The experience of environmental hazards and the way that hazards become disasters tell us as much about the human geography of class and power as they do about the physical geography of weather and landforms.

3.4 The Anthropocene: Living in a Modified Earth and Socially Constructed Nature

A popular text by Andrew Goudie that in 2013 summarized much of the research on people–nature themes was simply titled *The Human Impact on the Natural Environment*. It provides a good survey of material and case studies. But look again at the title. It assumes that people are impacting a natural environment. The increasing amount of human impact raises the question: Is there such a thing as the natural environment, or have we as humans so transformed the Earth that most of us now live in human-modified environments? Climate change brought about by our carbon base economy is creating permanent drought,

rapid melting of the polar ice caps, an expansion of the tropical region by as much as two degrees of latitude north and south, and increased ocean acidity. These are not short-term changes that will soon revert to a more natural state. We live more and more in a world literally of our own making.

We inhabit a modified Earth and a socially constructed nature. We are now in a new geological age, the **Anthropocene**, a name used to indicate the profound and fundamental restructuring caused by human agency. The term was first coined by Nobel Prize winner Paul Crutzen in 2002. Erle Ellis and colleagues have mapped the world's anthromes, biological areas that are shaped by human impacts.

This profound transformation comes with costs and benefits. The loss of more natural ecosystems, while providing benefit in the short term, may have longer-term costs. A major report by the United Nations, *The Global Biodiversity Outlook*, published in 2020 suggests that some ecosystems are close to the tipping point at which they will become less useful to humans. One indicator of the changes is the decline in biodiversity. More than one in five of all mammal species and close to one in three of all amphibian species are close to extinction. And yet it is only recently that we have started to put a specific value on nature and develop an economic metric for measuring biodiversity.

There is also an emerging movement known as **rewilding** that aims to create landscapes and ecosystems similar to those that existed in Paleolithic times. This ecological strategy consists of the reintroduction of predators such as wolves and cougars into core reserves, with migratory corridors linking the cores. It is often summarized as "cores, corridors, and carnivores." A 15,000-acre reserve in the center of the Netherlands known as Oostvaardersplassen was created in the 1980s and stocked with a variety of animals in an attempt to recreate a European wilderness at the end of the last Ice Age.

In response to the significant transformations humans have had on the environment, there has emerged a new discourse around sustainability.

Sustainability and Sustainable Development

Sustainability and sustainable development are often used interchangeably, but there are nuanced distinctions. Both concepts speak to the impacts and consequences of consuming or degrading resources faster than they can be replenished. The idea of sustainability emerged in the 1960s primarily out of concern about environmental degradation. It is a broad term that describes managing resources without depleting them for future generations.

3.2 The Tragedy of the Commons?

In a paper published in 1968, Garret Hardin described the tragedy of the commons. The setting for the tragedy occurs, according to Hardin, when individuals use a collective resource. He gives the example of a common pasture used by individual herders. The tragedy happens because each individual, seeking to maximize their personal interest, tends to overuse the commons, and the result is deterioration of the common space. The solution lies either in some form of government regulation or in a privatization of the common property.

The image invoked by Hardin is compelling, yet problems remain with the concept. Hardin describes the case of open-access resources, whereas in many traditional societies the commons were, and sometimes still are, either a common pool resource or, more often, common property. The tragedy of the commons occurs in traditional societies less frequently than Hardin suggests because they were regulated through collective property agreements.

In contrast to Hardin, political scientist Elinor Ostrom has offered a counterargument. Her research of common resources in Maine, Indonesia, and Nepal revealed that individuals and communities could, in fact, manage their own collective resources very effectively. However, Ostrom notes that common resources are well managed when those who benefit from them the most are in close proximity to that resource. For her, the tragedy occurs when external groups exert their power (politically, economically, or socially) to gain a personal advantage. Her work also highlights that government intervention is often ineffective unless it is supported by individuals and communities. Ostrom's work has led to the development of a set of design principles that have supported effective mobilization for local management of common pool resources in a variety of areas.

The environment can be considered a common-pool resource that is susceptible to overuse and deterioration. The depletion of deep-sea ocean fisheries, for example, is a case where users, acting rationally on the individual level to harvest as many fish as possible, collectively undermine the resource base. Global climate change, to take another example, results from the action of carbon polluters who do not pay the full cost of their actions. If we see the environment as a common-pool resource, a very valuable and fragile one, then we can imagine a range of policy responses along the line suggested by Ostrom.

REFERENCES

Frishmann, B., A. Marciano, and G. Ramello. 2019. "Retrospectives: Tragedy of the Commons after 50 Years." *Journal of Economic Perspectives* 33 (4): 211–228.

Hardin, G. 1968. "The Tragedy of the Commons." *Science* 162:1243–1248.

Lant, C. L., J. B. Ruhl, and S. E. Kraft. 2008. "The Tragedy of Ecosystem Services." *BioScience* 10:969–974.

Ostrom, E. 1990. *Governing the Commons: The Evolution of Institutions for Collective Action.* Cambridge: Cambridge University Press.

Ostrom, E. 2009. "A Polycentric Approach for Coping with Climate Change." *Social Science Research Network.* http://papers.ssrn.com/sol3/papers.cfm?abstract_id=1934353.

Sustainable development, while closely related, is the process for improving long-term economic well-being and quality of life for people now and in the future. Sustainable development emerged in the 1980s out of discussions about the persistence of poverty, poor health, and resource scarcity. In 1987 the United Nations World Commission on Environment and Development issued a report called *Our Common Future.* The report defined sustainable development as "development that meets the needs of the present without compromising the ability of future generations to meet their own needs." This definition is a fairly abstract concept, but its core is about finding a way to support human populations and their economies without ultimately threatening the health of humans, animals, and plants. It also takes a long-range perspective, noting that current decisions should not impair the prospects for maintaining or improving future living standards.

As concepts, sustainability and sustainable development consist of three pillars: the environment, the economy, and equity. These are often referred to as the "three E's." The argument is that sustainability can only be achieved when all three pillars are balanced and it rejects the notion that achievement in one pillar can be accomplished by sacrificing either of the other two.

Sustainability and sustainable development have become global buzzwords. Political leaders speak of a green economy or community resilience; corporations pledge to become carbon neutral or to enact corporate social responsibility; academics research ecological footprints, carrying capacity, and systems thinking; and the international development community advances programs to reduce poverty and end hunger.

Sustainability and sustainable development are integrative concepts. Hayden Washington called sustainability the destination, while sustainable development is the journey to get there.

3.3 The United Nations Sustainable Development Goals

In 2015, after several years of negotiations and debate, 193 countries agreed to adopt the United Nations' (UN) Sustainable Development Goals (SDGs), officially known as *Transforming Our World: The 2030 Agenda for Sustainable Development*. The seventeen SDGs set out to tackle a diverse range of issues, including ending hunger, improving health, combating climate change, and protecting oceans and forests. They recognize that ending poverty must go hand in hand with strategies that build economic growth and address a range of social needs, including education, health, social protection, and job opportunities, while tackling climate change and environmental protection.

Within each SDG is a set of targets. Combined, there are 169 targets to achieve by 2030. These targets will direct efforts to implement the three E's of sustainability—economy, ecology, and equity—and are the guiding agenda for the protection of both the local and the global commons. The SDGs set an agenda for investment in advancing sustainability; as a result, it is estimated that there has been a $200 billion increase in development investment efforts since the SDGs were launched.

The SDGs can be viewed as an attempt to address the profound changes caused by human action. But will they be enough, and will we transform in time?

REFERENCES

Brinkman, R. 2016. *Introduction to Sustainability.* Oxford: Wiley/Blackwell.

Macekura, S. J. 2020. *The Mismeasure of Progress: Economic Growth and Its Critics.* Chicago: University of Chicago.

Sachs, J. 2015. *The Age of Sustainable Development.* New York: Columbia University Press.

United Nations General Assembly. 2015. *Transforming Our World: The 2030 Agenda for Sustainable Development.* https://sdgs.un.org/gsdr/gsdr2023

PHOTO 3.6 United Nations Sustainable Development Goals.

Cited References

Aspinall, R., ed. 2010. "Climate Change." Special issue, *Annals of Association of American Geographers* 100:715–1045.

Blaikie, P. M. 1985. *The Political Economy of Soil Erosion in Developing Countries.* London: Pearson.

Campbell, J. 1983. *Historical Atlas of World Mythology.* San Francisco: Harper & Row.

Castree, N., and B. Braun, eds. 2001. *Social Nature: Theory, Practice, and Politics.* Malden, MA: Blackwell.

Conzen, M., ed. 2010. *The Making of the American Landscape.* New York: Routledge.

Crutzen, P. 2002. "Geology of Mankind." *Nature* 415:23.

Ellis, E. E., K. K. Goldewijk, S. Siebert, D. Lightman, and N. Ramankutty. 2010. "Anthropogenic Transformation of the Biomes, 1700 to 2000." *Global Ecology and Biogeography* 19:589–606.

Emel, J., and J. Wolch. 1998. *Animal Geographies: Places, Politics and Identity in the Nature-Culture Borderlands.* New York: Verso.

Etter, A., C. McAlpine, and H. Possingham. 2008. "Historical Patterns and Drivers of Landscape Change in Colombia since 1500: A Regionalized Spatial Approach." *Annals of Association of American Geographers* 98:2–23.

Goudie, A. S. 2013. *The Human Impact on the Natural Environment: Past, Present and Future.* 7th ed. Oxford: Blackwell.

Hoskins, W. G. 1955. *The Making of the English Landscape.* London: Hodder & Stoughton.

Marsh, G. P. 1864. *Man and Nature.* New York: Charles Scribner.

Martin, L. J., B. Blossy, and E. Ellis. 2012. "Mapping Where Ecologists Work: Biases in the Global Distribution of Terrestrial Ecological Observations." *Frontiers in Ecology and the Environment* 10:195–201.

McKibben, B. 2010. *Eaarth: Making a Life on a Tough New Planet.* New York: Time Books.

Mitchell, D. 1996. *The Lie of the Land: Migrant Workers and the California Landscape.* Minneapolis: University of Minnesota Press.

Neumayer, E., and T. Plümper. 2008. "The Gendered Nature of Natural Disasters: The Impact of Catastrophic Events on the Gender Gap in Life Expectancy, 1981–2002." *Annals of Association of American Geographers* 97:551–566.

Shiva, V. 2016. *Staying Alive: Women, Ecology, and Development.* Berkeley, CA: North Atlantic Books.

Summers, S. 2011. *Situations Matter: Understanding How Context Transforms Your World.* New York: Riverhead.

Thomas, W. L., ed. 1955. *Man's Role in Changing the Face of the Earth.* Chicago: University of Chicago Press.

Tyrrell, J. B. ed. 1916. *David Thompson's Narrative of His Explorations in Western America.* Toronto: Champlain Society.

Washington, H. 2015. "Is 'Sustainability' the Same as 'Sustainable Development'?" In *Sustainability: Key Issues*, 359–376, edited by H. Kopnina and E. Shoreman-Ouimet. New York: Routledge.

Watts, M. J. 1983. *Silent Violence: Food, Famine and Peasantry in Northern Nigeria.* Berkeley: University of California Press.

Select Guide to Further Reading

Arevalo, P., E. L. Bullock, C. E. Woodstock, and P. Oloffson. 2020. "A Suite of Tools for Continuous Land Change Monitoring in Google Earth." *Frontiers in Climate.* https://www.frontiersin.org/articles/10.3389/fclim.2020.576740/full.

Benton, L. M., and J. R. Short. 1999. *Environmental Discourse and Practice.* Oxford: Blackwell.

Benton, L. M., and J. R. Short. 2000. *Environmental Discourse and Practice: A Reader.* Oxford: Blackwell.

Cosgrove, D. 1984. *Social Formation and Symbolic Landscape.* London: Croom Helm.

Cosgrove, D. 2008. *Geography and Vision: Seeing, Imagining and Representing the World.* London: Tauris.

Dorn, F. M., and C. Huber. 2020. "Global Production Networks and Natural Resource Extraction: Adding a Political Ecology Perspective." *Geographica Helvetica* 75:183–193.

Foley, S. F., et al. 2013. "The Palaeoanthropocene—the Beginnings of Anthropogenic Environmental Change." *Anthropocene.* http://dx.doi.org/10.1016/j.ancene.2013.11.002.

Hartman, C., and G. D. Squires, eds. 2006. *There Is No Such Thing as a Natural Disaster: Race, Class, and Hurricane Katrina.* New York: Routledge.

Head, L. 2017. *Cultural Landscapes and Environmental Change.* London: Routledge.

Hecht, S. B., and A. Cockburn. 2011. *The Fate of the Forest: Developers, Destroyers, and Defenders of the Amazon.* Chicago: University of Chicago Press.

Howard, P., I. H. Thompson, and E. Waterton, eds. 2012. *The Routledge Companion to Landscape Studies.* New York: Routledge.

Lewis, S. L., and M. A. Maslin. 2015. "Defining the Anthropocene." *Nature* 519:171–180.

Mitchell, D. 2012. *They Saved the Crops: Labor, Landscape, and the Struggle over Industrial Farming in Bracero-Era California.* Athens, GA: University of Georgia Press.

Montz, B. E., and G. A. Tobin. 2011. "Natural Hazards: An Evolving Tradition in Applied Geography." *Applied Geography* 3:1–4.

Mulligan, M. 2018. *Introduction to Sustainability.* New York: Earthscan/Routledge.

Olwig, K., and D. Mitchell, eds. 2008. *Justice, Power and the Political Landscape.* New York: Routledge.

Peet, R., P. Robbins, and R. Watts, eds. 2010. *Global Political Ecology.* New York: Routledge.

Portney, K. *Sustainability.* Cambridge, MA: MIT Press, 2015.

Pulido, L. 2017. "Geographies of Race and Ethnicity II: Environmental Racism, Racial Capitalism and State-Sanctioned Violence." *Progress in Human Geography* 41(4): 524–533.

Robbins, P., J. Hinz, and V. Gaia. 2014. *Adventures in the Anthropocene: A Journey to the Heart of the Planet We Made.* Minnesota: Milkweed Editions.

White, L., Jr. 1967. "The Historical Roots of Our Ecological Crisis." *Science* 155:1203–1207.

Wisner, B., P. Blaikie, T. Cannon, and I. Davis. 2004. *At Risk: Natural Hazards, People's Vulnerability and Disasters.* London: Routledge.

4

People and Resources

LEARNING OBJECTIVES

4.1 Explain the concept of commodification of resources by discussing the example of oil.

4.2 Describe the history of global coal consumption and production, along with its contemporary uses and their consequences.

4.3 Discuss the sustainability of fossil fuels.

4.4 Identify the five laws of resource use and relate them to specific resource examples.

4.5 Describe the twentieth-century history of oil production and consumption, followed by present-day changes affecting the industry.

4.6 Explain at least two different environmental impacts and growing costs associated with the continued use of oil and gas.

Economic development is dependent on resource use. How nature is turned into a resource depends on many factors, including changing technologies and modes of activity. Coal and oil provide examples of the changing relationship between population and resources.

In this chapter, we will consider some of the general relationships between people and resources, using the nonrenewable energy resources of coal and oil as primary examples. Energy resources impact our everyday lives in many ways. We need energy to turn on a light or cook our meals, to power the tractors that plough the soil and the trucks that deliver the food, and to commute to jobs. Energy resources are complex and interconnected; energy is a crucial factor in climate change, rising prices for commodities, political instability in the Middle East, and environmental degradation in many parts of the world. Energy is used for two broad purposes: transportation (including cars, trucks, planes, and trains) and electricity generation, which powers industry, homes, and commercial buildings. Today, oil is the predominant energy source for transportation, and coal and natural gas provide most of the electricity. We need to better understand these relationships.

Let us begin with some definitions. In the *Oxford English Dictionary*, among the many definitions of "resource," the primary one is "a means of supplying a deficiency or need." This sense emerged first in the seventeenth century and is linked with the development of words and ideas associated with markets, money, buying, and selling. From its origin, then, the term "resource" is embedded in social considerations of money and capital. The definition also hints at the relational nature of resources; things are resources because they are linked to scarcity and desire.

Resources are socially constructed, which does not necessarily mean that humans make resources; oil and coal, for example, are the result of eons of geologic processes. "Socially constructed" in this case means that a resource is a function of social, technical, and economic considerations. For centuries, oil was just a messy black substance that oozed from the surface of the Earth. It became a valuable resource only with the creation of a **carbon economy**. The history and geography of resources are the unfolding tale of how "nature" becomes a commodity, with all the social and political consequences involved in this valuation and subsequent revaluation.

The translation from inert material to resource occurs through technical change and commodification. **Commodification** means that the material is valued and traded, it has a price equivalent, and it enters the arena of things bought and sold. For things to become commodities, technical change is often required. The movement of oil from sticky liquid to valuable commodity requires the design and construction of a vast technical apparatus to transform the raw material of oil into fuel and to produce its valuable byproducts, such as plastics. Technical change and commodification are not external to each other but are inexorably interconnected. Technical developments occur through the increased valuation of the resource. Let us further consider the case of two primary resources, coal and oil.

4.1 The Case of Coal

When the Romans first invaded what they called Britannia around 2,000 years ago, they found coal being used as raw material for jewelry. When carved and polished, it adorned the bodies of affluent Britano-Romans. Some local tribes in south Wales also used the material to cremate their dead leaders, but it was only in the twelfth century that people in Britain began using it as fuel. Before then, and for five centuries more, wood was the principal fuel for cooking and heating. As the population expanded, the forest reserves became depleted and, as so often happens in the history of resources, an alternative resource had to be exploited. Coal that was close to the surface was readily available, and soon there was a vigorous trade in coal mined in northeast England, around Newcastle, then shipped to the large urban centers, such as London. The burning of coal became such a common domestic practice by 1600 that there was rising concern with air quality. In 1661 John Evelyn described London as similar to an active volcano, as smoke and noxious fumes belched from the city's chimneys to create unhealthy air. The smoke blocked out the Sun, and sooty particles covered buildings. Despite its side effects, coal was preferred because wood was two to five times more expensive. And here we come to a second characteristic of resource use: the difference between private costs and benefits and social costs and benefits. For individual London households, it was cheaper to use coal than wood. As more households adopted this strategy to cope with private costs, they imposed **social costs** on their neighbors by degrading London's air quality.

As the demand for coal increased, it became profitable to make larger outlays to dig deeper into the Earth. There was a problem, however. The deeper the mines were sunk, the more vulnerable they were to flooding. Scientists and inventors tackled the problem with attempts to build a pumping machine using piston action fueled by the steam from burning coal. It worked, but it used tremendous quantities of coal; mines could easily access this fuel, but it was not suitable for wider usage. Then James Watt devised an improvement to the pumping machine to make it more energy efficient. He added a condenser that enabled the piston's cylinder to remain hot and thus able to pump more efficiently. His machine did not need such large quantities of coal; the innovation freed the machine from the coal mine. In 1776, Watt created two coal-driven, steam-powered engines: one to pump water from a coal mine, the other to power an iron foundry. It was a momentous year: the Industrial Revolution was inaugurated, Adam Smith published *The Wealth of Nations*, and the fledgling American Republic declared independence.

Coal and the Industrial Revolution

Coal powered the **Industrial Revolution** and would prove an important ingredient in the future manufacture of iron and steel. In 1830 Britain produced four-fifths of the world's coal, and by the middle of the nineteenth century, it was the world's leading industrial power and later the world's unrivaled superpower. Coal was not the only factor in the stunning national success, but it was an important ingredient. Coal mining was also at the center of a nexus of capital–labor relations. Coal-mining areas, with their large pools of workers living side by side, produced not only coal but also an organized force of working-class politics. Political radicalism as well as coal came from the coalfields of south Wales, northern England, and central Scotland.

By 1900, the United States had become the world's largest coal producer, and, as in Britain, coal was a vital resource in the country's rise to industrial prominence. Coal powered the growing industrial nation and was an essential part of the subsequent creation of a mighty iron and steel industry. Emissions from coal-burning power plants also figured in the nation's history of **environmental legislation**. There is no simple path between growing environmental health hazards and the resultant environmental legislation.

Smoke-belching factories also implied economic growth to such an extent that political scientist Matthew Crenson writes of the "un-politics of air pollution" in many industrial cities in the United States. Factory chimneys were seen as a sign of progress and full employment, so in many industrial cities air quality did not become a political issue; attempts to restrict emissions were often resisted and disputed, and the problems of polluted air were downplayed. In 1970, in part as a result of Rachel Carson's 1962 book, *Silent Spring*, and a series of widely publicized environmental disasters, the landmark Clean Air Act established new standards of air quality; this was just the opening salvo in the ongoing struggle to control and restrict air pollution. Coal producers and owners of coal-fired power stations continue to lobby to restrict further legislation. And here we come to another element of resource use: the producers always trumpet the benefits and try to ignore or downplay the costs.

Coal Production

The United States remains a major coal producer (Map 4.1). In 2023 the United States produced over 594 million tons of coal, almost 40 percent from the state of Wyoming and

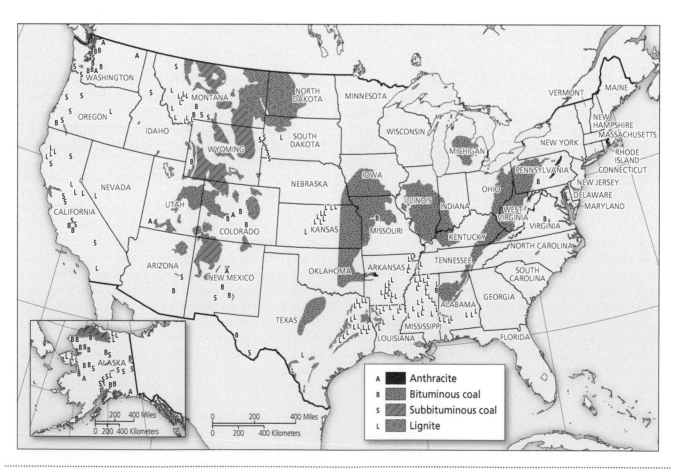

MAP 4.1 Coal regions in the United States.

12 percent from West Virginia. Since the 1990s, there has been a rapid growth in shallow mining (close to the surface), as opposed to deep mining, and more of this mining has occurred in the states west of the Mississippi, a rapid expansion caused by more efficient and powerful mechanical digging equipment and the need for lower-sulfur coal. In 1970, almost 70 percent of US coal came from Appalachia; now that figure is just 14 percent and falling. Coalfields in the Illinois basin in the central United States and the Powder River basin in the western states are now the principal sources of US coal production. Coal is dug out close to the surface by giant machines and then transported by train. Day and night, the coal trains move the fuel from the coal mines to the power plants. Almost 90 percent of coal produced in the United States is destined for power plants that generate electricity, which is used to power industry and residences. The country's coal reserves are difficult to estimate with accuracy. Although there may

4.1 The Hubbert Curve

The **Hubbert curve** is named after a geophysicist who first employed it in 1956 to predict the decline of US oil production. Essentially a symmetrical bell-shaped curve, it suggests that nonrenewable resource use rises quickly from zero to a peak and then falls off; at the top are peak production levels (Figure 4.1). Hubbert used the curve to predict that oil production in the continental United States would peak in 1965. The curve is perfectly bell shaped; in reality, resource use is rarely so symmetrical. And there is always the possibility that new technologies will increase the efficiency of extraction. However, it raises the issue of immediate and rapid falloff from peak production. The curve has been used to model historical resource exploitation, such as coal production in the United Kingdom. It is also part of the considerable debate about the existence and timing of a global oil peak. Figure 4.2 shows the gradual tailing off of most sources of global oil production.

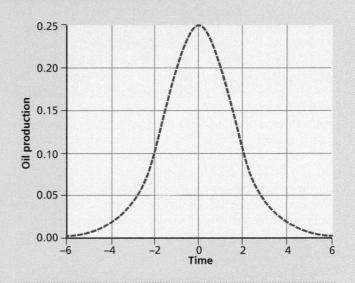

FIGURE 4.1 The Hubbert curve: ideal.

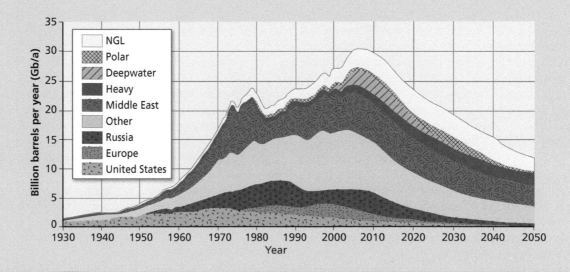

FIGURE 4.2 The Hubbert curve: empirical (NGL = natural gas condensates).

be 480 billion tons, largely in western states, probably only 10 percent of that is recoverable. The rest may be too expensive to mine in the current economic circumstances. Resources only become recoverable resources when their exploitation is economically feasible.

In 2022, the top five coal producers in the world were, in order, China, India, Indonesia, the United States, and Australia. Coal production in China, as in Britain and the United States before it, literally has fueled the country's very rapid economic growth. In 2022, China produced 3.7 billion tons of coal, almost eight times more than the United States. But even this is not enough to meet its domestic demand: China is also the largest importer of coal, after Japan. China's substantial and accelerating industrial growth draws in imports from all over the world. Coal from the United States' Powder River basin is now being sold to China, and the large US coal companies are seeking to build and expand more coal terminals in Washington and Oregon. Australia has experienced a commodity boom as it exports vast quantities of coal and iron ore to China.

However, since 2012, coal production has leveled off, and in some cases declined. Coal production fell from 640 million tons in 2019 to 487 million by 2020. There are two main reasons for this: The first is that cheaper power is now available from underground sources of gas and oil. As power costs decline, more expensive coalfields, such as those in Appalachia, become less profitable. In 2019, for example, the US energy consumption from renewable resources exceeded coal consumption for the first time in 130 years; this is because of the declining costs of renewables, which makes them competitive against coal. The second reason is because China's economic growth has slowed after three decades. The headlong growth was literally fueled with coal from domestic and overseas coalfields. The age of coal is not coming to an end, but it is being eclipsed as other sources of fuel provide cheaper and sometimes cleaner sources of energy.

Problems with Coal

Coal is still responsible for 42 percent of the world's electricity. Every time we recharge our cell phone, plug a device into a socket, or turn on a light, a coal-fired generator probably produces the power. Coal production continues to increase. Although the basic supply of coal is not yet an issue, the wider implications of its burning and exploitation remain problematic. There are proponents of clean coal, but that remains more a fantasy than an empirical fact.

Coal has two major problems. First, burning coal produces noxious fumes that pollute the skies and cause global climate change (Photo 4.1). Two particular gases, sulfur dioxide (SO_2) and carbon dioxide (CO_2), cause these problems. SO_2 is a major byproduct of burning high-sulfur coal, the kind of coal that is found in the eastern United States and throughout much of China. It is the main ingredient in acid rain, as the SO_2 combines with elements in the atmosphere to become sulfuric acid. The term "acid rain" was first used in 1872 by an English inspector of chemical works, Robert

---

4.2 The Unsustainability of Fossil Fuels

Energy resources can be categorized into two broad groups: nonrenewable and renewable. In the early twenty-first century, about 80 percent of the world's energy needs are met using nonrenewable energy resources. Nonrenewable energy resources are energy resources that, once converted for use by people, are gone forever. These include coal, oil, natural gas, and nuclear energy. In contrast, renewable energy resources derive energy from ongoing natural processes that are constantly being replenished. These include biomass, hydropower, wind, geothermal, and solar energy.

Oil, coal, and natural gas are called fossil fuels because all three are derived from plant materials and living organisms that are converted into hydrocarbons over millions of years. The type of fossil fuel formed depends on the nature of the organic material buried, the specific environmental conditions (types of sediment), and the time it takes for conversion. Natural gas is trapped methane, crude oil is the residual sludge of marine phytoplankton, and coal is highly compressed terrestrial organic matter—mostly plants—that decomposed less than the others. Today, the three major fossil fuels—petroleum, natural gas, and coal—account for most energy production.

The story of fossil fuels is complex. On the one hand, fossil fuels have served us well, allowing economic development of the Global North. On the other hand, this prosperity and wealth has come at a steep cost everywhere: it has exacerbated the concentration of wealth, accelerated resource exploitation, and generated air and water pollution that have led to respiratory diseases. Further, fossil fuels are driving climate change, creating disasters from leaks and spills, encouraging mountaintop removal, and subjecting economies to market volatility. The list goes on, making fossil fuels unsustainable energy resources.

REFERENCE

Reisser, W., and C. Reisser. 2019. *Energy Resources: From Science to Society.* New York: Oxford University Press.

PHOTO 4.1 Coal-fired power plant in Denmark.

Smith, who noticed how the poisonous fumes from British factories turned into acidic rain that eroded buildings and polluted the water. The term and the underlying process were little used or discussed apart from a paper published in 1955 by Eville Gorham that identified its significance in the Lake District of England. Then, as SO_2 levels mounted, a flood of papers began to note its significance. A Swedish researcher, Svante Odén, noted its effects in Scandinavia in destroying forests and making lakes and rivers more acidic. Much of the acid smoke that fell as acid rain in Scandinavia came from coal burned in Britain and Germany. Airborne pollutants do not respect international boundaries. And here we come to another element in resource use: while the benefits may accrue to an individual company, region, or country, the costs are often borne by other companies, regions, and countries. There are social costs to private enterprise. While the benefits may be privatized, the costs are often socialized.

SO_2 became a major problem in the United States, especially because of the burning of high-sulfur coals. In the upland areas of the Adirondacks, it was discovered that by the 1980s, 80 percent of lakes had suffered from acidification that killed fish and microorganisms. Because of the region's geology, mainly igneous and metamorphic rocks with limited soil cover, there were few carbonate rocks or soils to buffer the effects. The small lakes at higher elevations were particularly vulnerable because they had less buffering soil; the higher elevation also created more rain and snowmelt, which led to more acid rain. The very small lakes that received great runoff were being doused in acid baths that killed the fish stock. Things got so bad that lime was dumped into the lakes to counter the rising acidity. The Adirondack lakes are far from the immediate polluting sources. The principal source was the power stations along the Ohio River. In the wake of the Clean Air Act, they were forced to build taller chimneys to reduce the local impact of emissions. Increasing the height of the chimneys did take the SO_2 from the immediate area, but pushed it higher into the atmosphere, where it drifted eastward and eventually fell as acid rain on the mountains of upstate New York. Resource use often has unforeseen consequences.

While acidity levels in some lakes remain high, SO_2 levels across the United States have decreased since the beginning of the twenty-first century. An acid rain program was adopted in 1990 that required power plants to halve their SO_2 emissions by 2010. The process was aided by the greater use of low-sulfur coal from the western states. The net effect: SO_2 was much reduced. In China, by contrast, because it has lower pollution standards and uses sulfur-rich coal, the problems of acid rain continue to be a major side effect of the country's economic growth, poisoning the air, polluting water, and contaminating soils. China is now on the path previously followed first by the United Kingdom and then the United States.

The burning of coal also produces CO_2. Coal burning around the world is responsible for one-third of all CO_2 emissions, and coal-fired power stations are the single biggest source of CO_2 in the Earth's atmosphere. CO_2 is a "greenhouse" gas, so named because it produces a greenhouse effect of trapping heat in the atmosphere.

The link between CO_2 and global climate change is now well established by science. The steady rise of CO_2 in the atmosphere is recorded in the **Keeling curve**, which plots the concentration of the gas in the atmosphere from 1958 to the present day (Figure 4.3). The curve is named after Charles David Keeling, who initiated the measurements from the Mauna Loa Observatory in Hawaii. The data from this site, free from many local sources of pollution, are a good indicator of global patterns. The yearly variation occurs because more coal is burned in the Northern Hemisphere during winter. The curve is on a seemingly inexorable upward swing. When James Watt invented the steam engine, CO_2 levels were roughly 280 parts per million (ppm). At the start of Keeling's measurements in 1959, the figure had risen to 315 ppm. In 2023 it reached 425 ppm. (The very latest figure is available at http://co2now.org.) The consensus among most climate scientists is that climate change becomes significant at 350 ppm.

A second problem is that coal mining often involves environmental destruction as the land is poked and prodded, vegetation is destroyed, habitats are threatened, and waste material pollutes the soil and the air. One of the most environmentally damaging forms of coal mining in operation in the Appalachia region of the United States is **mountaintop removal**; more than 500 mountains have been destroyed and 1.2 million acres of mountaintop removal mining have occurred. The process involves removing an entire peak to gain access to coal seams and dumping the material into the neighboring valleys. The environmental damage includes

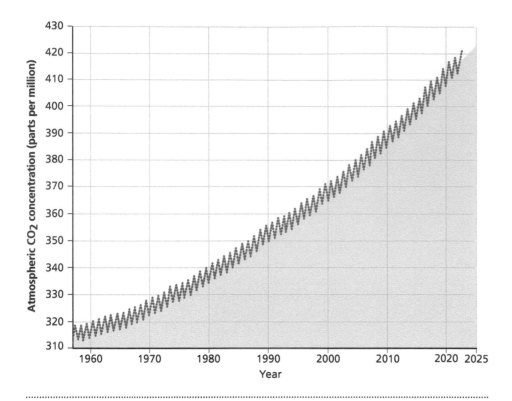

FIGURE 4.3 The Keeling curve shows the upward trajectory of CO_2 emissions in the atmosphere.

disruption of local **ecosystems**, loss of **biodiversity**, and polluted air, soil, and water. A study by geographers Bard Woods and Jason Gordon showed that while the negative environmental consequences were obvious, the positive economic benefits to local communities, often touted by supporters of mountaintop removal, were less apparent.

In some cases, reclamation is possible. Across the old industrial landscapes of Europe and North America, land reclamation projects are turning mining wasteland into recreational opportunities. In Leipzig, Germany, for example, a former coal mining area has been turned into a theme park; it opened in 2005, and by 2017 a recreational area of seventeen interlinked lakes had been created. The old mining area has been reimagined and remade as a place of recreation and pleasure rather than pollution and work.

4.2 Laws of Resource Use

We can summarize some of the themes we have considered so far as five laws of resource use:

1. Resources become resources because previous resources become exhausted, unavailable, depleted, or too expensive. Coal was used in Britain because wood was becoming more expensive. Nuclear power was developed in countries such as Japan and France because they had few coal reserves or oil supplies. Wind and waves are being exploited as energy sources because of the rising cost of nonrenewable sources and as their costs decline.

2. Resource use creates new geographies, whether it be in the eighteenth-century creation of plantation economies in the West Indies to feed the sugar habit of Europeans, the nineteenth-century creation of coalfield landscapes across Europe and North America, or the late-twentieth- and early-twenty-first-century creation of vast mining operations in Australia to feed China's demand for energy and raw materials.

3. All resource usage tends to have wider social and political consequences. Nineteenth-century coal production, for example, was also the setting for organized labor movements and political radicalism. The national competition for resources is an important element in the interactions between states. The United States' insatiable demand for oil makes it curry favor with oil-rich countries such as Saudi Arabia, even though it has vastly different cultural and political values.

4. All resource use has implications for public spaces and the wider community, as well as for private enterprises. The use of fossil fuels, for example, has many positive consequences, but also pollutes the air we breathe and is a cause of climate change that affects us all. The greater use of fracking has the very real possibility of polluting public water supplies.

5. Initiatives to regulate and diminish the negative public consequences of resource use and exploitation always come up against powerful interests that initially deny the problem and then minimize it. In 1980, the head of the US National Coal Association stated that the negative effects of acid rain were only in the imaginations of those who wanted to promote regulation. When major oil companies fund those who deny climate change, they are taking part in a long and constant story of resistance to regulation.

4.3 Energy Inequalities

Access to energy resources varies around the world and is highly connected to income. A 2020 World Bank report found that 4 billion people around the world still lack access to clean, efficient, convenient, safe, reliable, and affordable cooking energy. Most of the 4 billion without clean cooking energy are low-income or very-low-income populations. The report also notes that the lack of clean cooking is costing the world more than $2.4 trillion each year, driven by adverse impacts on health, climate, and gender equality. For example, household air pollution, mostly from cooking smoke, is linked to around 4 million premature deaths annually, with half of those deaths being children under the age of five. Women and children are disproportionately affected by household air pollution, because they are most exposed as they often spend a significant part of their day collecting the fuel—firewood or charcoal, for instance—needed to cook a meal.

Families who switch to clean cookstoves significantly reduce time spent collecting wood, money spent on solid fuels, and smoke output, significantly improving the health of women and children. In many rural areas and urban slums, clean cookstoves can have positive impacts. A clean cookstove can save a family around $200, which can pay for a whole year of schooling for one child. It may not seem like a lot, but this is a huge cost saving for the average family.

While clean cooking fuels and technologies are now more available, consumer awareness, accessibility, and affordability remain significant challenges. The provision of clean cooking solutions does not guarantee that rural and urban communities will stop using traditional cooking methods. A study of cookstoves in Kenya by Nzengya and colleagues found that although many of the proposed cookstoves allowed faster and more efficient cooking, they were much less flexible in adapting to women's needs than traditional three-stone fireplaces. Many poor and rural recipients of clean cookstove programs thus continue to use traditional fuels and solutions for sociocultural, economic, and pragmatic reasons. These results suggest that programs to replace traditional but unsustainable fuel use will likely only succeed if they prioritize user preferences and local cooking contexts.

REFERENCES

Nzengya, D., P. Mwari, and C. Njeru. 2021. "Barriers to the Adoption of Improved Cooking Stoves for Rural Resilience and Climate Change Adaption and Mitigation in Kenya." In *African Handbook of Climate Change Adaptation*, edited by N. Oguge, D. Ayal, L. Adeleke, and I. da Silva, 1641–1658. Cham, Switzerland: Springer. https://doi.org/10.1007/978-3-030-45106-6_133.

U.S. International Energy. 2019. "SDG7: Data and Projections." https://www.iea.org/reports/sdg7-data-and-projections/access-to-electricity.

World Bank. 2020. "The State of Access to Modern Energy Cooking Services." https://www.worldbank.org/en/topic/energy/publication/the-state-of-access-to-modern-energy-cooking-services.

4.3 The Limits to Growth?

The relationship between population and resources, like the relationship between population and food, is dominated by the notion of **limits to growth**. In the middle of the twentieth century, biological concepts such as **carrying capacity** were employed to indicate that perhaps we were coming up against the limits of growth. The basic idea was that the world could carry only so much population and sustain so much growth. There was a limit to the carrying capacity of the Earth. For neo-Malthusians such as Garret Hardin and Paul Ehrlich, the limits had been reached by the 1970s; the only response was to severely limit population growth.

In 1972, an influential book entitled *The Limits to Growth* was published. Its central argument, summarized in its title, was that there were limits to continuous economic growth. The supply of nonrenewable resources, such as iron and oil, was not infinite, and on current trends, the limits to supply would soon be reached (Photo 4.2). The book's argument was strengthened the next year when oil prices increased fourfold. However, the 1973 **oil shock** led

PHOTO 4.2 Oil derricks in Los Angeles. This oil field, like many in the United States and around the world, is coming to the end of its life.

PHOTO 4.3 Nuclear power plant construction near Chinon, France. Without oil reserves and with coal reserves long since depleted, France relies heavily on nuclear power as a source of electricity. Fifty nuclear power plants produce 70 percent of France's power.

to much more efficient use of oil resources and a renewed search for alternative sources of supply (Photos 4.3 and 4.4). The limits-to-growth argument also highlights another law of resource use: there is always the tendency to exaggerate the risks and preach a form of resource Armageddon.

Meadows's book was dismissed by critics for its neo-Malthusian pessimism compared to the reality of human ingenuity. Resources did not run out. Critics such as the economist J. L. Simon argued that as we come close to "limits," price changes stimulate new forms of resource use, alternative supplies, and new ways of doing things. Rather than simply walking over the edge of fixed supply, we negotiate an increasing price curve.

However, Meadows's book did stimulate a debate about longer-term prospects and strengthened the notion of sustainability. The arguments have shifted and morphed since the early 1970s. Rather than mechanistic limits to growth, there is an emerging notion of the desirability of more sustainable resource use.

While the idea of limits to growth has its weaknesses and major resource scarcities have yet to materialize, there is a more general point to be made. If there is a heavy reliance on a particular resource, then supply issues can become problematic. We will illuminate this with reference to the case of oil.

4.4 The Case of Oil

The Never-Ending Demand for Oil

Demand for oil is insatiable. In 1970, around 45 million barrels of oil were consumed worldwide every day. By 2022, the figure was closer to 100 million. Even with energy efficiencies, the raw demand keeps growing. Rapid industrialization is based on oil. In 1970, China consumed only half a million barrels of oil a day. By 2020, it was 14.2 million. The United States has the world's biggest thirst for oil; it consumed 17.1 million barrels a day in 2020. Our modern world runs on oil.

Despite the increase in demand, the price of oil remained stable for decades as cheap oil was exploited around the world. From 1881 to 1973, the price of oil, in real terms, remained flat. Oil prices were low and, more important, stable. Economic growth in the first three-quarters of the twentieth century was literally lubricated by the steady supply of cheap, easily available oil. However, in 1973 the **Organization of the Petroleum Exporting Countries (OPEC)** used its power as an oil cartel to increase the price of oil dramatically, causing an oil shock.

The 1973 oil shock had a number of rippling consequences. The era of cheap fuel was over, and energy had

PHOTO 4.4 Windmills off the coast of Denmark, a leader in wind power generation. Wind power from turbines such as these provides almost one-fifth of all electricity production in the country, which now has a technological lead in the provision of this alternative energy source. About half of all wind turbines used around the world are made in Denmark.

to be used more efficiently. Although US gas prices have always been much lower compared to those of other industrialized countries, car companies in Japan, where oil prices were always much higher, already had experience in fuel efficiency. In the United States, a generation of cheap oil meant that car companies had few fuel-efficient models and were slow to realize the seismic shift in oil prices and associated consumer demand. This opened the door to the much more fuel-efficient Japanese cars. The long decline of the Detroit car industry began with the 1973 oil price shock.

As the demand for oil increased, even after the OPEC price hike, oil companies sought out new reserves. Higher prices also meant that oil patches previously considered only marginally feasible now became potentially profitable. The oil companies went long and they went deep in their search for oil, developing new techniques and technologies, such as deepwater drilling and hydraulic fracturing.

The Changing Supply of Oil

The supply of oil on the world market is no longer dominated by the Middle East. In 2021 Saudi Arabia remained the largest single exporter of oil (15 percent of global oil exports), followed by Russia (11 percent), the United States (8.1 percent), Canada (7.3 percent), and Iraq (7.0 percent).

Although demand seems infinite, the supply is fixed. Estimating oil reserves is a difficult business. There is always the possibility of discovering new reserves. And even the knowledge of existing reserves is more an estimate than a fine calibration. Yet the long-term prospect does not look good. We may be at **peak oil**, the high point in the Hubbert curve. The total world supply is estimated at 1.65 trillion barrels of oil. But with oil consumption rising and most of the world now surveyed for oil reserves, it is possible that supply soon may peak. The rate of extraction is now double that of new discoveries. Estimates of oil's peak years vary from 2010–2015, a view shared by US military intelligence, to a more optimistic 2050. It is estimated that the world's oil reserves will run out by 2052.

Even if we take the most optimistic estimates and put the peaking of oil into the distant future, we are still in the era of tough oil. Most of the **easy oil** reserves close to the surface have been tapped. Many of the big fields are in steep decline. Mexico and Indonesia used to be major oil exporters but are now importers. More than one-third of the world's oil comes from large fields producing more than 5 million barrels a day. Discoveries of these fields declined from 130 in the 1960s to only 34 in the 2000s.

Even if the exact timing of global peak oil is difficult to assess accurately, it is clear to see that we are in an era of **tough oil**, not only in terms of the cost and risks of deep drilling into ever more dangerous places, but also in the search for immediate alternatives. The tar sands of Alberta, Canada, are being exploited for their oil at the cost of large amounts of water and heat; elevated pollution rates, because the tar sand oil produces between 5 and 15 percent more CO_2 emissions than other oil; and the need for a 1,700-mile pipeline through much of North America.

We have moved from an era of easy oil, one of large, shallow reserves that were easily exploitable, to an age of tough oil, in which reserves are deeper, farther offshore, and more expensive to exploit. The new fields are smaller and not as long-lasting; they peak fast, then decline sharply, so there is a continuous search for new fields. Companies are persistently at the furthest extent of their experience and technical capabilities as they drill deeper in ever more inaccessible places.

There is considerable debate about peak oil. There are those who focus on the narrow debate of estimating the precise date of the oil peak. This is a discussion that sheds more heat than light, since the oil supply business is one of estimates rather than solid empirical facts. Then there is the debate that widens to look at the consequences of oil supply peaking sometime in the future. We can consider three of them.

The first is the search for alternate sources of supply, from alternative sources of oil and gas to **renewable energy** sources such as biofuels, wind, waves, and solar. A list of **biofuels** is given in Table 4.1. Although Brazil has huge oil supplies, it also uses half of its entire sugar cane crop to produce fuel ethanol. Skylines across the world are sprouting metal windmills and solar panels are being installed on roofs in numerous cities. For example, the city of Chicago may be known as the Windy City, but the city is also well suited to produce solar energy throughout the year. To promote more adoption of solar energy, the city plans to reduce the time and paperwork involved in the solar permit approval process for small-scale solar installations. In many cases, the costs of alternative energies are still high, but as more efficient renewable energy technologies are introduced, we may be on the cusp of a major shift in energy supply. Solar energy costs in the United States are fast approaching par with power from the electric grid—so-called grid parity.

A second consequence is the long-term future of countries that derive most of their revenue from oil. In oil-based regimes such as Venezuela, whose peak production was in 1996, the decline of oil export revenue, as global demand declines and prices fall, has been the source of political conflict and social unrest. Saudi Arabia derives all its wealth from oil exports, yet oil production peaked in 2005. When the major and almost only source of revenue begins to decline, national economic stability is put at risk. The possibility of moving past peak production into a world of possibly declining revenue raises issues of social stability and regime longevity. In places such as Dubai and Abu Dhabi, the frenetic pace of building and development is in

4.4 The Geopolitics of Oil

Oil is a vital resource. The need for energy security, for ensuring long-term and reliable sources of oil, is an increasingly vital national security consideration. Oil demand creates a strong nexus of mutual interest between major oil importers and large oil companies. It also dictates foreign policy. The United States, for example, must take into account relations with major oil-producing nations, sometimes creating alliances with dictatorial and authoritarian regimes and other times being actively involved in regime change and direct military interventions. The crucial test is not that foreign oil suppliers treat their own people well, but that the oil keeps coming at a reasonable price. Securing oil supplies was long a vital part of the operation of US global power.

China, as a leading oil importer, also tailors its foreign policy to its energy needs, but in the process may get dragged into regional disputes such as that between Sudan and South Sudan. The new pipeline that takes oil from Russia to China represents a recent reversal of old enmities. Linking the supply and demand for oil can turn centuries-old enemies into today's trading partners.

The money from Russian oil and gas exports funded Putin's ambitious goal for great power status and the invasion of Ukraine. Germany was placed in a difficult position. It joined the European support for Ukraine while still deeply dependent on Russian oil and gas.

Large and powerful oil importers need to ensure the supply of oil. Treaties, interventions, and wars, in some cases supporting regimes and in others undermining them, all become part of the changing tactics devised to meet the need for energy security. The geopolitics of oil links importers and exporters in complex, ever-changing relationships. The suppliers band together in cartels to regulate the supply and price, while the importers draw on a wide variety of tactics and use their political, economic, and sometimes military power.

REFERENCES

El-Gamal, M. A., and A. M. Jaffe. 2018. "The Coupled Cycles of Geopolitics and Oil Prices." *Economics of Energy and Environmental Policy* 7:1–14.

Yergin, D. 2011. *The Quest: Energy, Security, and the Remaking of the Modern World.* New York: Penguin.

TABLE 4.1 Biofuels

CROP	FUEL SOURCE	MAIN PRODUCING COUNTRIES
Sugar cane	Ethanol (from sugar)	Brazil, India, China, Thailand
Sugar beet	Ethanol (from sugar)	France, United States, Germany, Russia
Cassava	Ethanol (from starch)	Nigeria, Brazil, Thailand, Indonesia
Maize	Ethanol (from starch)	United States, China
Oil palm	Biodiesel	Malaysia, Indonesia, Nigeria, Thailand
Rapeseed	Biodiesel	China, Canada, India, Germany
Soybean	Biodiesel	United States, Brazil, Argentina, China

produced. They use less power and can navigate dense urban streets (Photo 4.5). Much of the United States' suburbanization relies on low and stable oil prices. All those part an attempt to build an alternative economic future before the oil runs out.

The third consequence is the implication for economies and socio-spatial assemblages that are predicated on relatively cheap oil. In Italy more very small cars are being

PHOTO 4.5 This tiny car in Florence, Italy, is ideal for managing the narrow streets of the old city. It also has the advantage of very high gas mileage, a necessity in a country with high oil prices. Small cars, hybrid vehicles, and electric cars tend to become more popular as gas prices increase.

low-density family homes, out-of-town shopping centers, and long-distance commutes arose at a time of dependably cheap gas prices. But as prices rise and become unstable, the suburbs, with their heavy reliance on private automobiles, now look like a landscape predicated on oil prices that may never return to their former stability. The reliance of a built form precariously balanced on one fossil fuel with high and fluctuating costs raises issues of long-term sustainability. The suburbs were built on gasoline from oil priced on the order of $27 a barrel (at 2007 prices). In one month in the summer of 2008, the price reached over $140. Prices will remain deflated during recessions but will then tend to rise. There are few large oil reserves left, and the price will inevitably rise when the global economy ticks upward. Where does that leave low-density suburban sprawl, which is so reliant on large-scale private car usage? The general answer: in a very precarious position. This same question has been asked before, in 1984:

> The growth of the suburbs was based upon a number of conditions which are now disappearing into recent history.... If these changes are long term and large scale this may lead to lack of demand for suburban housing and a rapid fall off in the house prices of suburban districts. Those who live in the suburbs may find it difficult to either sell their houses or recoup their investment.
>
> (Short 1984, 23)

The long-term sustainability of low-density, energy-profligate suburbs with a heavy ecological footprint is now a matter of serious consideration. Suburbs may become a thing of the past.

4.5 Fracking in the United States

For decades the United States was a classic case study of a country with a very heavy reliance on oil but a limited supply. Oil production of the traditional sort in the United States peaked at 10.5 million barrels in 1985. By 2008 it was just over 7.8 million barrels. Consumption increased in the same period from 15.7 million to 19.6 million barrels. Because oil is literally an essential lubricant of economic growth, the United States had to import an increasing proportion of the oil it consumed. The oil predicament also made the country susceptible to geopolitical as well as economic considerations. US foreign policy had to take into account the sensibilities of such oil-rich countries as Saudi Arabia while also positioning itself to secure access to oil. Its desperate need for oil meant sidling up to some dictatorships while undermining others. The one constant in US foreign policy in the Middle East was the need to secure long-term access to oil. This need for a steady supply was an important shaper and determining factor in US foreign policy.

The gap between demand and domestic supply entailed not only importing from overseas but also exploring tougher sources of oil, such as the expensive-to-extract oil in the Arctic and tar sands in Alberta, Canada. Such extraction is costly and environmentally damaging to produce. The oil is extracted either through strip mining, which tears away the surface vegetation, releases heavy metals, and creates air pollution; or by large injections of water and solvents, which leads to water pollution.

The **oil crunch**, the large and growing gap between the national supply and demand, led to a vigorous search for domestic oil reserves. All the easy shallow reserves in places like Pennsylvania and California had long since dried up. The search then turned to places that only decades before were considered too difficult or too expensive to extract oil from.

First there was offshore drilling. The history of oil extraction in the Gulf of Mexico highlights the shift. Drilling first started in 1937, just off the Louisiana coast. The first well was located in 14 feet of water less than a mile from the shoreline. By 1993, approximately 12 percent of all oil produced was from **deepwater wells**, defined as 5,000 feet below the water's surface or deeper. Only 2.5 percent of new wells, forty in total, were deepwater. By 2009, 80 percent of oil came from deepwater wells and over a third of all new wells sunk were deepwater. Between 2006 and April 2010, the number of deepwater rigs grew by 43 percent. As they ventured further offshore and drilled deeper and deeper, oil companies were always at the ever-extending limits of their technical capabilities. Massive and potentially dangerous machinery must be placed on small platforms far out to sea. At the deeper levels, the oil is subject to up to 10,000 pounds of pressure per square inch, making the safe and steady release of the oil a difficult proposition. It is easy to fracture the surrounding geologic formations. Oil is a hazardous, highly flammable liquid. Pumping millions of gallons of it at high pressure though a small pipe more than 3 miles down is a high-risk activity. On the ocean floor, where the wellheads are located, temperatures are low and visibility is hazy. Adjustments must be made using remotely guided machinery. Blowout preventers on the ocean floor are connected to the surface platforms by long pipelines exposed to strong undersea currents. In summary, it is a difficult and dangerous operation to extract deep-sea oil from a small, floating platform so far above the wellhead. The very recent shift to deep offshore drilling, made urgent by the need for oil, means that the technology is continually evolving, which generates new risks and accidents. On April 20, 2010, an explosion aboard

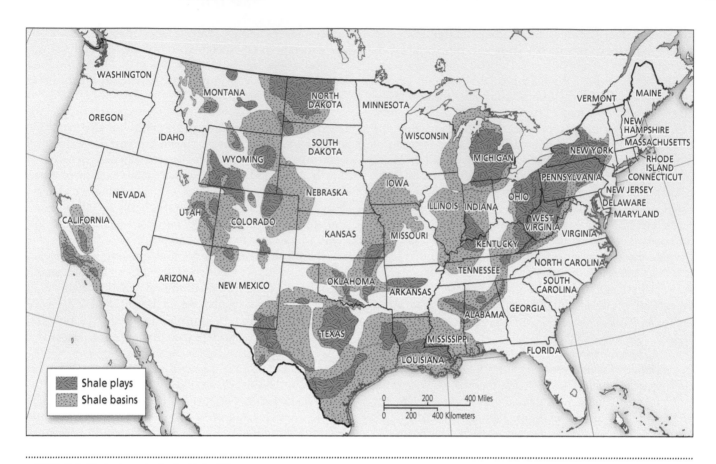

MAP 4.2 Shale deposits in the United States.

the *Deepwater Horizon* drilling rig led to oil gushing uncontrollably from the underground reservoir. For almost two months, 53,000 barrels a day, 5 million barrels in all, gushed from the well into the Gulf of Mexico, poisoning the water, contaminating the beaches, and killing sea creatures and animal life.

Second was the **fracking** revolution that involved the exploitation of oil and natural gas locked up in shale deposits. The distribution of major shale deposits in the United States is shown in Map 4.2. Two of the biggest sites are the Marcellus field in the Northeast and the Bakken field in North Dakota (Photo 4.6). The oil and gas locked up in these shale deposits are released through hydraulic fracturing—high-pressure application of water, sand, and chemicals that essentially crack the rocks and release the oil and gas. This injection technique was developed more than fifty years ago but another innovation made it more powerful: the development of horizontal drilling in the 1990s that allowed drillers to go deep and then sideways to exploit widely dispersed reserves. By 2010 fracking had transformed the US energy sector.

The basic fracking technique is to drill down up to two miles and then drill horizontally as far as a mile. Fracking fluid, a mix of water and chemicals, is then pumped down

PHOTO 4.6 Flex drilling rig in North Dakota.

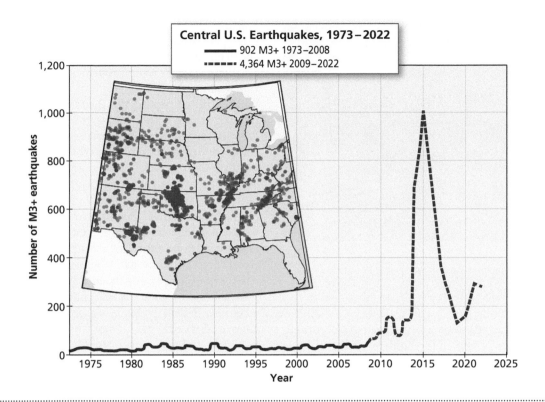

FIGURE 4.4 Central US earthquakes.

at very high pressures to shatter rock formations to release the fossil fuels. The oil and natural gas are then pumped back to the surface along with the millions of gallons of flow-back. In 2000 there were 276,000 gas wells in the United States. By 2010 the number had doubled. Approximately 13,000 new wells are now drilled each year.

The impact of the fracking revolution was dramatic. In 2008, the United States produced 7.8 million barrels of oil a day and ranked third in the world. By 2021, it was producing 18.6 million barrels of oil a day and was the largest oil producer in the world, with 20 percent of the world's total.

There are downsides to the fracking revolution. Fracking pollutes the local water supply. The fluid that is used to fracture the rock contains acids, detergents, salts, lubricants, and disinfectant, as well as other chemical additives, to ensure that the oil and gas are released. A number of studies have identified benzene in local water supplies where fracking occurs. Fracking causes air pollution because some of the natural gas leaks into the atmosphere. There is also the strange case of fracking-induced earthquakes. In Ohio, Oklahoma, and Texas, the number of earthquakes has increased in direct association with the increase in fracking (Figure 4.4). The culprit is the injection deep underground of the millions of gallons of waste-

water byproducts. Before 2009 and large-scale fracking in the state, Oklahoma experienced 2 earthquakes a year. Since then, it experiences 2 a day. The number of earthquakes of magnitude 3 or higher have increased from 20 to 700 a year.

4.6 The Limits to Growth Revisited

The initial limits-to-growth argument was misplaced. The real issue is not that supplies of resources will run out, but that their continued supply may create other costs and negative consequences. What we must ask is not "Is it going to run out?," but "Is continuing its supply worth the rising costs?" The ultimate law of resource use is that we always pay a price for what we get. The cost can be shifted from private to public, from local to national, or from one country to another, but the world is a closed system. The cost and benefits are always exacted, although they may be shifted across space. There is no free lunch.

Despite these challenges, there are signs we are moving away from fossil fuel dependency. As prices for renewable

PHOTO 4.7 Informal housing and renewable energy in Vietnam.

energies continue to fall, they are increasingly becoming an economical energy choice for homeowners and businesses. Many homes and businesses are installing solar energy, and in many of the low-income countries in the Global South, solar is making a significant difference to the quality of life. Photo 4.7, for example, shows informal housing just outside Hanoi in Vietnam. Notice both the solar panels and the small wind machines that generate cheap electricity.

There have been improvements in energy conservation and efficiency as a result of policies and incentives. In the United States, the number of jobs in the solar industry exceeds the number of jobs in coal mining. A growing number of countries, businesses, cities, and states have pledged to reach carbon neutrality by the year 2050 or sooner. Dell, Google, and PepsiCo are Fortune 500 corporations that have pledged carbon-neutral initiatives. Finally, the United Nations Sustainable Development Goals (SDG) have various targets that focus on increasing renewable energy. Targets in SDG 7, Affordable and Clean Energy, for example, aim to provide clean electricity to the almost 1 billion people who currently live without it, many of them in Africa. It is possible that Africa could become the first major region to develop its economy primarily by using energy efficiency, renewables, and natural gas—all of which offer huge untapped potential and economic benefits. In addition, SDG 13, Climate Change, has targets that encourage the transition from fossil fuels to renewables.

Ending the dependency on fossil fuels will not be easy, but it must be done.

Cited References

Crenson, M. A. 1971. *The Un-politics of Air Pollution: A Study of Non–Decision Making.* Baltimore: Johns Hopkins University Press.

Gorham, E. 1955. "On the Acidity and Salinity of Rain." *Geochimica et Cosmochimica Acta* 7:231–239.

Meadows, D. H. 1972. *The Limits to Growth.* New York: Universe.

Odén, S. 1976. "The Acidity Problem: An Outline of Concepts." *Water, Air and Soil Pollution* 6:137–166.

Short, J. R. 1984. *An Introduction to Urban Geography.* Henley, UK: Routledge.

Woods, B. R., and J. S. Gordon. 2011. "Mountaintop Removal and Job Creation: Exploring the Relationship Using Spatial Regression." *Annals of the Association of American Geographers* 101:806–815.

Select Guide to Further Reading

Burns, S. S. 2007. *Bringing down the Mountains: The Impact of Mountaintop Removal on Southern West Virginia.* Morgantown: West Virginia University Press.

Freese, B. 2003. *Coal: A Human History.* Cambridge, UK: Perseus.

Holzman, D. C. 2011. "Mountaintop Removal Mining: Digging into Community Health Concerns." *Environmental Health Perspectives* 119:a476–a483. http://dx.doi.org/10.1289/ehp.119-a476.

Jacobson, M. 2021. *100% Clean, Renewable Energy and Storage for Everything.* New York: Cambridge University Press.

Klare, M. 2010. *Rising Powers, Shrinking Planet.* New York: Holt.

Kolbert, E. 2006. *Field Notes from a Catastrophe: Man, Nature and Climate Change.* New York: Bloomsbury.

Mehany, M. S. H. M., and A. Guggemos. 2015. "A Literature Survey of the Fracking Economic and Environmental

Implications in the United States." *Procedia Engineering* 118:169–176.

Raiml, D. 2018. *The Fracking Debate*. New York: Columbia University Press.

Stokstad, E. 2014. "Will Fracking Put Too Much Fizz in Your Water?" *Science* 344:1468–1471.

Van de Graaf, T., and B. K. Sovacool. 2020. *Global Energy Politics*. Medford, MA: Polity Press.

Willow, A., and S. Wylie. 2014. "Politics, Ecology, and the New Anthropology of Energy: Exploring the Emerging Frontiers of Hydraulic Fracking." *Journal of Political Ecology* 21:222–236.

Zimmerer, K. S., ed. 2011. "Geographies of Energy." Special issue, *Annals of Association of American Geographers* 101:705–980.

The Human Population

Part 2 introduces the general themes of population and resources. Chapter 5 outlines the basic trends of population growth and decline and examines the dynamics of the demographic transition. Chapter 6 considers the geography of population. The relations between population and food supply are detailed in Chapter 7.

OUTLINE

Population Dynamics

LEARNING OBJECTIVES

5.1 Describe the historical factors that lead to major population changes.

5.2 Describe the main trends of world population change since 1800.

5.3 Identify the four phases of the demographic transition and **explain** the social and economic concerns that accompany each stage.

5.4 Explain how cultural and political factors may influence contemporary population geographies.

5.5 Explain the benefits of the demographic dividend.

The world population is now over 8 billion. On average, there are 60 babies born each minute of each day. In the United States, a country of 332 million, there is a baby born every nine seconds and a death every twelve seconds. World population growth is calculated by adding the number of babies born and subtracting the number of people who have died in a given year. For a country or city, population growth is calculated by births minus deaths, plus the total of in- and outmigration.

The so-called population explosion of the twentieth century did not occur because women's total fertility rates increased, but because rapidly declining mortality rates far outpaced declines in birth rates.

World population grew from a very small base. The number of humans who left Africa to populate the world probably ranged from 3,000 to 5,000 people. This tiny group was the basis for the spread and growth of the human population around the globe. Global population remained small and steady until 4,000 years ago, when there was a slow, steady rise, then an explosive upward growth beginning around 1800 (Figure 5.1). Population growth has accelerated since the early twentieth century. It took thirty years, from 1930 to 1959, for an extra billion people to be added. It took only another thirteen years, from 1998 to 2011, for yet another billion to be added.

Throughout most of human history, the world's population was relatively small, with moments when the human population came precariously close to extinction. Around 70,000 years ago, the eruption of the volcano Toba in Sumatra caused a volcanic winter of a long stretch of colder, cloudier weather. The total human population may have been reduced to only thousands. The result was a very small pool of genetic diversity. The huge population of today is actually genetically very close. We share a common ancestry, and despite the cultural production of racial and ethnic differences, we are biologically just part of one large extended family.

Despite such setbacks, humans prospered. They were adaptable, smart, and mobile. They transformed their surroundings, trapping animals, creating gardens, and efficiently developing gender divisions of labor between male hunting and female gathering and foraging. Men and women accumulated a long-term, deep, and sophisticated knowledge of the environment.

5.1 Population Declines

The long-term trend is one of global population growth. The increase in the **world population** masks significant examples of population decline. While wars and social conflict can reduce population in specific regions at specific times, longer-term declines are mainly the result of disease and climate change.

Disease

In Athens between 430 and 427 BCE, almost a third of the population of the city-state died from a disease as yet undetermined, likely a form of typhus. Yet disease has sometimes reached pandemic proportions. A **pandemic** is an epidemic on a wide geographic scale. We can consider three examples.

The first was the **bubonic plague** that swept in a series of waves across the world from 541 to 747 CE. It was especially pronounced in the Byzantine Empire. The first wave occurred from 541 to 544 and is named after the ruling emperor of the time, Justinian. It spread quickly along trade routes. The total death toll is estimated at 25 million people. In Constantinople, at the height of the first wave, almost 5,000 people were dying every day, ultimately halving the city's population. The disease spread as far west as England and as far east as Persia. The mass death created social confusion, caused economic dislocations, and prompted political upheaval. Villages were abandoned. Arable land lay fallow, and labor was in short supply. The lack of labor stimulated technological improvements such as improved plow design and new crop rotations. The pandemic impeded the growth of the Byzantine Empire and led to what some describe as the birth of Europe, because there was a localization of economies and political systems rather than unification under an empire. It was the pivot of large-scale social change, including the shift from antiquity to the Middle Ages and the opening stage of the rise of western Europe to continental and then global dominance.

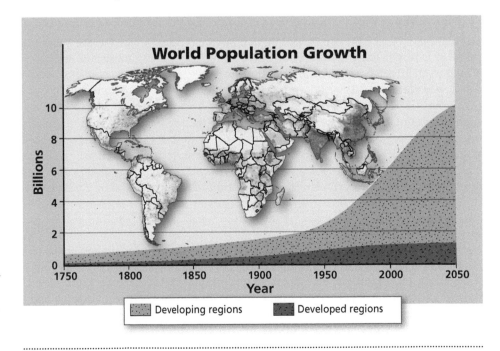

FIGURE 5.1 World population growth.

5.1 The Reproductive Revolution

The demographic transition is less a natural fact than a social revolution. It is a result of changes in public health and nutrition that extended the average life span. It is also the consequence of a reproductive revolution. The reproductive capacity of women is potentially very high. In theory, each fertile woman can have up to, and in some cases has had more than, fifteen children. Before the demographic transition, the reproductive efficiency was low, with few women living to puberty, high **infant mortality rates**, and low life expectancy. Fewer than one in ten girls reached the age of puberty. The result was a relatively low reproductive capacity. Each child born had limited chances of reaching adulthood. Emphasis was thus placed on each and every woman having as many children as possible. This largely defined the role of women and is the fundamental basis of **patriarchy**, the belief system and practice that men are the authority figures and women are restricted to home, hearth, and the bearing and rearing of children. Patriarchy—although mediated through different cultural and religious filters—is an attribute of low reproductive efficiency.

After the demographic transition, more women survived into their fertile years, and infant mortality rates fell. This higher reproductive capacity means that women can have fewer children. As the risk of each child dying declines, women do not need to have so many children to ensure that some survive into adulthood. The increased life expectancy also extends more women's lives beyond childbearing ages. The reproductive revolution calls into question the basis of patriarchy.

There are other reasons behind the fall in birth rates, including innovations in birth control, the rise of feminism, and the emergence of gender equality. In many countries women have more control over their own bodies and the decision to have children and how many. John McInnes and Julio Diaz argue that the origin of other social changes, including the deregulation and privatization of sexuality and the rise of the importance of individual identity, also ultimately resides in the increased efficiency of reproduction. Their case is suggestive and intriguing, yet like all ambitious arguments it is open to debate and discussion, criticism, improvement, and refinement.

REFERENCE

McInnes, J., and J. P. Diaz. 2009. "The Reproductive Revolution." *Sociological Review* 57:262–284.

A second pandemic, a recurrence of bubonic plague, swept throughout Asia, North Africa, and Europe: the "**Black Death**," first mentioned in written records in 1346, by 1353 had killed between 30 and 40 percent of the population, with some estimates of up to 60 percent mortality in Europe. The population of Europe probably declined from 80 million to 30 million. The effects of such a rapid population loss were immense. Forests regrew throughout Europe. Societies dominated by low wages, high rents, and high prices were quickly transformed into ones of high wages, low rents, and low prices. The lack of population forced the development of more efficient, less labor-intensive technologies. Landowners switched from the strenuous **arable agriculture** of grain production to the less taxing **pastoral farming** of producing meat, wool, and dairy products. The switch "from corn to horn" also gave more employment opportunities to women, who were traditionally employed in the pastoral sector. There were also social upheavals. In England the 1381 **Peasants' Revolt** was the organized resistance to attempts by authorities to reduce wages to pre–Black Death levels. The wider significance of the Black Death is debatable. There are some who argue for continuities before and after, while others see a more profound shift from the medieval world to the modern world—a world where religious devotion was undermined for some and strengthened for others; a world where minorities, especially Jews, could become scapegoats for disasters; a world where the authorities were made responsible for public health; a world where the relationship between rich and poor was dramatically revealed as based less on unchanging custom and tradition than on the brute facts of labor supply and demand.

Perhaps the most dramatic relationship between disease and population declines involves the **Columbian Encounter** that took place when Europeans discovered and colonized the Americas. The impact on the Indigenous population was devastating. The pre-Columbian population of the Americas is difficult to gauge precisely, but estimates put the figure between 50 million and 100 million. The Indigenous population was descended from peoples who had come from Eurasia before the development of agriculture. There was thus a major difference between the Eurasian and the American human populations. In Eurasia the agricultural revolution involved a closer association between animals and people. Over time, diseases that passed from animals to humans, such as influenza and measles, were less fatal, because those who were fatally susceptible died out, leaving behind a more resilient human population. This process did not occur in the New World, making

5.2 Climate Change and Population Decline

The development of agriculture made human population more dependent on **agricultural productivity**. A decline in agricultural output could increase mortality and lead to fewer births. The decline in food supply occurred especially when climate change reduced the length of the growing season. Patrick Galloway noted a connection between population change and fluctuations in climate change for the mid-latitude regions of the world. Across these regions there was a warming trend, peaking around 1200 CE, followed by cooling until 1450; then warming until 1600, followed by a rapid cooling until around 1690. These changing temperatures caused a decline in agricultural productivity—colder weather reduced harvests and increased the risk of hunger and starvation—that suppressed population growth. In England, for example, during the cooling period of 1600 to 1650, population levels declined because, with lower food supplies, women had fewer children.

Harry Lee and colleagues, testing Galloway's model, examined the data for China over the past thousand years. They found that five major population contractions occurred during cooler temperatures. Long-term cooling led to falling harvest yields, fewer births, and increasing mortality rates.

Historian Geoffrey Parker extended the analysis to the global level. He pulled together a vast array of material to consider the political implications of global climate change in the seventeenth century. From around 1640 to 1690 the climate was appreciably colder, at a time when most people were dependent on agriculture. This **Little Ice Age** led to harvest failures, food shortages, and hunger. A third of the world's population died and the unfolding consequences rocked political foundations. From China to England there was an unparalleled outbreak of revolts and revolutions as people rebelled against and resisted an existing order unable to cope with the crisis.

REFERENCES

Galloway, P. R. 1986. "Long-Term Fluctuations in Climate and Population in the Preindustrial Era." *Population and Development Review* 12:1–24.

Lee, H. F., and D. D. Zhang. 2010. "Changes in Climate and Secular Population Cycles in China, 1000 CE to 1911." *Climate Research* 42:235–246.

Parker, G. 2013. *Global Crisis: War, Climate Change and Catastrophe in the Seventeenth Century*. New Haven, CT: Yale University Press.

its Indigenous inhabitants fatally susceptible to everyday diseases from the Old World. Diseases such as influenza, measles, and smallpox proved fatal to the people of the Americas. Soon after contact, Indigenous populations collapsed as fatal diseases spread quickly, with devastating effect. "Great was the stench of the dead," noted a 1571 report of an epidemic in Guatemala. In the northwest region of Guatemala, the population fell from 260,000 in 1520 to 47,000 in 1575. Across the entire continent, from 1492 to 1640, the Indigenous population was reduced by close to 90 percent. This has been called the demographic holocaust, and its consequences were wide and long-lasting. Later European colonists arrived in a land largely emptied of its original inhabitants. The wilderness that nineteenth-century observers noted—the vast forests, empty plains, and abundant animal life—was not a pristine landscape but an environment of much reduced human impact. Geographer William Denevan deconstructs the **pristine myth** that North America was a vast wilderness before the coming of the Europeans. The later colonists encountered not what they thought of and described as a "natural" wilderness but what was in fact a regenerated wilderness, the product of a dramatic decline in the Indigenous population and its environmental impact. The American landscape of the eighteenth and nineteenth centuries was not an unchanging wilderness but the result of the population collapse of the sixteenth and seventeenth

centuries. The Columbian Encounter did not reveal the American wilderness, but (re-)created it.

Societies develop equilibrium when demographic trends stay the same or exhibit only a slight increase or decrease. Change, even large change, if it develops slowly, can be easily incorporated into social institutions. But when population levels fall drastically, the resultant ruptures and tears in social norms and social relationships produce dramatic changes and long-term unfolding consequences.

5.2 The Demographic Transition

Population dynamics are complex and connected to social, economic, political, and environmental factors. The demographic transition is a shift from high birth and death rates, through declining death rates and high birth rates, to declining birth and death rates. Different parts of the world are at different stages of this transition. Some have high growth rates, while others are experiencing net population decline.

Until 1800, global demographics were characterized by short life expectancy, high fertility, a young population, and slow population growth. In 1800, the average life span was only twenty-seven years, most fertile women gave birth to six children, one in three people were younger than fifteen years old, and the annual population growth rate was around 0.51 percent. People's lives were short, women spent much of their time having and raising children, and there were very few old people.

Around 1800, changes in mortality and fertility, known as the **demographic transition**, altered this global pattern. The transition has two dominant trends. First, there was the decline in **mortality rates**, initially in the industrial economies of the world, such as western Europe and North America, brought about by improved personal hygiene, public health measures, and increased affluence, which led to better diets. These all combined to reduce death rates and lengthen **life expectancy**. More people in the core countries of the global economy lived longer. While the peoples of the peripheral and colonial areas saw little improvement, a benign cycle was created in the West, where increased affluence led to more food being available, which led to longer and more productive lives, which in turn led to increased economic development. In 1950 the average life expectancy in Japan was 63.5 years; by 2022 it had reached 84.9. Life expectancies have taken longer to increase in lower-income countries. Zimbabwe, for example, still has an average life expectancy of only 61.8 years. Malaria, largely eradicated elsewhere, annually sickens more than 200 million people and kills close to 800,000 children in Africa, where it is the leading cause of death for those under five years old. There are still global disparities in healthcare provision that have led to marked spatial differences in child mortality and overall life expectancy.

Second, the increase in life expectancy was soon followed by a decline in **fertility**, at first only in the highest-income countries and then later more widespread. In 1800 there was an average **total fertility rate (TFR)** of 6 births for every woman; by 2022 the figure was 2.4. This TFR is very close to the **replacement rate**, or the number of births needed to keep the population stable. Generally, when the TFR is greater than 2.1, population in a given area will increase, and when it is less than 2.1 it will eventually decrease. Today, the replacement rate is 2.1 in higher-income countries and 2.3 in lower-income countries. The difference is the result of higher juvenile mortality rates in poorer countries.

Many countries moved toward lower birth rates between 1890 and 1920. Today, in high-income countries like Japan, the TFR rate is 1.3. In the United States, it is 1.7. Many factors were involved in lowering TFR, including access to more reliable birth control methods, changes in household economies such that children were not so much workers as dependents, and increased investment in an individual child's caring and rearing. Family size declined. However, in the lower-income countries the fertility rates

began to drop later, beginning in the 1960s. They are still higher than in the highest-income countries—5.5 births per woman in Mali, for example—but across the world, there is a transition toward lower birth rates.

Three important trends in fertility have affected the demographic transition. First, fertility has declined everywhere, but not at the same rate in all places. As a result, there is an uneven pattern of fertility, with the highest-income countries having the lowest TFR and the lowest-income countries having the highest TFR. The good news is that many lower-income countries are likely to continue fertility declines and achieve lower fertility (and hence lower population growth rates) in the coming decades. Second, although fertility has declined around the world, it has lagged behind the more rapid declines in mortality rates. This has resulted in the so-called population explosion of the twentieth century. Third, unlike mortality, declines in fertility can be subject to more complex social, cultural, technical, economic, and political factors that can inhibit or accelerate fertility decline. For example, within families and communities there may be pressures on women to maintain larger family sizes. Governments may discourage family planning by making it illegal or unaffordable. Religious institutions may consider family planning immoral. The marginalization of women can further impede declines in fertility; if a woman is only valued for her role as a mother, higher fertility rates will likely be maintained. However, increasing educational attainment for females lowers fertility, as does increasing economic opportunities for women.

The demographic transition has transformed the human population. Life is now longer as life expectancy increases and mortality rates decline. There is also a steady graying of the population. In 1950 only 4 percent of the total global population was over age sixty-five; by 2000 this had increased to almost 6 percent. By 2100 it is estimated to be 21 percent. The world's population is aging, a trend especially marked in the highest-income countries of the world.

These demographic changes have cultural consequences. Consider, for example, gender roles and representations. Masculinity was often constructed around the image of the strong, virile body. But what of masculinity when men live long past their peak physical strength and virility? Similarly, as birth rates drop, women are freed from constant childrearing. As their lives extend beyond their fertility, we need new models of what it means to be a woman. Family size declines, so that large families are now replaced by smaller, nuclear families. The family is less a large extended group of people than a tight nexus of a few individuals. Family life takes on a different hue from how it was long imagined and lived. And what of the elderly? When they were a small group, they were revered and often subsidized.

The preference for a son over a daughter continues to be the social norm in many societies. When boys are valued more highly than girls, pressure to have a son is intense. The preference for sons over daughters may be so pronounced that couples will go to great lengths to avoid giving birth to a girl or will fail to care for the health and well-being of a daughter they already have. Data, such as differences in mortality rates and excess deaths among female infants and young girls, confirm that this is still occurring.

The preference for sons has also led to imbalances in the number of men and women, distorting the sex ratio balance of countries' populations. Women, on average, live longer than men. This means that all else being equal, females should account for slightly more than half of the total population. In 2020, the global sex ratio is 101 males to 100 females, but in some countries the sex ratio is skewed beyond what is expected. Because of its three-decade-old one-child policy in operation until 2016, China's male-to-female ratio in 2021 was 114 males to 100 females, which translates into nearly 34 million more males than females. Parents restricted to having only one child often chose to abort female embryos. India, another country with a deeply held preference for sons and

male heirs, reported a ratio of 108 males to 100 females, an excess of 37 million males. Vietnam, Pakistan, and Azerbaijan also have skewed sex ratios. The term "missing women" refers to the shortfall in the number of women relative to the expected number of women in a country. It is estimated that there are over 130 million missing women in the world as a result of selective abortion and excess female deaths.

The sex ratio imbalances can have far-reaching consequences: they entrench gender inequality within a society, distort labor markets, and exacerbate human trafficking. In countries where there is a strong son preference, girls' mortality rates are much higher than would be expected. Sex ratio imbalances also have consequences for males. Large numbers of men in India and China may be unable to find partners and have children.

REFERENCES

Aksan, A. 2021. "Son Preference and the Fertility Squeeze in India." *Journal of Demographic Economics* 87:67–106.

Nguyen, M., and K. Le. 2022. "Maternal Education and Son Preference." *International Journal of Educational Development*. 89: 102552. https://doi.org/10.1016/j.ijedudev.2022.102552.

But when this group grows in size and lives long past its economic productivity, it can become a fiscal liability as well as an important resource of accumulated knowledge and wealth. With a rapidly aging population, **intergenerational inequities** and conflicts can become more pronounced, just as intergenerational transfers of wealth can become more important.

5.3 Stages of the Demographic Transition

There is a chronology and a geography to the demographic transition. In the richer countries, the transition is almost complete. Life expectancy has increased, and birth rates have dropped. In poorer countries, although life expectancy has increased, the birth rate has declined more slowly. The result is a complex pattern of different demographic regimes in countries at different phases of the transition.

Figure 5.2 highlights five distinct stages of the demographic transition. Figure 5.3 depicts the population pyramid of four individual countries at different stages in this

transition. The pyramids are the graphic representation of the national population along the vertical axis in five-year age cohorts, with males to the left and females to the right of the vertical line through the middle of the pyramid.

The First Stage

The first phase is one of high birth and death rates and stable, slow population increase. This was the pattern across most of the world prior to 1800.

The Second Stage

The second phase is associated with high birth rates, falling death rates, and—because death rates are falling faster than birth rates—a very rapid increase in population. Countries with this profile include present-day Afghanistan, Uganda, and Zambia. In Uganda, for example, with a population of 38.8 million, population growth is around 3.4 percent, driven by the high average birth rate of 5.9 births for every woman. The top left population pyramid depicted in Figure 5.3 shows the typical pattern of a large youthful base because of the high birth rates, tapering off quickly because of the relatively low life expectancies. Birth rates are high because there is little effective contraception

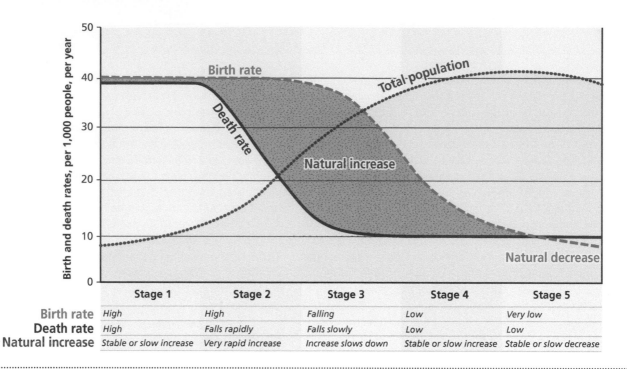

FIGURE 5.2 Phases of the demographic transition.

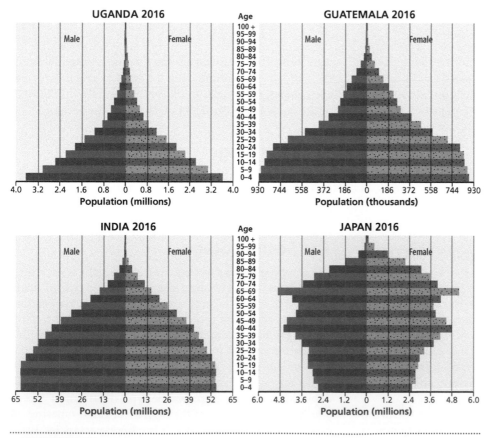

FIGURE 5.3 Population pyramids.

and cultural mores inhibit its use. High infant and child mortality rates may encourage parents to have more children so that at least one or more survive into adulthood. The net result is a very young population with high growth rates. Almost half of the population of Uganda is younger than fifteen. Similarly in Kenya, 43 percent of the population is under the age of fifteen (Photo 5.1). There are also considerable gender inequalities in income and job opportunities, since a woman's role is largely defined as giving birth to many children. The challenge for Uganda, as for all second-stage countries (which now constitute 9 percent of the world's population), is how to slow down the rate of growth to provide enough jobs and opportunities for a young population coming into the job market and to extend the economic opportunities for women.

PHOTO 5.1 Young children in rural Kenya. Almost 43 percent of the population in Kenya is under the age of fifteen and 75 percent is under the age of thirty.

The Third Stage

In stage 3, birth and death rates fall and population increase begins to slow. Countries in stage 3 constitute 7 percent of the world's total population and include middle-income countries such as Ghana, Guatemala, and Iraq. In Guatemala, because of improved public health, life expectancy is now seventy-one years. The birth rate has declined to three births per woman. The decline is the result of an increased use of modern family planning methods and the desire for smaller families. Almost half of all women in Guatemala use modern family-planning methods. By 2022 the annual population growth rate was 1.8 percent. As the population pyramid on the top right of Figure 5.3 shows, more than half of the population is under the age of nineteen, so Guatemala, like Uganda, shares the problems of finding opportunities for its many young people. A distinctive feature in Guatemala is the difference between the Indigenous Mayan and non-Mayan populations. The Indigenous Mayan population has a birth rate of over five per woman, whereas it is less than two for richer, non-Mayan women. National population statistics can hide marked differences in income, race, and ethnicity (Photo 5.2). In the case of Guatemala, there is a difference between the Indigenous Mayan population, which is more engaged in peasant agriculture, where children are important contributors to family income, and the non-Mayan population, which tends to be more affluent and less reliant on children's labor.

The Fourth Stage

Stage 4 of the transition is associated with low birth and death rates and a stable or only very slow increase in population. Countries in phase 4 now constitute 38 percent of

the world's population and are characterized by sharply falling birth rates of below 3 per woman. Figure 5.3 shows the population pyramid of India on the bottom left. Compared to Uganda and Guatemala, there is a widening in the middle cohorts, from ages twenty to fifty, and less marked tapering as the population ages. India has seen a remarkable decline in birth rates, from over 5 births per woman in the 1950s to around 2.1 by 2022. There are marked regional variations. Birth rates per woman in the southern state of Kerala have dropped below 2, while they are still high in the poorer states of the north. In Uttar Pradesh, with a population of 200 million, the rate is 2.7 births per woman. With 1.4 billion people, India is one of the most populous countries in the world. In ten years' time, it may overtake China as the most populous country on the planet.

The Demographic Dividend

For countries in stage 4, there is the possibility of a **demographic dividend**. The dividend occurs when birth rates fall substantially, requiring less investment in the very young, and before the population starts to age dramatically, requiring more money spent on caring for the elderly. As the cohorts from previous growth spurts enter the job market, there are proportionately more people of working age compared to the very young (as birth rates decline) or the very old (as life expectancy for the older groups is still relatively low). The net effect is a relative and absolute

PHOTO 5.2 An Embera family in Colón, Panama. Throughout Central and South America the birth rate for Indigenous peoples is much higher than for non-Indigenous peoples.

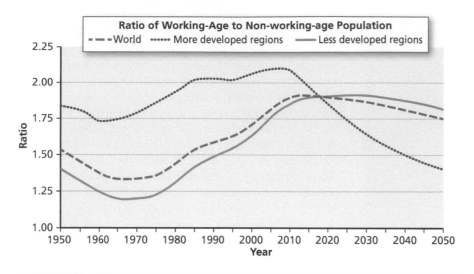

FIGURE 5.4 Ratio of working-age to non-working-age population.

increase of younger, more productive workers. Between one-third and one-half of economic growth in states such as India may be a result of the demographic dividend.

One way to highlight the demographic dividend is to estimate the **dependency ratio**. One measure is calculated by dividing the working-age population, those aged fifteen to sixty-five, by the non-working-age population, those aged zero to fourteen and those older than sixty-five. The higher the ratio, the larger the demographic dividend. Figure 5.4 shows how the ratio and the dividend decline as societies pass through to the later stages of the demographic transition.

The dividend pays out in several ways. It increases the labor supply and reduces the relative size of the **dependent population** of the very young and the very old. More public and private investment can thus be devoted to increasing productive capacity because less public money is spent supporting the very young and very old. Working-age adults tend to save more, and this can create a larger pool of capital available for investment. The human capital of the country is increased, because with many fewer children, women are able to join the workforce, and because family income is spread across fewer children, more can be invested in each child's education.

Part of the economic growth of countries such as the United States in the 1950s and 1960s and, later, South Korea was based on this demographic dividend. Even more recently, the rapid economic growth of countries such as Brazil and China in the past twenty years is based on the demographic engine of the absolute and relative increase in its working population. In these two countries, the proportion of the population aged between fifteen and sixty-five—generally considered the most economically active age range—is 69 and 74 percent, respectively.

However, China is now facing a decline in terms of both decline in total population and the size of its working population. After its brief demographic dividend, China is now rapidly graying. In poorer countries such as Afghanistan and Cameroon, the respective figures are 52 and 54 percent. Both these countries have also seen significant emigration as people search for economic opportunities beyond the national borders.

The opportunity to realize this demographic dividend is relatively short, because increasing life expectancy soon ages the population, quickly reducing the ratio of workers to non-workers. There is a short window of time when there is a rising share of people of working age and the demographic profile is weighted toward the most productive. During this time, holding everything else constant, average per capita income rises, both reflecting and prompting enhanced economic growth. A full dividend is only possible if female **participation rates** are high. If societies continue to restrict the employment opportunities of women, the full benefits of the dividend are unrealized. One reason behind China's spectacular growth is not simply the demographic dividend but also its full implementation. Female participation rates are very high, at 94 percent for women aged twenty-five to thirty-four. The dividend is fully realized if the workforce is educated and skilled and the benefits of the growth are spread across the population rather than garnered by only a few. Rapid economic growth in Japan and South Korea from 1960 to 1990 and China since 1990 capitalized on their demographic dividend by increasing employment opportunities fast enough to maintain labor productivity growth.

The demographic dividend may provide a supply of relatively cheap labor, but only for a limited period. The **Lewis turning point**, named after economist William Arthur Lewis, occurs when the supply of cheap labor runs out. The increased demand for labor pushes up wages, and benefits are increased. Wages rise, profits may fall, and the economy loses its cheap labor-cost advantages. A vital question for China, with its below-replacement birth rate, is whether it is reaching or passing through a Lewis turning point.

From 1980 to 2010, China's spectacular growth was in part a function of the high proportion of working-age citizens between twenty and sixty-five years old balanced between lower skill and a more educated workforce. However, with declining birth rates and extended life spans, from 2020 to 2050 the proportion of people aged sixty and over will increase from 17 to 40 percent. By 2040 China will have more people aged over the age of sixty-five than the

5.4 Brazil and the Demographic Dividend

At the time of its first official census in 1872, the population of Brazil was only 10 million. In 2022 it was estimated at 215 million. After a century of rapid demographic growth, the country is undergoing a major demographic transition. Birth rates are dropping; the number of births per woman declined from 6.3 in 1960 to 1.6 in 2022. This remarkable fall is a product of many things, including rapid urbanization, the secularization of Brazilians, and the increasing use of contraception. The religious codes of the Catholic Church no longer exercise much influence. Popular soap operas with their emphasis on modernity play a hugely influential part in creating new gender roles and generating alternative models of family life and size. Better access to reproductive health and increasing female participation are part cause and part effect of the rapid fall in birth rates. It is not only the affluent who want to limit family size. The causes and effects feed off each other. Economic growth makes it possible for poorer families to have fewer children, which means that more can be spent on each child's education, which in turn means children are better educated and more able to find employment.

Brazil has experienced a demographic dividend since 1970 as the rapid fall in fertility has reduced the absolute and relative proportion of dependents. From 1970 to 2022, more than half of all economic growth could be explained by the dividend, which will be in operation until 2025. On the basis of this economic growth and increased democratization, recent decades have witnessed the creation of relatively generous social welfare programs of social security and a public education system. However, as the increasing life expectancy grays the population, the number of beneficiaries to workers will increase. Mortality rates have also fallen dramatically, to the extent that life expectancy has increased from 53.7 in 1960 to 76.3 in 2022. While the social welfare programs reduce inequality and promote economic growth, they also will be a significant source of political debate as the population ages. How to balance old age security and income equality with sustaining economic growth will become a significant policy issue for Brazil and the many other countries moving into the latter stages of the transition.

United States, and the decades of rapid growth may end as the demographic dividend fades.

The Youth Bulge

In the early and middle stages of the demographic transition, a **youth bulge** occurs (Photo 5.3). This is the rapid increase in the number of people aged between fifteen and twenty-four. It occurs when a rapid reduction in child mortality occurs before a rapid falloff in fertility. In ten years' time, the increase in the number of children under the age of fourteen becomes the youth bulge of fifteen to twenty-four. When this bulge constitutes more than 20 percent of the population, it has been linked to an increase in political instability and the possibility of increased political violence. Two political geographers, Gary Fuller and Forrest Pitts, looked at the emergence of this youth cohort in South Korea in the 1970s and 1980s and its correlation with the rise of political unrest.

A number of social scientists have sought to test the connection between youth bulges and **social conflict** in more detail. Using data from 1998 to 2005 covering 127 countries, Alfred Marcus and colleagues found that youth bulges are correlated with violent conflict. They also found that violence did fall off when the youth bulge was followed by a decline in the youth cohort. Henrik Urdal also tested this idea against empirical data. He found that an increase in the youth bulge of 1 percent increases the likelihood of domestic armed conflict in a society by 7 percent points. Countries with particularly large youth bulges are three times more likely to experience serious armed conflict. Economic stagnation plays an important role in turning a youth bulge into a tinderbox of armed conflict. In a more detailed study, Urdal and Hoelscher looked at

PHOTO 5.3 Young people in Tunis, Tunisia. A youth bulge can occur at the middle stages of the demographic transition.

5.5 Afghanistan's Population Trends

Afghanistan has an average gross domestic product per capita of $570, making it among the lowest-income countries. It began the demographic transition relatively late in the twentieth century and has seen fertility decline more slowly compared to mortality. In 2000, Afghanistan still had a TFR of 8 children per woman. Not surprisingly, the population pyramid for Afghanistan in 2022 resembles the typical shape of a country in stage 2 or early stage 3 that has high TFRs, high population growth, and a very young population (see Map 5.1). Currently Afghanistan's TFR is 4.5, with an average annual growth rate of 2.3 percent and a life expectancy of 53 years.

But a snapshot of Afghanistan today fails to show the underlying changes occurring. Fertility and population growth rates are declining. According to the 2023 US Census, which maintains an international database, the projected population pyramid for Afghanistan by 2055 will show a decline in fertility and an aging of the population, with more people over age forty than in 2022. By 2090, the projected population pyramid resembles a stage 4 country, with a steady TFR at replacement level and a much older population.

REFERENCE

United States Census International Database. "International Database: World Population Estimates and Projections: Afghanistan, July 12, 2023. https://www.census.gov/programs-surveys/international-programs/about/idb.html. And https://www.census.gov/data-tools/demo/idb/#/dashboard?COUNTRY_YEAR=2023&COUNTRY_YR_ANIM=2023&CCODE_SINGLE=AF&CCODE=AF

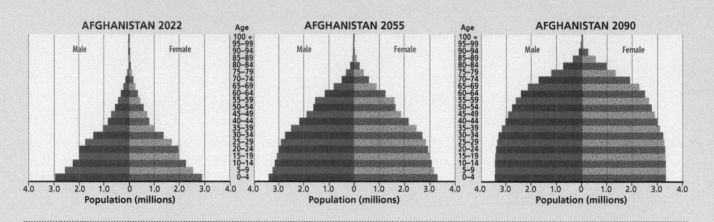

MAP 5.1 Population pyramids for Afghanistan.

fifty-five large urban centers in Asia and sub-Saharan Africa for the 1960–2006 period. The male youth population bulge did not increase levels of social disorder. A more important factor was youth exclusion from economic opportunity, even in the absence of extraordinarily large urban youth population bulges. It is not simply the number of youth but also their range of opportunity for active and meaningful participation in economic and political life that influences political conflict.

Across the Middle East and North Africa, the proportion of fifteen- to twenty-four-year-olds is 21 percent of the total population and 34 percent of the working-age population. The unemployment rate among this group is a very high, at 26 percent. The Arab Spring had many causes, but a disenfranchised youth bulge was certainly a significant factor.

The relative and absolute size of specific cohorts does have wider social implications, even for the more affluent countries. In an argument that runs counter to the demographic dividend hypothesis, the **Easterlin hypothesis**, named after economist Richard Easterlin, suggests that the fortunes of a cohort depend on the size of that cohort relative to the total population. There are luckier times to be born than others. If you were born in the United States in the 1930s, you experienced a demographic trough—only 18.7 babies were born for every 1,000 people, compared to 29.5 in 1915 or 24.1 in 1950. Holding everything else constant, you had smaller class sizes, easier access to college, and more employment opportunities. Although we imagine our economic success or failure to be a result of personal characteristics, it is also a matter of demographic luck.

Demographic troughs also have wider social effects. Crime rates, for example, showed a marked decline across several countries, including Brazil, Canada, and the United States, as the percentage of the population aged fifteen to twenty-four declined. This occurred in Canada and the United States in the 1990s and in Brazil in the early 2000s. Very different types of societies at different times all show the falloff in crime rates as the percentage of the fifteen- to twenty-four-year-old age group declined. Demography is behind the fall in crime rates, as the cohort more likely to commit crime diminishes in numbers and relative weight.

The Fifth Stage

The final stage of the demographic transition is marked by very low birth and death rates and stable or decreasing population. As countries become more affluent, they tend to move to the fifth stage.

In fact, it is more accurate to see the transition as a continuum along which countries are moving, sometimes slowly, sometimes very quickly. Rather than a strict categorization, we can think of the phases as way stations where countries may be only temporary residents. Take the case of the most populous nation in the world, China, which is rapidly shifting from stage 4 to stage 5. Since the Communist takeover, death rates have declined markedly and birth rates were tightly controlled by a one-child policy that lasted from 1980 to 2016. The birth rate is now 1.4 per woman, below replacement and much less than that of the United States. Population growth has declined to 0.7 percent per annum, and there are now problems with labor shortages. The working population, those aged between fifteen and sixty-five, has declined each year since 2012. Fear of long-term labor shortages forced the government in 2015 to lift its decades-old one-child policy.

Countries in this phase constitute 46 percent of the world's population. The population pyramids of these countries, like those of Japan shown in Figure 5.3, have a narrower base because of low birth rates and less obvious tapering because of longer life expectancies. There are fewer young people compared to the increasing proportion of old people. Japan has one of the highest life expectancies and lowest birth rates in the world. The world's average life expectancy is seventy, while in Japan it is eighty-four. With only 1.3 births per woman, Japan is falling below replacement levels. A similar trend can be noted for other East Asian countries, including South Korea (0.9 births per woman), Taiwan (1.2), Singapore (1.2), and Hong Kong (1.3) (Photo 5.4). In Japan the result is a declining population, a trend that is particularly marked in rural areas of the country. Much of rural and small-town Japan has been effectively depopulated as young people move to the larger cities for jobs and services. Across the nation, the population is aging. Almost one in four Japanese are over the age of sixty-five. Each year 1.3 million die in Japan, but only 1 million babies are born. With limited immigration, the population total has started to fall. By 2100, Japan is expected to lose 34 percent of its population as the total drops from 125 million in 2022 to 83 million.

PHOTO 5.4 Schoolchildren in South Korea. In higher-income societies with low birth rates, children are often the recipients of significant public and private investment.

One major reason behind the decrease in fertility in Japan, and throughout East Asia and elsewhere, is that women of reproductive age are not getting married and not having children. As a result of job discrimination, only about one-third of Japanese women remain in the workforce after having a child, compared to two-thirds of women in the United States. Because of the cultural mores against out-of-wedlock children and the heavy pressure to look after their husbands, for most Japanese women the choice is a stark one: stay single, and have a career and financial independence, or get married and devote oneself to husband and child/children. The percent of Japanese women who have opted to remain single has doubled over the past twenty years. Over the coming decades, the proportion of working-age Japanese will continue to dwindle. This situation is the opposite of the demographic dividend: it is the demographic deficit of a declining workforce and an increasingly dependent population. The Japanese workforce will decline by 70 percent by 2050. In countries with a similar age profile, immigration from outside the country fills some of the job vacancies. In Japan, there are strict restrictions on foreign immigration. Less than 2 percent of the population is foreign-born, compared to the United States, where the figure is close to 14 percent, or Australia, where it is 22 percent. Japan will experience a major crisis when its anti-immigration posture crashes against the reality of its demographic deficit. Another solution to this impending crisis would be to make it easier for women with children to enter the workforce. Again, however, this policy goes against traditional cultural norms. In 2011 a government think tank in South Korea, the Korea Development Institute, made the case for

5.6 Russia's Changing Population

In 2010 the Russian government offered the equivalent of $9,000 in cash to have a second child, evidence of a pronounced demographic shift. At the end of the Soviet era, the population of Russia was around 148 million, having risen steadily from a 1950 population of around 102 million. It fell to around 138 million in 2011 and then recovered to 145 million by 2022. What is behind this remarkable collapse and more recent resurgence?

Russia experienced increased mortality rates because of a decline in life expectancy for males; it was sixty-five in 1989 but fell to a staggering fifty-nine by 2005. Life expectancy for females is seventy-two, the disparity a direct effect of heavier alcoholism rates among men compared to women. The decline of the old Soviet system, with its extensive welfare programs, and its replacement, especially in the earlier years of reform, by a harsh market-driven approach initially led to a decline in general public health provision, increased unemployment, and lower living standards. This economic shock therapy led very quickly to a decline in general social welfare and an increase in mortality. The number of deaths increased from 1.58 million in 1989 to 2.36 million in 2003 as the death rate rose from 10.7 to 16.4 per thousand. There was also a decline in birth rates from 1.8 in 1989 to a historic low of 1.1 in 1999, well below replacement level. There were 2.1 million births in 1989 but only 1.4 million in 2003. More Russian women, with very easy access to abortion, simply limited the number of children they were willing to have. Economic uncertainty and social dislocation made having children a more difficult choice, and with fewer males there were not enough partners with whom to start or raise a family.

The population decline occurred because more people died and fewer children were born. Immigration was tightly controlled and kept low, so the foreign-born did not contribute much to population increase.

REFERENCE

Gontmakher, E. 2022. "Russia's Demographic Setback." https://www.gisreportsonline.com/r/russia-demographic-decline/.

reducing the social taboos in the country associated with out-of-wedlock pregnancies as one way to counter the declining fertility. With a fertility rate of 0.9 births per woman, South Korea ranks 222 of 222 countries in fertility. As the number of single women increases, easing legal safeguards of children born outside marriage is one way to counter a decreasing population. In many East Asian countries, "foreign brides," largely from Southeast Asia, are a recent but significant phenomenon.

At current rates, the world's over-sixty-five population is likely to double in fifty years. Tod Fishman uses the phrase "shock of grey" to refer to the reality of rapidly aging societies (Photo 5.5). The old will become an increasingly large part of the population in both absolute and relative terms. As people live longer, what it is to be old is redefined. The over-sixty-five cohort is growing so large and so quickly that new categories are being developed: the young-old, aged sixty-five to seventy-four; the old-old, aged seventy-five to eighty-four; and the oldest, aged eighty-five and above. We can also reconceptualize old age in a more positive light as an age of personal fulfillment rather than one of degeneration and dependency. People previously considered old now live longer and more engaged lives. And with more elderly, there are new market niches, such as retirement communities and caring for elders. Some residential developments in the United States, for example, have minimum age requirements as the affluent elderly sequester themselves into homogeneous communities (Photo 5.6).

PHOTO 5.5 An elderly lady in Inchon, South Korea. The elderly population will double over the next fifty years, creating increased demands for elder care and healthcare.

PHOTO 5.6 As the elderly become a larger proportion of the population, retirement communities like this one will become more important features of the urban landscape.

TABLE 5.1 Demographic Differences

COUNTRY	POPULATION AGED <15 (%)	WORKING POPULATION (%)
Uganda	46	52
Guatemala	33	61
India	26	67
Japan	12	59

Another form of demographic dividend is the increase in the number of older persons who are able to share their life assets with the younger generation. This is an important phenomenon in countries with strong extended family ties on one hand and limited social welfare programs on the other. Intergenerational transfers of wealth and income from an older generation to a younger generation are an important part of economic growth in East and Southeast Asia and Latin America. In South Africa, Monde Makiwane conducted fieldwork on these transfers and concluded, "Most of the elderly in South Africa use their meager old age pensions to support unemployed children and orphaned and vulnerable children." Intergenerational transfer as a form of wealth accumulation is very uneven in terms of race and class as existing inequalities tend to be reinforced and inherited.

This fifth stage of the demographic transition has developed in some richer countries, where it is associated with very low birth and death rates, more pronounced aging, more reliance on immigration to fill job vacancies (especially at the lower wage levels), and a disconnect between marriage and children. More children are born out of wedlock and more couples do not have children. A variety of household living arrangements occur as traditional marriage is no longer such a dominant model of household formation. These trends are especially marked in the countries of northwest and eastern Europe. In Norway and Sweden, 54 and 55 percent of all births, respectively, are to unmarried women. The figures reflect increasing cohabiting between couples rather than marriage and more single-parent households, in part the result of choices of women and their families. These choices may have negative economic consequences in some countries. In the United States, almost 41 percent of all births are to unmarried women. One in three households that experience child poverty are single-parent, female-headed households, compared to much lower rates in married, two-parent families. In many northern European countries, in contrast, children born out of wedlock are generally born to stable, cohabiting couples or to those with access to more generous social welfare and public health programs.

Summary of Transition

By way of a summary, let us return to the four countries that we considered earlier in the chapter when we examined population pyramids. Some of their basic demographic differences are summarized in Table 5.1. In terms of the relative distribution of the population, note how countries at the earlier stages of the transition have a higher proportion of those aged younger than fifteen and a corresponding young median age. In contrast, a country such as Japan, at the later stages of the transition, has fewer young people and more older people. Countries such as India are in the middle of the demographic dividend, with the highest percentage of people of working age.

Median age is the age that divides the country's population into two equal halves. Working population is defined as the percentage of the population older than fifteen and younger than sixty-four.

5.4 Problems and Opportunities of the Demographic Transition

At a very fundamental level, the demographic transition is an enormously life-enhancing phenomenon. People now live longer, healthier lives. The crushing tragedy of infant mortality and the lost opportunity of short lives, if not eradicated, are much reduced. People get to enjoy the gift of life for longer and in a healthier state, and they lead more productive lives.

Each stage of the demographic transition creates both problems and opportunities. In stage 2 and early in stage 3, as death rates decline, more people live longer, healthier lives. However, the still-high birth rate means that a society at this stage must cope with a very youthful population with a limited number of people in the most economically productive cohorts. There is the potential instability of a youth bulge.

In the middle stages of the demographic dividend, there is an increasing proportion of the economically active population. This demographic dividend can turn into a demographic time bomb, however, if economic growth does not keep up with the number of people coming into the job market. An increase in the working-age population can produce a demographic dividend, but also raises the possibility of social unrest if the army of young people fails to find gainful and rewarding employment. It is too simple to assign the Arab Spring, for example, only to the large number of young people in societies without job prospects—there are other countries with similar conditions that did not experience social uprisings—but it was an important demographic component in the content of long-term totalitarian regimes, marked inequality, and stagnant economic opportunities for the majority. When there are a large number of young people—the median ages in Egypt and Tunisia are twenty-four and thirty-two, respectively, compared to forty-seven in Germany—there is the possibility, in association with anemic economic growth and sclerotic political systems, of social unrest. Other things being equal, younger people more easily take to the streets than older people. And if the demography is weighted toward the young while the economy is geared toward the more elderly, the possibility of unrest rises.

In the later stages of the transition, birth rates drop and life expectancies stretch out the length of the average life. There are more people with the advantage of a long life experience. A society with more of these people should be wiser and more able to take the sage, long-term perspective. Again, opportunity and crises are possibilities at the later stages of the transition. On the one hand, there is the possibility of a second demographic dividend as the elderly, able to amass a long lifetime of assets, pass on their wealth to their children and younger family members. This ability depends on race and class, however. On the other hand, there may be little dividend if the elderly hold onto their assets and continue to receive and demand public benefits.

At the later stages of the demographic transition, there is an increase in the elderly nonworking population. We can follow one of the consequences by looking at Social Security in the United States, a social insurance fund that current workers pay into to support retired workers. When it was first introduced, there were many more workers paying into the system than recipients

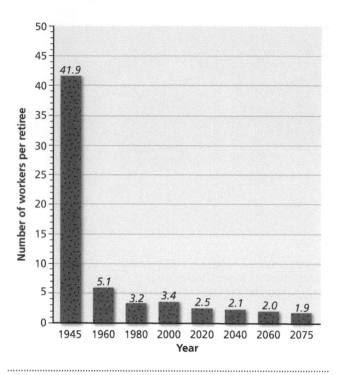

FIGURE 5.5 Number of workers to retirees in the US Social Security system.

receiving benefits. In 1950, for example, the number of workers per beneficiary was 16.4. In 1960, the number of workers per beneficiary was 5.1, and that number is likely to fall to below 2 by the end of the twenty-first century. Figure 5.5 highlights the dramatic shift. As more retirees live longer, they are supported by fewer workers. Over the longer term, this may be fiscally unsustainable. In social democracies with generous social insurance programs, the aging of the population creates a fiscal crisis for the welfare state. It can lead to generational inequalities because fewer workers have to support more retirees.

Generational inequity really kicks in if those paying for the elderly are unlikely to see the same benefits. In the United States, for example, older adults are advantaged because they have publicly provided pensions and healthcare that are unlikely be so generous to future generations. The net worth of a typical US household member aged over sixty-five is forty-seven times that of one younger than thirty-five. The disparity was ten to one just twenty-five years ago. Over a quarter of a century, there has been a relative shift of wealth from the young to the elderly and the creation of an intergenerational divide in life chances and benefits. Around the world, in countries with social welfare programs, the aging of the population creates fiscal tensions and the possibility of intergenerational conflicts. Class, race, and gender have long been identified as sources of difference, advantage, and disadvantage. We need to add to that list both age and the cohort's position in the demographic transition.

The shift toward a more elderly population can have major implications for continued economic growth, because without younger workers the number of dependents will exceed the number of productive workers and there will be increasing pressure on generous social welfare systems. In some rich countries, there is now a greater reliance on foreign immigrants to fill job vacancies. However, this can create tensions as the issue of citizenship, for example, becomes more problematic with more "foreign" workers. As welfare costs mount, there is increasing concern with noncitizen access to generous welfare arrangements. The economic need for more immigration can sometimes clash with a cultural debate about national identity and emerging political debates on citizenship.

There is a distinct global geography to the demographic transition, and it interconnects with levels of economic development and rates of growth. We can identify three types of countries:

1. Graying countries, such as Russia and most East Asian and European Union countries, with aging populations.

2. Demographic and economic growth countries, such as Brazil and Thailand, with rapidly growing populations but developing economies, with the possibility of a positive demographic dividend.

3. Rapid demographic growth with limited economic growth countries, such as Haiti and Afghanistan.

Youth revolts, crime levels, culture wars about the family, discussions about welfare reform, and national identity: there is a demographic context to some of these interesting social changes. Demographic changes are at the heart of some of the most profound cultural transformations and intense political debates occurring around the world.

Cited References

Denevan, W., ed. 1992. *The Native Population of the Americas in 1492*. Madison: University of Wisconsin Press.

Fishman, T. C. 2010. *Shock of Grey*. New York: Scribner.

Fuller, G., and F. R. Pitts. 1990. "Youth Cohorts and Political Unrest in South Korea." *Political Geography Quarterly* 9:9–22.

Makiwane. M. 2011. "Intergenerational Relations in Africa with a Focus on South Africa." Paper presented at the United Nations Expert Group Meeting on Adolescents, Youth and Employment. Population Division, United Nations Secretariat. http://www.un.org/esa/population/meetings/egmadolescents/pp09_makiwane.pdf. Accessed November 7, 2011.

Marcus, A. A., M. Islam, and J. Moloney. 2008. "Youth Bulges, Busts, and Doing Business in Violence-Prone Nations." *Business and Politics* 10 (3): 1-40. https://doi.org/10.2202/1469-3569.1227.

Urdal, H. 2004. *The Devil in the Demographics: The Effect of Youth Bulges on Domestic Armed Conflict 1950–2000*. Social Development Paper No. 14. Washington, DC: World Bank.

Urdal, H., and K. Hoelscher. 2009. *Urban Youth Bulges and Urban Social Disorder*. Washington, DC: World Bank.

Select Guide to Further Reading

Benedictow. O. J. 2004. *The Black Death, 1346–1353: The Complete History*. Woodbridge, Suffolk, UK: Boydell.

Bongaarts, J. 2009. "Human Population Growth and the Demographic Transition." *Philosophical Transactions of the Royal Society B: Biological Sciences*. 364:2985–2990.

Cai, F. 2010. "Demographic Transition, Demographic Dividend, and Lewis Turning Point in China." *China Economic Review* 3:107–119.

Demeny, P. 2011. "Population Policy and the Demographic Transition: Performance, Prospects, and Options." *Population and Development Review* 37 (Suppl.): 249–274.

Dyson, T. 2010. *Population and Development: The Demographic Transition*. New York: Zed Books.

Goldscheider, F., E. Bernhardt, and T. Lappegård. 2015. "The Gender Revolution: A Framework for Understanding Changing Family and Demographic Behavior." *Population and Development Review* 41:207–239.

Handwerker, W. P., ed. 2019. *Culture and Reproduction: An Anthropological Critique of Demographic Transition Theory*. London: Routledge.

Inayatullah, S. 2016. "Youth Bulge: Demographic Dividend, Time Bomb, and Other Futures." *Journal of Futures Studies* 21:21–34.

Laslett, P. 1989. *A Fresh Map of Life: The Emergence of the Third Age*. London: Weidenfeld & Nicolson.

Lee, R. 2003. "The Demographic Transition: Three Centuries of Fundamental Change." *Journal of Economic Perspectives* 17:167–190.

Lee, R., and D. Reher. 2011. "Introduction: The Landscape of Demographic Transition and Its Aftermath." *Population and Development Review* 3:1–7.

Lerch, M. 2018. "Urban and Rural Fertility Transitions in the Developing World: A Cohort Perspective." *Population and Development Review* 45:301–320.

Lesthaeghe, R. 2014. "The Second Demographic Transition: A Concise Overview of Its Development." *Proceedings of the National Academy of Sciences* 111:18112–18115.

Little, L. L., ed. 2006. *Plague and the End of Antiquity: The Pandemic of 541–750*. Cambridge: Cambridge University Press.

Lovell, W. G. 1992. "'Heavy Shadows and Black Night': Disease and Depopulation in Colonial Spanish America." *Annals of Association of American Geographers* 82:426–443.

Newbold, B. K. 2021. *Population Geography: Tools and Issues Fourth Edition*. New York: Rowman & Littlefield.

Nordås, R., and C. Davenport. 2013. "Fight the Youth: Youth Bulges and State Repression." *American Journal of Political Science* 57:926–940.

Pablos-Mendez, A., S. Radloff, K. Khajavi, and S. A. Dunst. 2015. "The Demographic Stretch of the Arc of Life: Social and Cultural Changes That Follow the Demographic Transition." *Global Health: Science and Practice* 3:341–351.

Pearce, F. 2013. "Why the "Youth Bulge" Can Make or Break a Country." *New Scientist* 219:42–45.

Rosen, W. 2007. *Justinian's Flea: Plague, Empire and the Birth of Europe*. New York: Viking.

Shenk, M. K., H. S. Kaplan, and P. L. Hooper. 2016. "Status Competition, Inequality, and Fertility: Implications for the Demographic Transition." *Philosophical Transactions of the Royal Society B: Biological Sciences* 371 (1692): 20150150.

Strulik, H., and S. Vollmer. 2015. "The Fertility Transition around the World." *Journal of Population Economics* 28:31–44.

Thompson, V., and M. C. Roberge. 2015. "An Alternative Visualization of the Demographic Transition Model." *Journal of Geography* 114:254–259.

Tuchman, B. W. 1978. *The Distant Mirror: The Calamitous 14th Century*. New York: Knopf.

Zaidi, B., and S. P. Morgan. 2017. "The Second Demographic Transition Theory: A Review and Appraisal." *Annual Review of Sociology* 43:473–492.

The Geography
of Population

This chapter considers the size, distribution, and movement of population. The population is composed of different categories of gender, age, race, ethnicity, and other factors of social difference. The social construction and spatial distribution of these categories reveal how social and cultural geographies are created and maintained.

The world population is now just over 8 billion people. The basic distribution across the world is depicted in Figure 6.1. Notice how the population is concentrated in certain regions of the world. This is largely a function of the distribution of landmass, but there are also significant geographical macro trends. There is a concentration in the northern mid-latitudes. Nine of every ten people live in the Northern Hemisphere. The human population is located in a relatively narrow terrestrial band. Our inhabited world is centered on the Tropic of Cancer at approximately 24 degrees north. Farther south from this favored location is the historically unhealthy region of the equatorial tropics, while farther north it is too cold to support large populations.

The world's population distribution reflects our historical reliance on agriculture. The centuries-old agricultural regions of East, South, and Southeast Asia continue to have heavy concentrations of people. Population has also

LEARNING OBJECTIVES

6.1 Identify the current population of the world and **describe** its geographic distribution and associated attributes.

6.2 Compare population distributions and differences within and between countries.

6.3 Explain why race and ethnicity are social constructions.

6.4 Explain the forms and reasons for movement of populations in a variety of scales and contexts.

6.5 Explain the difference between forced and voluntary migration and describe trends in both.

6.6 Identify and **explain** the values and limitations in using different models of population movement.

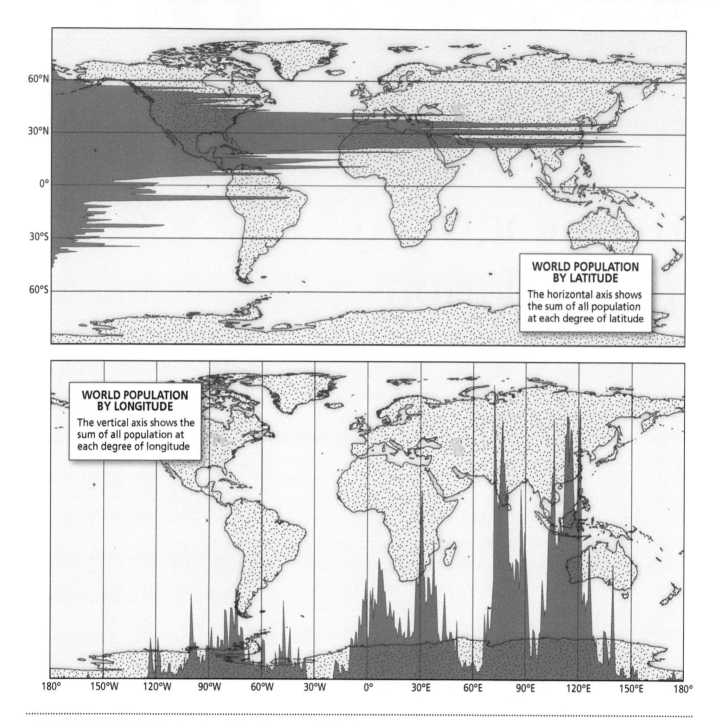

FIGURE 6.1 Global distribution of population (courtesy of Bill Rankin, www.radicalcartography.net).

concentrated in the urban industrial regions of the world: first in western Europe, then in North America, and more recently in East Asia. The result is a distinct population clustering in long-established agricultural and industrial areas in the Northern Hemisphere and, more recently, in large metropolitan regions across the globe.

6.1 The Distribution of Population

The world's population is distributed and demarcated in cultural, political, and economic space. Human populations are culturally specific; they speak particular languages and have different beliefs, and part of this difference is a function of where they are. A major geographic truth and variant of Tobler's law is that who you are depends on where you are.

International Differences

Figure 6.2 depicts population distribution by countries. More than a third of the world's population lives in just two countries, China and India. Half of the world's total population lives in just six countries. And the ten largest countries combined contain almost 58 percent of the world population.

Since the 1970s, the fastest rates of growth were in the Global South. High birth rates and declining mortality levels resulted in rapid population growth in countries such as India and China, leading to their present-day huge populations. Meanwhile, growth slowed in the Global North. Today, growth has slowed even in countries such as India and China, but remains higher in sub-Saharan Africa.

Population size is an important component of national power. Very small countries tend to wield less power than larger countries. A large population allows for a bigger economy, a larger military, and a greater ability to project economic, cultural, and political power beyond national boundaries. However, population size is not everything. Switzerland, with a population of just over 8.7 million, has a total gross domestic product of $850 billion, while that of Uganda, with 48.4 million, is only $34 billion. The difference is caused by the stark difference in gross national income per capita: $83,400 in Switzerland but only $910 in Uganda. It is not only the total population that defines national power but also the wealth and productivity of the population.

The most powerful countries are those with a large population and a successful economy. The United States wields enormous power and influence not only because of its population size (the world's third largest), but also because of its economy (the world's largest). Superpowers have demographic, economic, and military power.

We may be witnessing a change in geopolitics and geo-economics as some countries, such as Brazil, China, and India, have not only large populations but also growing economies. They combine a demographic dividend with economic growth. Their growth marks a shift in the relative balance of power in the world.

Population size has advantages and disadvantages. Large countries have greater relative weight in the world. China's global status is reinforced not only by its impressive economic growth since 2000 but also by the basic fact that this growth occurred in the world's most populous country. But a populous country may also have possible problems maintaining national coherence because very large populations may be diverse. In the case of India, for example, the sheer size of the Muslim minority means there is a significant resistance to Hindu nationalism.

Small countries may have the benefit of greater homogeneity and coherence, but may also be more vulnerable to external shocks. A hurricane landing in the Caribbean island of Dominica, with a population of just over 72,000, has a greater national impact than a hurricane landing in the United States. A single hurricane can paralyze Dominica's entire tourist trade and negatively affect its entire

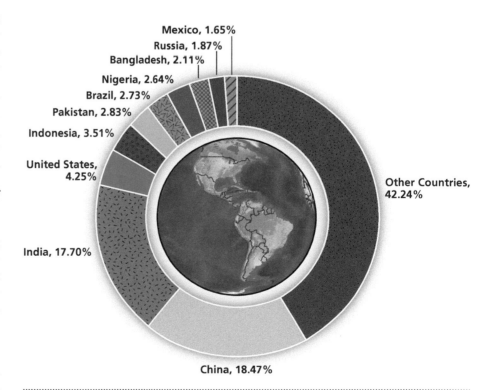

FIGURE 6.2 Population distribution by country.

national economy. By contrast, although a hurricane striking America might harm the tourist trade in Florida or the Carolinas, these states are only a small proportion of the US economy.

National Imaginaries and Population Distribution

Not only is the global population unevenly distributed, so too are national populations. Consider the case of the United States. Map 6.1 highlights the uneven nature of population density. Notice the towers of high density in the major metropolitan areas, the columns of adjacent suburbia, and the flat nature of much of the rural United States.

Population distribution often connects with the national imaginaries. **National imaginaries** are a distinct form of geographical imaginary that refer to the spatial nature and representation of the nation-state. Map 6.1 highlights the metropolitan nature of US society, but many of its cultural representations focus on the American West and its rural regions as the quintessential United States. The notion of a US heartland, the region of the rural interior United States, may be culturally significant, but it is of minor and declining demographic significance. Although

wilderness is a major theme in how the United States sees itself, the demographic reality is a nation of big metro areas.

And here we come to an interesting phenomenon of the differences between demographic realities and political representation. In the United States, states have equal representation in the Senate. Both California and Wyoming have two senators in Congress, yet they have very different populations. California is the most populous state, with around 39 million people, while Wyoming has closer to 582,000. Wyoming is vastly overrepresented in the US Senate while California is vastly underrepresented.

Because the Senate makes important independent decisions, such as confirming Supreme Court nominations and foreign policy agreements, these different weightings given to the population in different parts of the country have significant political effects. Table 6.1 shows the disparity between the topmost populous states. California, Texas, Florida, New York, and Illinois combined hold more than a third of the entire US population but only have 10 percent of seats in the US Senate. The five least populous states, Wyoming, Vermont, Alaska, North Dakota, and South Dakota, have just 1 percent of the nation's population but also hold 10 percent of Senate seats. Rural conservative states are thus given a greater political representation in the Senate than the larger, more metropolitan states.

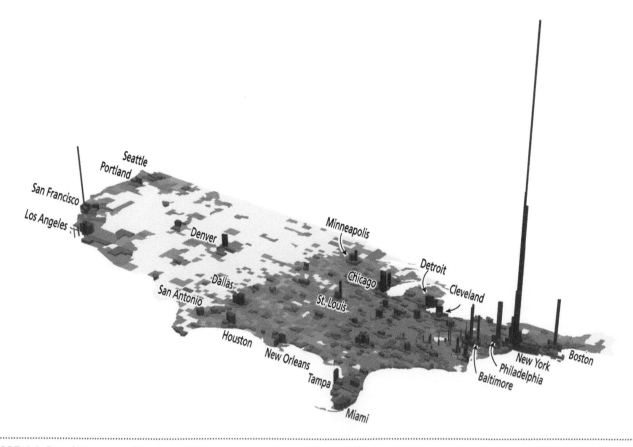

MAP 6.1 Population map of the United States.

TABLE 6.1 Population and Political Representation in the US Senate		
	FIVE LARGEST STATES	**FIVE SMALLEST STATES**
Population	123 million	3.6 million
% of national population	37.1	0.9
% of US senators	10	10

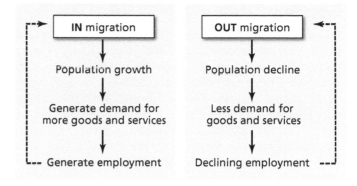

FIGURE 6.3 Demographic cumulative causation.

There are disparities between population distribution and political representation in other countries. In Japan, for example, rural areas are losing population yet retain significant electoral representation. The electoral boundaries and number of representatives were decided decades ago, before the massive rural-to-urban migration. The result is that rural areas are overrepresented in the national parliament while the cities are underrepresented. Rural residents often have political power way beyond their demographic number or economic importance. Voters in rural areas have on average 250,000 voters per national parliament seat, compared to over 1 million voters per seat in the big cities. These rural and agricultural areas and their predominantly elderly populations have an outsized role in the national politics of Japan because they are granted disproportionate electoral power.

Demographic Cumulative Causation

Across the world we find regions of population growth and areas of population decline. This uneven distribution is a function of changing demographics; regions with higher birth rates increase population, while regions such as rural Japan witness more deaths than births and so lose population.

There is also a redistribution as people move to areas of greater economic opportunity. We can identify a **demographic cumulative causation**. As people move from a declining region to growing regions, profound changes occur that can reinforce the regional differences (Figure 6.3). It is the most able and active who make the move, adding to the labor power of the receiving region and detracting from the labor stock of the sending region. The receiving regions gain extra sources of labor and effective economic demand, leading to greater regional growth. This may be offset by greater pressure on goods and services in the receiving region, increased overcrowding, and congestion. Meanwhile, the sending region loses labor power and purchasing power. While sending back remittances may offset this, the net effect is often a further exacerbation in economic development disparities between the richer receiving region and the poorer sending region. Continued

population movement may exacerbate regional differences. Governments can iron out regional differences through regional polices such as establishing growth poles in or guiding government investment and jobs to declining regions.

6.2 Population Differences: Gender, Age, Race, and Ethnicity

"Population" is an aggregate term that covers many different types of people. The sources of difference are many and include age, gender, sexual orientation, race, and ethnicity. We should be careful of imputing biological origins to sources of social difference. While male and female may be biological categories, what it signifies to be man or what it implies to be woman is socially produced and reproduced as well as celebrated and challenged. Our definition of what it means to be old, and when to be old, to take another example, is less a biological fact than a constantly shifting and changing social categorization. It is important to remember that differences are a product of social relations, not unchanging biological verities.

Social differences are codified, reinforced, and policed. They are contested and embraced as they are imposed and adopted, resisted and celebrated, becoming sources of constraint as well as platforms for creativity. They are also subject to change—even the most rigid divisions of gender or sexual orientation. We are in the middle of a period of flux when traditional categories are being redefined. In terms of sexual identity, some people identify themselves under the broad category of LGBTQIA+?. There are many sources of difference and identity. Here we will limit ourselves to exploring just three major sources of difference: gender, age, and race/ethnicity.

Gender

Space and place are gendered. The **gendering of space and place** is an important topic in human geography. Gender differences have spatial dimensions because men and women occupy space differently, certain places are gendered in that they are associated more with one gender than the other, sexism influences the shape and location of places, and the migration movements of men and women can and do differ. Across much of the world, women often have a form of second-class citizenship and are denied equal status with men.

Gender distinctions are often an integral part of cultural and social identity. In Aboriginal Australia, for example, there were separate sacred sites for men and women. Authority and control over these sites were highly gendered. In today's Catholic Church, women are allowed to be members of the church but are not allowed to become priests, bishops, or popes.

Different cultures assign different roles for men and women in the public and private spheres. Economic activities are often highly gendered. In areas of peasant agriculture, women are responsible for much of the agricultural labor. Traditionally, women were not involved in heavy industry such as coal mining and steel making, but the advent of mass assembly work meant more jobs for women. For example, in central Scotland in the 1960s the men worked in the mines and many of the young unmarried and older married women worked in the textile mills. Today, women and especially young women are a major source of labor in regions of manufacturing growth, such as along Mexico's border with the United States or in the booming coastal cities of China.

Gender roles in the economy vary over time. During the Second World War, for example, women were actively encouraged to enter the manufacturing labor force to help in the war effort. Rosie the Riveter, an image of female muscular power, was used to recruit women (Photo 6.1). After the war ended, women were encouraged to return to the domestic sphere.

In peasant economies and in economies shifting away from manufacturing toward services, women often play a more active role in the public realm. The **female participation rate** is the percentage of women aged fifteen to sixty-four employed in the labor force. The highest rates of around 85 percent are found in countries such as Cambodia, Burundi, and Zambia, where women work in the fields and as traders in the towns. In other regions of the world, women are restricted to the domestic sphere. The lowest figures are recorded in traditional Islamic countries such as Afghanistan, where it is 16 percent. Map 6.2 shows that the traditional Islamic countries have the lowest rates of female participation. In the United States the figure is 56 percent.

Women also do a great deal of work in the domestic sphere, and writers have referred to the double bind that

PHOTO 6.1 Rosie the Riveter.

exists for many women who work hard in the formal economy as well as in the domestic economy.

Space is an arena where gender roles are inscribed and transgressed, codified and undermined. The ability to move across space is also gendered. Many **patriarchal societies** limit women's mobility. An extreme example is the practice of foot binding of young girls in traditional China. This was the ultimate form of limiting women's freedom to move. Even today, women in Saudi Arabia are banned from driving, thus severely limiting their ability to move freely through urban space without a male presence.

There are also the more informal mechanisms. Women's use of urban space, for example, is more constrained than men's because of the fear of male violence. Women are more sensitive to the fear of sexual violence, and this structures their behavior in many cities. Strategies of individual safety include avoiding certain places at certain times, going to others only when accompanied, or not participating in an entire repertoire of activity, especially at night.

Gendered identities are rarely stable and never fixed. Amy Hanser looks at changing conceptions of gender and sexuality in her survey of new service work regimes in China. Under the Maoist regime, strong, physically robust, hard-working women were positively portrayed, but in the

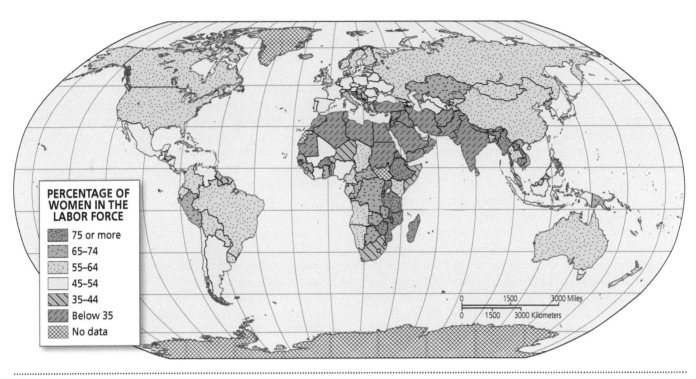

MAP 6.2 Female labor force participation rates. The participation of women in the labor force varies and can be affected by social, cultural, and political factors.

new, market-driven China other forms of femininity are employed. Hanser surveyed three retail outlets in the Chinese city of Harbin. In the state-owned department store, there was little attempt at gendered display of the women workers. Uniforms were bulky and unisex. Older rather than younger women worked in the store. The atmosphere echoed the Maoist era. In a wholesale/retail sector selling cheap clothes to recent rural migrants, the selling was done by young, sexily dressed women. There was a hypersexualized, undisciplined atmosphere. In the high-end, privately owned department store, the apex of the retail sector, the emphasis was on a youthful but disciplined femininity. The young women were trained in physical deportment and demeanor. They were presented for the discriminating **male gaze**.

Gender is a site of resistance and contestation as well as a setting for subordination and oppression. Elizabeth Wilson suggests that much early feminist writing on the city underscores the construction of women as passive victims. She argues for a wider debate that recognizes women's more active agency with cities as platforms for liberation where women widen horizons and escape traditional expectations.

There are tensions, connections, and intersections between race, class, ethnicity, age, and gender. Thomas Abowd, for example, examines the gendered politics of residential space in Jerusalem and looks at how unmarried Palestinian women are deemed "out of place." In this case, gendered relations intersect with other relations of power based on race, class, nationality, and religion.

Age

According to William Shakespeare, there are seven stages in the life cycle, from infant through schoolchildren going "unwilling to school" through adulthood to the final stages of a "second childishness."

What it means to be at the different stages varies across space and through time. In traditional peasant societies, young children were an important form of labor. They were prized for their muscles, not their brains. As society becomes more modern, people tend to have fewer children and devote more resources to each child. Young children are seen less as a source of domestic labor force and more as a receptacle of a formal education.

Biology determines when a woman can have a child, but the socially acceptable age of when to have a child varies. The social norm of when to have a child is a result of economic and social changes in what it means to be a woman of a certain age in society. In the United States, for example, the average age of a woman giving birth to her first child is now around 26.3 years, and across much of Europe it is closer to 29.4. Women are living longer and thus are able to delay childbirth. More women are spending more time in formal education and in the formal economy, although the result of delaying childbirth is a greater risk of birth defects and infertility. By contrast, in Bangladesh, where more traditional roles hold sway, the average age of women having a first child is 18. In Honduras the average age of women having a first child is 20, and before they are 18, 34 percent of women are married and 26 percent have given birth. In Hong Kong, where women feel that having children reduces their employment prospects, the average age of women having their first child is closer to 30. In Italy and Japan, it is 31.

Mary Robinson, the first woman president of Ireland and a former UN High Commissioner for Human Rights, wrote that "it is a sad irony at a time when the world has more older people than ever before—living longer with even greater wisdom and experience to offer—that they are often not respected as they have been in the past."

The practice of discriminating against a person because of their age, or ageism, is widespread in the world, and many have been marginalized because of their age. Older people are often regarded as victims of declining mental and physical capacity or threats to the opportunities of younger people, outdated but persistent stereotypes.

In most countries, it is still considered acceptable to deny people work, access to healthcare, education, or the right to participate in government because of their age. Recent studies have shown that ageism holds back more older women and men in the world from living well and with dignity than any other single factor. Other recent studies have shown that households with older heads or members tend to be poorer than other households and that ageism marginalizes and excludes older people in their communities.

The World Health Organization notes that ageism is the most socially "normalized" of any prejudice and is not widely countered—like racism or sexism. These attitudes lead to the marginalization of older people within our communities and have negative impacts on their health and well-being.

We could examine each of the stages of the life cycle and show how what they mean and how they are considered varies across the world, because societies have different religious beliefs and levels of modernity. In higher-income countries where people tend to live longer, the age of being "old" has shifted upward, expressed in such phrases as "sixty is the new forty." Because people on average live longer than in previous eras when our attitudes were shaped and codified, what it means to be old, and how "old" people should conduct themselves, is now subject to revision and change.

Race and Ethnicity

We need to begin this section with a warning: Race and ethnicity are neither natural categories nor are they unchanging. The very definitions of race and ethnicity are social constructions, classifications that themselves are the result of socioeconomic, not biological, processes.

Race is not so much a fact of biology as a social construct, partly adopted, partly imposed, and always constructed and defined in relation to others. Black is defined in relation to white and Asian to non-Asian. There is no separate biological fact of race other than the relative, socially constructed definitions that we use. Race is the name we give to perceived social differences based on a shifting and unclear combination of half-understood genetics and half-remembered history. We can more accurately consider "race" as a fluid, relational concept rather than a fixed, eternal, biological fact.

Races are identified in relation to other races, and this categorization is not innocent of political and economic considerations. They embody and reflect power relations. We classify to count, to demarcate, and ultimately to control. The term **racialized groups** in reference to this social construction reminds us that races are artificial constructs that embody political power and exact a social cost on those in the least favored, nonhegemonic categories.

Ethnicity, on closer linguistic attention, is even more vague. Race and ethnicity slide into each other in common discourse. And there are some categories that combine both, such as "Hispanic" in the US Census. Ethnicity is a more general term for different collective identities that are in part imposed and in part adopted. We can distinguish between **imposed ethnicity**, when a group of people are named and treated as ethnically different, and **adopted ethnicity** or **symbolic ethnicity**, when individuals choose to express an ethnic identity.

Although race and ethnicity may be socially constructed, and largely abstract, this does not negate their potent real-world influence. It is not simply that we have racial categories, but that we have constructed racial hierarchies that create layers of inequalities.

People often prefer to live with people like themselves. **Racial/ethnic clustering** occurs around the world as these neighborhoods provide a support system and a platform to the new society. The degree of clustering is a function of the degree of difference between the group and the dominant population. Racial/ethnic clustering may also result from discrimination, as households are forced explicitly or implicitly to live in certain areas, through coercion or social sanctions. The policy of spatial demarcation by race was an integral part of the apartheid system in white supremacist South Africa.

Race and ethnicity are also created and maintained by acts of power. Kay Anderson provides a fascinating case study of the evolution of Chinatown in Vancouver, British Columbia. She shows how the racial category of Chinese was constructed and maintained by white politicians and local elites. Chinatown in Vancouver, for example, was a

In the United States, the 2020 protests over the police killings of George Floyd, Breonna Taylor, and others renewed national conversations on structural racism and racial inequity. A recent survey of US mayors reported that the four groups most discriminated against in their cities and across the country are immigrants, transgender individuals, Black people, and Muslims. Access to affordable healthcare, primary and secondary education, and safe and affordable housing continues to favor white people. More recently, there has been a rise of anti-Semitism, anti-Asian hate crimes have been directed at women, and some Southern states are enacting new laws restricting voting and anti-trans healthcare laws. In fact, a 2021 report by Aainna Lynch and colleagues noted that "every State in the Union would receive a failing grade of 'F' for their performance toward reducing inequalities for Black, Hispanic, Indigenous, Asian and Multiracial or 'Other' communities."

Ensuring equal access to justice for all will involve, among other things, promoting campaigns to enhance legal awareness and literacy, scaling up services to provide advice and assistance, developing alternative dispute resolution mechanisms, and, ultimately, improving the institutional framework for resolving disputes, conflicts, and crimes. Discrimination also challenges the ability of those affected to have their voices heard. A key step to promote their inclusion is to remove obstacles to political participation, including the right to vote. Finally, addressing the root causes of discrimination calls for structural reforms, of the justice system and other national institutions.

REFERENCES

Lynch, A., H. Bond, and J. Sachs. 2021. *In the Red: The US Failure to Deliver on a Promise of Racial Equality*. New York: Sustainable Development Solutions Network.

McFarland, C., Andrews, L., Wasserman, A., Jones, C., Einstein, K., Palmer, M., Glick, D., and K. Lusk. National League of Cities. 2018. "Mayoral Views on Racism and Discrimination." https://www.nlc.org/sites/default/files/2018-09/CSAR_BostonU_REAL_Report_FINAL_small.pdf.

racial and spatial category created by institutional power. As she notes,

"Chinatown" was a shared characterization—one constructed and distributed by and for Europeans, who, in arbitrarily conferring outsider status on these pioneers to British Columbia, were affirming their own identity and privilege. That they directed that purpose in large part through the medium of Chinatown attests to the importance of place in the making of a system of racial classification.

(Anderson 1987, p. 594)

Chinatown in Vancouver, like other Chinatowns in cities around the world, has evolved from a place of compulsion and restriction of ethnic difference to what is now a setting for performance and celebration of ethnic difference.

In 1963, Nathan Glazer and Daniel Patrick Moynihan published their classic work on ethnicity, *Beyond the Melting Pot*. It was primarily a study of New York City and the extent to which various ethnic groups sturdily maintained their distinctiveness. Their detailed review of the city's different immigrant groups showed that urban living did not diminish the ethnic identities, but reinforced them.

The **melting pot** notion was always something of a fiction, but so is the idea that identities are unchanging, singular. There is a complex social interaction as newcomers and natives interact with one another, changing each in turn. Rather than a multicultural society, a condition that implies unchanging groups remaining separate, there are different forms of **polyculturalism** as various groups interact with and change one another. Rather than thinking of a simple continuum of assimilation, measured from the initial immigrants being different to successive generations becoming more like the rest of the society, we can think of a complex picture with both the immigrant groups and the host society metamorphosing. Thus, immigrants moving from southern Italy in the early nineteenth century were not Italians until they came to the United States. The construction of an Italian American identity was not transposed from Italy but created in the United States. And the active creation of a new Italian American identity shaped the United States.

6.3 The Movement of Population

Maps of population distribution are enlightening, but can also be deceptive. They freeze the movement of a restless species into a static picture. In reality, people are always on the move. There are the daily rituals of journeying to and from school and work, the weekly movements to places of recreation, the seasonal and annual trips and outings. There are also the more radical ruptures when we change houses, jobs, or even countries. Movement takes place on a daily, weekly,

PHOTO 6.2 Space-time traveler. This photograph of a young woman walking through an installation by the artist Jesús Rafael Soto at the Hirshhorn Gallery in Washington, DC, highlights the sense of movement though space.

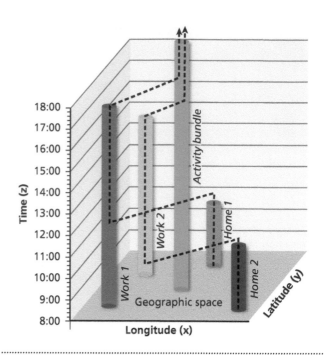

FIGURE 6.4 Space-time mappings.

seasonal, and annual basis following circular and noncircular patterns, sometimes involving permanent relocations.

Daily Space-Time Paths

Each day we make traces through space and time (Photo 6.2). Swedish geographer Torsten Hägerstrand realized that we move through both space and time, and he provided a way to map activity in each. Figure 6.4 traces the path of two people who leave their homes to go to work, then meet up in a cafe around 4:00 p.m. Such maps can highlight colocations in time and are enormously useful in tracing the paths we weave through space-time.

Our space-time paths are constrained. Much of our activity revolves around fixed pegs in space-time, such as home and work. A prism outlines the space that can be covered in the time available; those with a car will have a much larger prism, because they can cover a larger distance. They can therefore look much further afield for employment. What are the pegs and prisms that constitute your typical day?

Seasonal Movements

We move with the seasons. **Transhumance** is the traditional movement of mountain people with their animals from summer to winter pastures. In the Alps and the mountains of Scotland, for example, a traditional form of seasonal movement was to take cattle and sheep up the mountains to feed on the summer pasture. The herders and shepherds lived in modest shelters, known as sheilings

in Scotland. In winter, people moved back, with their flocks and herds, to the more protected and warmer valleys.

Rural-to-urban migration has swollen city populations across the globe. Sometimes the move is permanent. When it is seasonal, it is known as circular migration. In Delhi, for example, seasonal workers are known as *barsati mendaks*, the name of the frogs that appear during the monsoon. In one neighborhood of the city, Bara Tooti,

> The barsati mendaks, the rain frogs of Bara Tooti, are the seasonal workers from villages in Delhi's neighboring states of Uttar Pradesh, Haryana and Rajasthan. Most of them have land back home, a few acres that their fathers own, which will soon be divided among brothers. They come first in January after the winter crop has been harvested and the fields lie fallow, and return home for the sowing season in July. Once sowing is complete, they return to Bara Tooti for another few months of work before heading back to the village around Diwali [Hindu festival of light]. The barsati mendaks work frantically and live frugally to save as much money as they can. In the weeks leading up to Diwali they stop drinking or smoking, and save every last rupee so as to have something to show for the long absence of home. On the day before they leave, the mendaks hurriedly pay off their debts and pile into interstate buses headed homewards.
>
> (Sethi 2012, p. 16)

6.2 The Age of Distraction

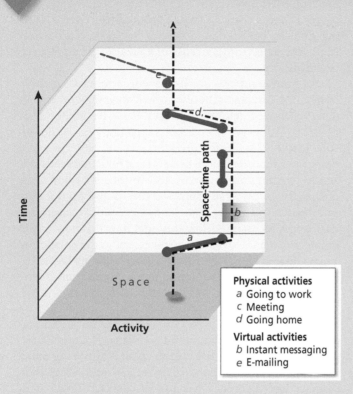

Physical activities
a Going to work
c Meeting
d Going home

Virtual activities
b Instant messaging
e E-mailing

In the world we live in, people occupy space while also communicating with people and things in other places. We need to amend the traditional space-time map to look more like Figure 6.7.

A common experience: you are walking down the street and someone is walking in the opposite direction toward you. You see him, but he does not see you. He is texting or looking at his cell phone. He is distracted, trying to do two things at the same time, walking and communicating. There is also the telltale recognition of a car driver on a phone; she is driving either too slow or too fast for the surrounding conditions, only partly connected to what is going on around her. Connected to someone else in another place, she is not present in the here and now.

These types of occurrences are now common enough that we can label our time as the **age of distraction**.

The age of distraction is dangerous. A recent report by the National Safety Council showed that walking while texting increases the risk of accidents. More than 11,000 people were injured in 2015 in the United States while walking and talking on their phones.

Distracted drivers have more fluctuating speed, change lanes fewer times than is necessary, and in general make driving less safe for everyone. Texting while driving resulted in 16,000 additional road fatalities from 2001 to 2007. More than 21 percent of vehicle accidents are now attributable to drivers talking on cell phones.

Multitasking is not so much an efficient use of our time as a suboptimal use of our skills. Student achievement improves when cell phones are banned, with the greatest improvements accruing to lower-achieving students, who gain the equivalent of an additional hour of learning a week. Students whose laptops are open learn less and recall less than students who have their laptops closed.

We are more efficient users of information when we concentrate on one task at a time. When we try to do more than one thing, we suffer from inattention blindness, which is failing to recognize other things, such as people walking toward us or other road users.

Media theorist Douglas Rushkoff asserts that our sense of time has been warped into a frenzied present tense of what he calls **digiphrenia**, the social media–created effect of being in multiple places and more than in one self all at once.

REFERENCES

Rushkoff, D. 2013. *Present Shock: When Everything Happens Now*. New York, NY: Penguin.

Short, J. R. 2015. "The Value of Unplugging in the Age of Distraction." *The Conversation*. http://theconversation.com/the-value-of-unplugging-in-the-age-of-distraction-43572.

All of the *barsati mendaks* are men. Their wives and sisters are at home caring for children, farming, and tending to household duties.

There are also the rituals for those who can afford vacations. College students across the United States look forward to spring break and Canadians to "reading week." Some concentrate on their studies, or continue to work or volunteer, while others take to the beaches and resorts in an annual rite of passage. There are also family vacations over the summer and, for some, ski vacations in the winter. We move to recreate and we recreate through moving.

Metropolitan Redistributions

There are various forms of movement between and within metropolitan areas. Urban-to-urban migration occurs as people tend to move from declining urban economies to expanding ones. Thus, people in Cleveland or Flint tend to move to better economic opportunities in the bigger cities and Sunbelt cities, such as Chicago or Dallas.

There is also **suburbanization**, as people move their permanent residence from the central city to the suburbs. This was the dominant form of metropolitan movement in the United States from the 1950s to 1990s. The move was selective, because those with more income were more able to afford the move. The poor were sometimes trapped in the central city neighborhoods.

More recently there has been a marked shift of population back to the city, especially of the more affluent, younger households in search of central city employment and urban lifestyle. Empty nesters whose children have left their suburban homes are now also leaving them to return to the city. As the number of single-person and nonchild households increases, a move to dense, walkable neighborhoods is also likely to continue. And again, the move is selective, with those who have more income better able to move into central city accommodations.

Rural to Urban

There are also relocations that involve a permanent change. The largest single permanent relocation of the past fifty years across the globe is rural-to-urban migration. The move to the cities is particularly pronounced in countries experiencing rapid population growth, industrialization, and modernization. Across the world, cities are growing in size as more people move in from the countryside in search of jobs, opportunities, and the prospect of a better life for them and their children.

In contemporary China, for example, there are approximately 280 million rural migrants to cities, especially to the burgeoning coastal cities. They constitute almost 20 percent of the total population. They face significant obstacles making their way. Under **hukou**, a system inaugurated in the 1950s to restrict internal movement, residents were classified as either urban or rural depending on where their parents were registered. Most were registered as rural and thus technically were not permitted to go to work and live in the cities. But from the 1980s, rural migrants have flooded into the city. Without the urban registration, they were denied access to equal housing and education. They are sometimes referred to as the "floating population." They now form a giant underclass.

Sometimes the rural-to-urban movement takes place within one country, as in the case of China. In other cases, people move from rural areas in one country to cities in another country. For instance, rural migrants from Mexico move to cities in the United States as well as to cities inside Mexico.

Rural-to-urban migration involves not only a change in location but also often a remaking of cultural identity as people move from the small-scale, intimate societies of the countryside to the larger, more heterogeneous mix of the big city. While rural-to-urban migration can bring economic benefits through remittances that can inject much-needed investment into the rural sending areas, it comes at some cost. To those uprooted from home, leaving spouses, children, and parents behind creates family trauma and, for some, a sense of alienation and precarity in the big city.

An immigrant city was the backdrop to what has been termed the **Chicago school** of urban sociology. One of its most celebrated figures is Robert Park; its other major figure, E. W. Burgess, is considered in Chapter 18. Robert Park (1864–1944) had a great love of the city. He was a newspaper reporter before he became a full-time academic, and his writings are infused with a journalist's acute sense of the city. In his 1916 essay, *The City: Suggestions for the Investigation of Human Behavior in the Urban Environment*, Park posits an **assimilationist model** in which new immigrants provide the dynamic raw material in the creation of a new urban society. New categories of people are created in the finer division of labor found in the city and the greater tolerance of difference. While the small community tolerates eccentricity, the great city rewards it.

Rural migrants to the city are often the most vulnerable when there are economic downturns. A study of migrant workers in China by geographer Kam Wing Chan and colleagues found that between 30 and 50 million workers lost their jobs because of the COVID-19 pandemic. More than 90 percent of rural-hukou workers could not find jobs, compared to only 42 percent of urban-hukou workers. The pandemic exacerbated preexisting inequalities for rural migrants to the city.

Forced Migrations

Forced migrations occur when people are forced to move. One of the most significant cases of forced migration was the **slave trade**.

The Slave Trade

From the fifteenth to the nineteenth century, people were torn from their communities in West and Central Africa and transported to the coast to be shipped off as slave labor for plantations, mines, and homes in the Americas. There was a **triangular trade** pattern that linked port cities such as Liverpool, England, with West Africa and then the colonies in the Caribbean and South and North America. Ships would carry trade goods to Africa, take on slaves there, sail across the Atlantic to the New World, and then return to Europe with commodities such as sugar and cotton for European markets and industries.

The human toll was enormous. Almost 4 million died during transportation within Africa and around 2 million died in the crossing as people were tightly packed together in appalling conditions. Around 11 million people arrived as slaves. Most went to South America; what is now Brazil received almost 40 percent. Around 7 percent ended up in the United States. The effects were enormous. In Africa the population stagnated and the once-small differences between the African and European economies widened. The profits from slavery were a basis for the accumulation of wealth in Europe and the Americas. The endowment of Brown University was built on the slave trade. The Catholic university of Georgetown made money by selling slaves. The evils of slavery permeated societies, permanently disfiguring their race relations and their very humanity. Slavery was the poisoned chalice that the Founding Fathers of the Republic such as Thomas Jefferson refused to challenge. It was only in the bloody Civil War of 1861–1865 that slavery was ended in the United States. Struggles like Black emancipation and equal rights continue even today. In 1999 the city of Liverpool offered an official apology for its role in three centuries of slave trading. In 2007 the state legislature of Virginia also offered an official apology. More states, including Alabama, Maryland, and North Carolina, have made similar declarations. In 2008, the US House of Representatives passed a resolution condemning American slavery and subsequent discriminatory laws. Throughout Africa there are calls for tribal chiefs to account for and apologize for their role in the slave trade.

The enforced migration of people from Africa changed both Africa and the New World. It is impossible to understand race relations in the United States without accounting for the history of slavery and its enduring legacy. Accumulated wealth in Europe and the Americas is built, in part, on the backs of unwilling migrants forced into slavery.

Forced Migration Today

Today, more common causes of forced migration include war, political conflict, discrimination, and social unrest. The United Nations defines a **refugee** as a person who, as a result of well-founded fear of being persecuted for reasons of race, religion, nationality, membership of a particular social group, or political opinion, is outside the country of his or her nationality. Refugees have crossed international borders and are entitled to protection and assistance from the states into which they move and from the international community through the United Nations and its specialist agencies.

Precise measurement is difficult, but estimates suggest that currently almost 20 million people have been forced to leave their country. They include people escaping civil war in Syria and families escaping gang violence in Honduras. Since 2015 almost 1 million Rohingya Muslims have fled Burma's Rakhine State to escape the military's large-scale campaign of ethnic cleansing. They have fled

PHOTO 6.3 Refugees leaving Ukraine and crossing into Poland.

mainly to Bangladesh, but also Malaysia, Thailand, India, and Indonesia. The vast majority of refugees are in the Global South. The largest refugee camps in the world are in the Horn of Africa, where drought, civil war, and social dislocation have displaced many people seeking food, shelter, and security and who end up in refugee camps. The largest refugee camp in the world, Dadaab in Kenya, houses close to 330,000 people.

More recently, the Russian invasion of Ukraine in 2022 caused a major population movement. It is estimated that within three months of the invasion, over 5 million people were forced to leave the country, the vast majority women, elderly people, and children (Photo 6.3). Over 7 million were displaced inside the country as people were forced to flee to escape Russia's unprovoked assault on a sovereign country.

We can also identify economic refugees who are not forced because of war or social conflict but feel compelled to move overseas in search of economic opportunity lacking at home. The mingling of political and economic refugees was evident during the million-strong wave of refugees to Europe in 2015. They were escaping the murderous civil war in Syria as well as the lack of economic opportunities in Africa, Iraq, and Afghanistan. The flood of refugees was so large and so sudden that it overwhelmed the traditionally welcoming nature of German society; in its wake, barriers to refugees were raised across Europe and the rhetoric of nationalism and anti-immigration became more strident throughout the continent.

While international refugees garner most public attention, millions of people are displaced inside their own country. The Internal Displacement Monitoring Centre estimated that in 2020–2021 there were 59.1 million

6.3 Climate Change, Inequality, and Migration

Climate change–related impacts and disasters are emerging as key drivers of human displacement and mass migrations. As climate change accelerates, drinking water may become scarcer, crops and livestock may struggle to survive, sea level rise may flood coastal areas, and desertification and other hazards from extreme weather events may force many people from their homes. There has been debate about whether to use the term "climate refugee." In 2018, the United Nations recognized that climate, environmental degradation, and disasters increasingly drive displacement, but it has not yet officially adopted the term, referring to those displaced by climate change as "environmental migrants." This may have more to do with the legal rights afforded a refugee and less to do with the reluctance to acknowledge that climate change is the underlying cause.

Climate change is predicted to worsen gender inequalities, for example, in cases where girls are the first to be withdrawn from schooling in response to drought or other climate-related shocks. Climate-related disasters could lead to increased vulnerability of women and girls to violence, for example, if they cause a shift in family power relations or lead to women and girls being vulnerably housed. Women's unequal access to economic resources can also compound their vulnerability to climate impacts.

In his book *Extreme Cities: The Peril and Promise of Urban Life in the Age of Climate Change*, Ashley Dawson examines the impacts of climate change on the world's megacities, including Jakarta, Delhi, São Paulo, and New York. Using a political ecology perspective, he argues that neoliberalism and racial capitalism have already made cities places of stark economic inequality (what he terms "extreme"); as a result, vulnerable populations such as people of color and those experiencing poverty will be even more vulnerable to floods, sea level rise, heat events, and so on. For example, in Indonesia the major brunt of climate change will be faced by the 26.5 million who live below the poverty line and have limited resources and capacity for resilience. Millions of these live in Jakarta. The climate shocks and stresses will also force the near-poor population hovering marginally above the national poverty line to fall into poverty. A political ecology perspective draws attention to the poverty–climate nexus and underscores that climate actions need to be carefully designed so that they explicitly benefit the poor and near poor and do not inadvertently increase vulnerability and inequality.

REFERENCE

Dawson, A. 2017. *Extreme Cities: The Peril and Promise of Urban Life in the Age of Climate Change*. New York: Verso.

internally displaced persons (IDP), 53.2 million as a result of conflict and violence and another 5.9 million because of disasters. That is up from a total of 26.4 million in 2012. The figure currently includes 6.6 million in Syria, 6 million in Colombia, 3.3 million in Iraq, 2.3 million in Sudan, 2.9 million in the Democratic Republic of Congo, and 2.6 million in Nigeria. Although many IDPs face the same difficulties as international refugees, they are not granted the same rights under international law. In addition, because IDPs stay within their own country, they remain under the protection of their government, even if that government is the reason for their displacement. IDPs still have rights, but some governments are unable or unwilling to honor those rights. So while this is the largest population of displaced people, it can also be one of the most vulnerable.

International Migration

Voluntary international migration is an important form of population movement. Almost 281 million people live outside the country of their birth, around 3 percent of the total world population. Half of all international migrants are women. If we include illegal and undocumented workers, the level of international migration across the globe is even higher. These immigrants are either temporary migrants, perhaps sending money back in the form of remittances, or permanent migrants making a life in a new country. International migrants are the raw material of cultural globalization because they bring their culture with them, adding a new cultural dimension to their destination. It is rare that cultures are mechanistically transported; there is creative reworking as cultural forms are adapted and changed.

Compared to the rest of the population, international migrants tend to be younger and better educated and are moving for employment-related reasons.

There are powerful push-and-pull factors at work as people search for better living conditions or religious and political freedoms in other countries. The level of migration depends on the range of factors pushing and pulling migrants, as well as the regulatory regime that controls migration. Countries vary in the severity of their immigration regimes, which can be defined as the set of rules that govern the entry, settlement, and assimilation of foreign migrants. Regimes tend to be more severe during

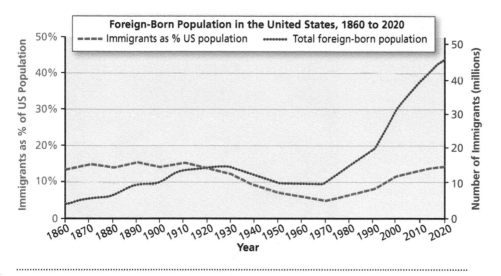

FIGURE 6.5 Foreign-born population in the United States, 1860 to 2020.

economic downturns and more relaxed when economic growth creates labor shortages. In the late nineteenth century, for example, the United States had an open-door policy that encouraged immigrants from Europe but severely restricted those from China. Today, most countries try to control immigration. Regimes vary from the more open, such as those in Australia and Canada, to the more closed and restricted, such as those in Japan and South Korea.

Migration affects both origin and destination countries. Origin countries may lose some of the most talented population but receive remittances. Destination countries may receive skilled migrants, but there may be a depression of wage levels in sectors where the often-cheaper immigrant labor force predominates.

International migration tends to increase during periods of increased globalization, such as in the late nineteenth and early twentieth centuries and over the last thirty years. Figure 6.5 plots the immigration history of the United States. High absolute and relative numbers of immigrants were recorded from 1890 to 1930. During this period, there were few restrictions on immigration from western Europe, although from 1924 there were restrictions on immigrants from east and southern Europe. People came from Ireland and Italy, Russia and Scandinavia, in the hundreds of thousands, changing the religious and ethnic composition of the country and leaving a permanent legacy. This period of large-scale immigration ended after 1930 as tighter immigration controls were launched at the start of the Great Depression. The number of foreign-born declined in absolute numbers and as a proportion of the total population. Between 1930 and 1960, the previous waves of immigrants were incorporated into the mainstream of US society. When the controls were lifted in the mid-1960s, a new wave of increased immigration ensued. The absolute and relative levels of

the foreign-born rose to levels not seen for over a hundred years. Between 2000 and 2005, 8 million immigrants entered the country, the largest five-year total in the nation's history. By 2020 there were almost 45 million foreign-born immigrants in the United States.

This second, more recent large wave of immigration has a wider source than the first, including Africa and Asia as well as Europe and Latin America, and gives the United States a new and different feel, making it more Hispanic, more Asian, and more cosmopolitan. While many immigrants in the first wave arrived in cities, in the second wave there has been a wider spread throughout greater metropolitan regions.

The current wave of international immigration across the world is now substantial. Resistance in destination countries is generated from perceptions of depression of living standards for native-born workers and feelings of local and national cultures being swamped by the foreign others. In the United States, for example, there is the added sense that between 12 million and 15 million people are in the country illegally. The description of this group is itself a battleground, with terms such as "illegals," which references the illegality of the people's permanent residence, competing with the terms "unauthorized" and "undocumented." Many countries are now significant sites of foreign immigration. The Gulf States, for example, with their combination of wealth, small population, and large demand for labor, have some of the highest percentage rates of foreign-born, including the United Arab Emirates (88 percent), Qatar (75 percent), and Kuwait (74 percent). However, most of the immigrants in these countries are temporary, with no path to citizenship provided. Among the larger developed nations, the highest rate is found in Australia (28 percent). In all these countries, the percentage of foreign-born has increased steadily since the 1990s, a result of increasing flows of migrant labor. In the United Kingdom, the percentage of foreign-born increased from 5.8 percent in 1971 to 14.0 percent in 2021—from over 3 million to just over 9.5 million in absolute terms. In the United States, the percentage of foreign-born has increased from 4.7 percent in 1970 to close to 13.7 percent in 2015. The United States has the largest absolute total of foreign-born, more than 45 million. Scale becomes important here as well. Although the foreign-born represent around 13.7 percent in the United States overall, the figure is higher for the major metropolitan areas. Nearly 40 percent of residents of New York are foreign-born and Miami has 55.8 percent foreign-born. Suburban cities in metropolitan areas such as Fremont, California (43 percent), and Irving, Texas (34 percent), have significant numbers of foreign-born individuals.

We should also note that there is a wide range of countries with very low numbers of foreign-born. They include rich countries such as Japan (2 percent), poor countries such as Sudan (1.7 percent), eastern European countries such as Poland (1.8 percent), authoritarian regimes such as China (0.1 percent), and South American countries such as Colombia (0.6 percent). A mixture of limited economic opportunities, more attractive alternatives, and restrictive immigration policies limits the number of foreign-born in these countries.

DIVERSE EXPERIENCES

There are multiple migrant experiences. Take the case of Dubai, where there are two very different streams of migrants. There is the encouragement of investors, technical experts, and those associated with **flight capital**, the practice of seeking a safe and opaque site for money made elsewhere. Then there are the construction migrants, often coming from rural communities in Pakistan and Bangladesh, who live many to a room and have difficult working conditions. Their entry is tightly controlled and monitored. Luxury homes and migrant camps, foreign investors and building laborers, welcome guests and barely tolerated foreign workers: these are the extremes of divergent migrant experiences. There are also the expatriate communities of technical workers and experts—English financial specialists in Singapore, Scottish engineers in Saudi Arabia, European aid workers in Nairobi, Indian doctors in the United States. And there are the lower-skilled and often exploited workers—the female Filipino domestic servants in Hong Kong, the Mexican gardeners in California, and the vast armies of undocumented and illegal workers in shadow urban economies around the world.

GATEWAY CITIES

International migration channels are often routed very largely, although not exclusively, through major metropolitan areas. In cities around the world, the number of the foreign-born has increased. In the United States, for example, a third of all the foreign-born are located in the five largest metropolitan areas. Marie Price and Lisa Benton-Short identify immigrant gateway cities around the world. They include

- Established gateways such as Amsterdam, Birmingham (United Kingdom), New York, Sydney, and Toronto.

PHOTO 6.4 Ethnic homogeneity in Seoul, South Korea.

- Emerging gateways such as Dublin, Johannesburg, Singapore, and Washington, DC.
- Exceptional gateways such as Riyadh and Tel Aviv.

Global and globalizing cities are now places of minority populations, **diasporic communities**, and **hybrid identities**. The metropolis is now the place of the Other, who, with their distinctive languages, cuisines, and cultures, now add to the cosmopolitan mix of the big cities. There is undeniable global mixing in the large metropolitan regions, but not in all big cities. In countries with restrictive regimes of tight regulations on the entry of foreign-born immigrants, even very large cities may remain remarkably homogeneous. Seoul, in South Korea, has a population of more than 10 million, but less than 1 percent of them are foreign-born. Seoul is a closed gateway city (Photo 6.4).

Immigrant places are where otherness is renewed and celebrated, performed and remembered in creative acts of representation and memory. And this representation adds to the cosmopolitan mix of the city.

6.4 Sustainability, Equity, and Migration

Migration is a powerful driver of sustainable development. Migrants can bring significant benefits in the form of skills, strengthen the labor force, increase cultural diversity, and contribute to improving the lives of communities in their countries of origin through the transfer of skills and remittances. Despite these benefits, if migration is poorly managed, it can negatively impact development. Migrants can be put at risk or may face hostility from their host countries, limiting them from reaching their full potential.

The International Convention on the Protection of the Rights of All Migrant Workers and Members of Their Families aspires to guarantee dignity and equality in an era of globalization. The convention aims to protect migrant workers and members of their families; its existence sets a moral standard and serves as a guide and stimulus for the promotion of migrant rights in each country.

The convention does not create new rights for migrants but aims at guaranteeing equality of treatment and the same working conditions, including in the case of temporary work for migrants and nationals. The convention relies on the fundamental notion that all migrants should have access to a minimum degree of protection.

Currently, only fifty-five countries have ratified the treaty (the United States has not, nor has Australia, both of which are major receivers of migrants). Countries that have ratified the convention are primarily countries of origin of migrants (such as Mexico, Morocco, and the Philippines). For these countries, the convention is an important vehicle to protect their citizens living abroad.

6.4 Models of Population Movement

There are numerous models of population movement. Each seeks to explain or predict who, how, where, and why migrants move.

Ravenstein

One of the earliest models was developed by E. G. Ravenstein (1834–1913), who was born in Germany but moved to Britain, eventually becoming a professor of geography. He arrived at a time of intense population movement associated with rapid urbanization and industrialization. Spurred by a comment that the migration process was random, he analyzed UK census data in 1871 and 1881 to propose several laws of migration. These included the following:

1. Most migrants move only a short distance, toward the nearest city.
2. Growing cities attract migrants from nearby rural areas that depopulate rapidly; in turn, migrants from more distant areas fill the "gaps" left in the rural population. Migration is stepped.
3. Each main current of migration produces a compensating countercurrent.
4. Long-distance migrants tend to move to major cities.
5. Rural people have a higher propensity to migrate than urban people.

6. Women have a higher propensity to migrate than men within countries, but more men tend to move in international migration.
7. Migration is driven in largest part by economic forces.

These laws are not physical laws, but generalizations made at a particular time in a specific context. While some of them (laws 3, 4, and 7) stand up in many different circumstances, the others can vary depending on the context. There are more female than male Filipino international immigrants, undermining law 6. In the United States, more than one-third of all Filipino migrants live in just three metro areas, tending to support law 4.

Gravity Models

Gravity models suggest that population movement between two areas is a function of their population size and an inverse function of the distance between them. It is sometimes presented as an equation,

$$\frac{P1 \times P2}{D^2}$$

where $P1$ is the population of city 1, $P2$ is the population of city 2, and D is the distance between them. It a is direct application of Newton's law of universal gravitation, which states that the pull between two bodies is a function of their combined mass divided by the square of the distance between them. While this model may work in a world of empty space, the distance between places in the real world is more complex; it is warped by space-time and distorted by political boundaries. Nevertheless, the model does provide some useful predictors such as how larger places attract more people and how places closer together generate more migrant flow.

The space between places is also the site of **intervening opportunities**; the gravity model thus needs to be tweaked to account for these.

Lee's Push–Pull Model

Moving voluntarily requires a comparative assessment of the origin area's characteristics with those of possible destinations. Everett Lee specifies a model with reference to specific push and pull forces between origins and destinations. Typical push forces include limited economic opportunities, social unrest, and political instability, while pull factors include readily available employment opportunities, presence of friends and family, and social and political stability.

Origin and possible destination areas have a range of positive and negative factors. For example, positive factors for origin areas include connection to family, existing social networks, and a feeling of belonging. Negative factors may include lack of jobs, restricted social opportunities, and poor health and educational facilities. The main factors that push and pull migrants are employment and social opportunities. Migrants assess or sometimes are forced to assess the relative weight of different factors. When the pull or push factors reach a threshold, or when the positive factors in the destination outweigh the positive factors in the origin areas, then people move.

Consider the case of the great migration of African Americans. More than 6 million African Americans moved between 1916 and 1970 from the rural South to the towns and cities of the North and the East. They were pushed out by racial discrimination, economic marginalization, and restricted educational and social opportunities. They were attracted by the jobs and less explicit racial discrimination outside the South. African Americans could triple their wages by leaving the rural South for the urban industrial North and be free from the injustice of institutionalized racism. The **Great Migration** changed the identity of African Americans, making them a predominantly urban rather than rural population and in the process changing the racial balance of cities across the county.

Lee also identified intervening difficulties of distance, cost, and accessibility. The model is summarized in Figure 6.6, where the pluses and minuses refer to the push and pull factors. Thus, a migrant leaving Syria may decide to head for Austria since it is closer and perhaps easier to get to than Sweden. And while many migrants in North Africa and the Middle East would like to get to the United States, the distance, cost, and difficulties are high, and the intervening opportunities are many.

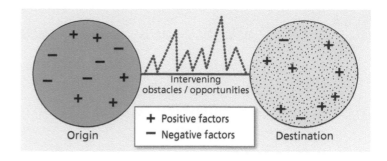

FIGURE 6.6 Lee's push–pull model.

The model assumes that people can rationally evaluate all the positive and negative factors. In most cases, people act in a suboptimal manner rather than optimally. They have limited knowledge, limited resources, and often too little time to make complex and difficult decisions. They are strongly affected by the information and advice from friends and relatives. This explains the existence of **chain migration**, where people move to a new destination and then convince friends and relatives to join them. This form of chain migration is a common characteristic of migration flows; people tend to move along channels and to destinations, followed by friends and families. One chain links domestic female workers from the Philippines to jobs in Hong Kong and the Gulf States, another links male manual workers from South Asia to work on construction sites in the Gulf States, and yet another links young men and women in Goa, India, with work on cruise ships. Once a strong link is established, a stable chain of migration links people with specific skills from specific origin areas with particular destinations.

There are many models of migration. For rural-to-urban migration, they include the Lewis model, the Fei–Ranis model, and the Harris–Todaro model. There are also the economic models of overseas migration that argue that households send workers abroad to improve their position relative to other households at home. In many countries, such as the Philippines, the significant flow of workers overseas generates remittances that are used to boost household income and spending power. Many new houses at home are financed with money earned abroad.

Like all models, population models are abstract representations of some portion of the real world. Migration models may help explain why some people are more likely to migrate than other people at specific times. Yet they are also limited in explaining the complex reasons why migrants move to certain places. In addition, models generally acknowledge that the timing of migration and the destination of migrants cannot be predicted with certainty.

6.5 The Zelinsky and Metz Models

We have already touched on the demographic transition model in Chapter 5. This model was the inspiration for the geographer Wilbur Zelinksy, who wanted to develop a transition model for mobility. The **Zelinsky model** consists of three phases. In the premodern traditional society, residential movement and personal mobility are very limited and circumscribed. In the advanced society, residential mobility increases; migrants are on the move, especially in the form of rural-to-urban migration; and circulatory movement such as tourism begins to increase. In the superadvanced state there is an increasing acceleration of movement in terms of both residential and personal mobility. In this final stage, Zelinsky suggests that there might be limits to increasing mobility; he gives the example of traffic lights that limit speed, despite the ability of vehicles to drive fast. Speed is reduced in the interests of traffic management and public safety.

Transport specialist David Metz employs a similar model, though he identifies four rather than three phases. In the first era of human travel, our hunter-gatherer ancestors walked out of Africa and populated the rest of the Earth. In the second era, they settled in agricultural communities and towns, where travel was generally limited to about an hour a day on foot. The third era began early in the nineteenth century with the coming of the railways, when the energy of fossil fuels could be harnessed to achieve faster travel through a succession of technological innovations, culminating in mass mobility made possible by the motorcar. There is now emerging evidence that we have passed peak car use. We are entering a fourth era, according to the **Metz model**, in which aggregate travel times, trip rates, and distance traveled are holding steady. The "peak car" phenomenon, whereby car mode share in cities like London reached a peak and then recently declined, marks the transition from the third to the fourth era.

REFERENCES

Metz, D. 2013. "Peak Car and Beyond: The Fourth Era of Travel." *Transport Reviews* 33:255–270.

Zelinsky, W. 1971. "Hypothesis of Mobility Transition." *Geographical Review* 6:219–249.

Cited References

Abowd, T. 2007. "National Boundaries, Colonized Space: The Gendered Politics of Residential Life in Contemporary Jerusalem." *Anthropological Quarterly* 80:997–1034.

Anderson, K. J. 1987. "The Idea of Chinatown: The Power of Place and Institutional Practice in the Making of a Racial Category." *Annals of the Association of American Geographers* 77:580–598.

Che, L., H. Du, and K. W. Chan. 2020. "Unequal Pain: A Sketch of the Impact of the COVID-19 Pandemic on Migrants' Employment in China." *Eurasian Geography and Economics*. 61: 4–5, 448–463: https://doi.org/10.1080/15387216.2020.1791726.

Glazer, N., and D. Moynihan. 1963. *Beyond the Melting Pot*. Boston: MIT Press.

Hägerstrand, T. 1970. "What About People in Regional Science?" *Papers of the Regional Science Association* 24:7–21.

Hanser, A. 2005. "The Gendered Rice Bowl: The Sexual Politics of Service Work in Urban China." *Gender and Society* 19:581–600.

Internal Displacement Monitoring Centre. 2022. *Global Report on Internal Displacement 2022*. https://www.internal-displacement.org/global-report/grid2022/.

Lee, E. S. 1966. "A Theory of Migration." *Demography* 3: 47–57.

National Safety Council. 2021. "Pedestrians." https://injuryfacts.nsc.org/motor-vehicle/road-users/pedestrians/data-details/.

Park, R. 1916. "The City: Suggestions for the Investigation of Human Behavior in the Urban Environment." *American Journal of Sociology* 20:577–612.

Price, M., and L. Benton-Short, eds. 2008. *Migrants to the Metropolis*. Syracuse, NY: Syracuse University Press.

Ravenstein, E. G. 1885. "The Laws of Migration." *Journal of the Statistical Society of London* 48:167–235.

Robinson, M. 2018. "Do We Respect Our Elders?" Age International. https://www.ageinternational.org.uk/policy-research/expert-voices/do-we-respect-our-elders/.

Rushkoff, D. 2013. *Present Shock: When Everything Happens Now*. New York, NY: Penguin.

Sethi, A. 2012. *A Free Man: A True Story of Life and Death in Delhi*. London: Jonathan Cape.

Wilson, E. 1991. *The Sphinx in the City*. London: Virago.

Select Guide to Further Reading

Anderson, K. 1991. *Vancouver's Chinatown: Racial Discourses in Canada. 1875–1980*. Montreal: McGill University Press.

Cordey-Hayes, M. 2012. "Migration and the Dynamics of Multiregional Population Systems." *Environment and Planning A* 7:793–814.

De Haas, H., S. Castles, and M. J. Miller. 2020. *The Age of Migration: International Population Movements in the Modern World*. 6th ed. London: Bloomsbury.

Goldscheider, F., E. Bernhardt, and T. Lappegård. 2015. "The Gender Revolution: A Framework for Understanding Changing Family and Demographic Behavior." *Population and Development Review* 41:207–239.

King, R., and R. Skeldon. 2010. "'Mind the Gap!': Integrating Approaches to Internal and International Migration." *Journal of Ethnic and Migration Studies* 36:1619–1646.

Laslett, P. 1989. *A Fresh Map of Life: The Emergence of the Third Age*. London: Weidenfeld & Nicolson.

Pablos-Mendez, A., S, Radloff, K. Khajavi, and S. A. Dunst. 2015. "The Demographic Stretch of the Arc of Life: Social and Cultural Changes That Follow the Demographic Transition." *Global Health: Science and Practice* 3:341–351.

Strulik, H., and S. Vollmer. 2015. "The Fertility Transition around the World." *Journal of Population Economics* 28:31–44.

Wong, A., ed. 2020. *Disability Visibility: First Person Stories from the Twenty-First Century*. New York: Vintage Books.

Population and Food

In this chapter, we will consider how population connects with one of the most important resources, food. In 2021, an estimated 950 million people across ninety-three countries faced food insecurity and hunger, up from 821 million in 2018. A World Food Programme report points to several reasons behind a recent rise in hunger, including conflict, climate variability, and economic slowdowns. Food and hunger are intricately linked to environmental and economic change. The relationship between population and food is complex and multidimensional.

7.1 Population and Agriculture

Before the advent of agriculture about 10,000 years ago, the total world population did not exceed 10 million people, fluctuating between 5 million and 15 million and spread thinly across the surface of the globe. **Settled agriculture** allowed an increase in population because the food resource base was increased and made more reliable. With a specialized division of labor, effective utilization of crops and animals, and centralized political control, population could increase beyond the limits of **hunting-gathering societies**. By 2,000 years ago, the population had increased to 200 million. Improvements in agricultural technology, such as the use of the plow and hybrid crops, enabled more food to be produced, sometimes by extending the amount of arable land and sometimes by increasing the productivity of existing agricultural land. The areas of greatest population density had rich, fertile soils and reliable sources of water, as well as centralized political control.

7.2 Malthusian Melancholy

The relationship between food and population has long interested social observers. One has left a permanent intellectual legacy. Thomas Robert Malthus (1766–1834) was one of eight children born to a prosperous middle-class family in Surrey, England. He is best known for his work *An Essay on the Principle of Population*. He worked on six successive editions of this text from 1798 to 1826.

The historical context is important. Writing in the convulsive times of the French Revolution, Malthus was one of the many English reactionaries who were wary of radical change, distrustful of social progress, and skeptical of massive improvement and the idea of upward progress. His basic argument is that while population increases at a geometric rate, food supply grows only at an arithmetic rate. The end result is that there are more people than the available food supply. When things are good, people have more children, but eventually this population growth exceeds the food supply needed to feed them. Families respond to this dire situation by having fewer children, and so population decreases until things are back in equilibrium, when the cycle starts all over again. Human history is one of population increase overshooting the available food supply. Misery, poverty, and famine, the so-called **Malthusian checks**, are "natural" elements that bring population growth back into alignment with the food supply.

This simple model has policy implications. As a social conservative, Malthus was against social policy that disrupted the existing social order and its workings. He was highly critical of policies aimed to help the poor. Social welfare provision, Malthus believed, allowed the poor to have children and, for him and his followers, that simply made things worse. He believed that the number of children born to the poor should be limited.

His ideas influenced contemporary British politicians, who introduced the census in 1801 as a means of counting the population; and the New Poor Law of 1834, which launched workhouses for the poor, so familiar to the readers of Charles Dickens. His ideas filtered out to social theorists such as Herbert Spencer (1820–1903), who espoused **social Darwinism**, based on the idea of the survival of the fittest. Malthus's works were roundly criticized in the radical working-class journals of the day. In 1820, William Godwin calculated that, using China as his model, the world's population could comfortably reach 9 billion. He was writing at a time when the population was little more than 1 billion and few at the time could even imagine today's population of 8 billion.

Malthus's work was socially specific, written at a time of intense political debates and just as the demographic transition and the Agricultural and Industrial Revolutions were taking off. Yet his ideas have proven very influential, echoing down through the subsequent years.

One continuing legacy is a pessimism regarding the ability to expand population beyond a fixed base because the food supply is considered relatively fixed or capable of only slow increase. Malthus presents a specter of population increasing faster than the supply of food to feed it. He was writing just as the agricultural revolution was transforming agricultural productivity. Since then, in the past 200 years, agricultural productivity has increased remarkably, creating a much larger food supply base. In one important way, Malthus's gloomy predictions proved incorrect. The global population has increased to 8 billion, and yet global hunger is not inevitable, even with a larger population. According to the Food and Agriculture Organization, the proportion of people in developing countries with food intakes below 2,200 calories per day fell from 57 percent in 1965 to just 10 percent in 1998. For those with access to food, the price has steadily declined by 60 percent since the 1980s. The Food and Agriculture Organization also estimates that the number of undernourished people in the world fell from over 1 billion in 1992 to 792 million in 2015, even as the population increased from 5.4 to 7.5 billion. However, after the COVID-19 pandemic in 2019/20 and the disruption of global supply chains and a decline in economic activity, hunger has increased. Since 2019 there has been an uptick in the global prevalence of undernourishment. In 2020, between 710 million and 811 million faced hunger, and this figure is likely to rise as the full effects of economic recession take their toll, especially on the world's poorest. The most recent estimate is as high as 950 million people.

We can also make a distinction between "undernourished," lacking enough calories to sustain a healthy state, and "malnutrition," a diet based on poor-quality food lacking necessary vitamins and minerals to sustain good health. Later, we will discuss the problems of malnutrition, a growing burden in many countries that nevertheless have significantly reduced the issue of undernourishment.

The Danish economist Ester Boserup reversed the Malthusian argument. Rather than agricultural technology determining population, it is population that determines agricultural technology. Necessity, in this case population pressure, is the mother of invention. In the history of agriculture, when population levels and densities are low, shifting agriculture can be employed, but when population levels and densities increase, field use becomes more permanent, which spurs agricultural innovations. These include irrigation, development of hybrid plants, weed control, specialization in specific crops, and fertilizer use. Agriculture intensifies as population pressure mounts.

In one sense, at least, Malthus's prophecy of doom has failed to materialize, while Boserup's more optimistic view is sustained. Population has increased, yet so has global food supply (Photos 7.1 and 7.2). All the evidence suggests that world agricultural production can grow even more, in line with increased demand. Charles Godfray and colleagues looked at the ability of the global food production system to meet the demand of 9 billion people by 2050. They suggest that it is possible through a variety of methods:

- Closing the **yield gap**—that is, the difference between possible and actual productivity. In Southeast Asia, for example, there is a 60 percent difference between average and maximum rice yields. Yield gaps are larger in failed or dysfunctional states or in poor countries with limited access to the world markets or to easy and cheap credit. Political stability and world trade can increase yields by making it profitable for farmers to invest more.

- Increasing production limits. The **Green Revolution** successfully increased agricultural production throughout the world from the 1960s and 1970s by developing hybrid varieties of maize, rice, and wheat and increasing the use of fertilizers and pesticides. Genetically modified crops can increase productivity. However, there is mounting public disquiet about the long-term implications of engineering our food supply and the unfolding consequences of synthetic pesticides and fertilizers in the food chain.

- Reducing food waste. Around 35 percent of food in the world is wasted. The reason varies from lack of cold storage to "supersized" portions. Food is so cheap in some places that there is little incentive to avoid waste. Each year in the United States, over 70 billion pounds of food is wasted, up 20 percent from a decade ago.

- Changing diet. As societies get richer, they tend to consume more meat. This involves growing grain to feed livestock. Reducing the consumption of meat allows grassland for animals to be turned into arable farming. Reducing meat consumption also has a positive health benefit, since it reduces heart disease rates.

- Expanding aquaculture. Fish farming increases the amount of relatively cheap protein. There are, however, issues of environmental pollution and genetic contamination with an increase in aquaculture.

Despite the tremendous increases in agricultural productivity that have rendered Malthus's original predictions obsolete, problems remain. Let us consider four of them: hunger, sustainability, nutrition, and ethics.

PHOTO 7.1 Grocery store in Budapest, Hungary. Local, national, and international food chains converge on this one store in Budapest to ensure fresh fruit and vegetables in January.

PHOTO 7.2 Food preparation and sale in Grenada. Women constitute the vast majority of food preparers and sellers at this weekly market in Grenada's capital city, St. George's, as they do in much of the world.

7.3 Hunger, Famine, and Food Insecurity

The declining levels of undernutrition are testament to the ability of agricultural production to surpass the Malthusian limits. Yet hunger remains. Map 7.1 shows the

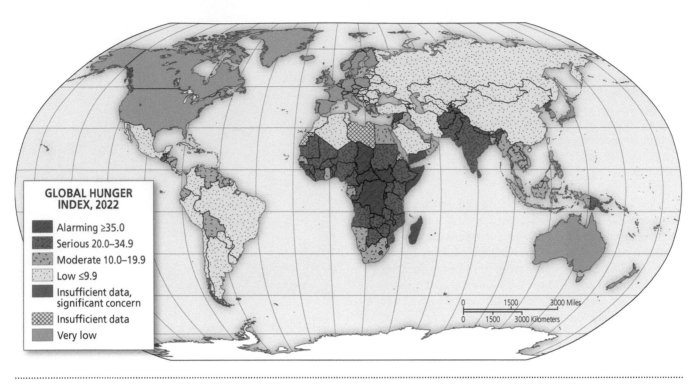

MAP 7.1 Global Hunger Index.

distribution of global hunger. Notice the extreme concentration in tropical Africa and South Asia.

The term "food security" is employed to refer to food's availability. Food security is when people have relatively easy access to safe and nutritious food. In contrast, food insecurity is a lack of access to enough safe and nutritious food. Food security can be conceptualized as an umbrella term for three interconnected concepts, often called the "three S's":

1. *Security*: Food must be available and accessible at all times, supplied through domestic production or imports (including food aid).

2. *Sovereignty*: Food is always accessible and people are empowered to make their own choices about the food they eat by purchasing food, growing food for their own consumption, or bartering for food.

3. *Safety*: Food must be safe and have a positive nutritional impact on the body.

Of particular note is food sovereignty, which began as a call for empowering farmers to make decisions about what to grow and when. In many countries in the Global South, the food sovereignty movement has become a bottom-up movement to protect the rights of local farmers and small-scale farmers, many of whom are Indigenous. Food sovereignty is part of the food justice movement and has the potential to substantially transform political economic relationships around food, land, and markets.

Food insecurity can be either chronic (long-term) or transitory; it leads to hunger and undernourishment, making people more susceptible to disease and illness. Worldwide, it is estimated that there are approximately 1 billion without enough calories and a further 1 billion who are malnourished in a form of hidden hunger because they do not consume enough nutrients for a healthy life. Almost 250 million preschool children lack vitamin A, and 500 million women have an iron deficiency that causes 60,000 deaths each year during pregnancy.

We can make a distinction between the demand for food—a universal need, since we all have to eat to remain alive—and the effective demand for food, which is the ability to pay for food. The issue is not the relationship between demand and supply, but that between effective demand and supply. Hunger occurs not only because there is not enough food but also because people are too poor to buy food. Fundamentally, hunger is caused by poverty.

Although hunger and food security are related, they are not the same. On the one hand, hunger is physiological and reflects the physical pain and discomfort an individual may experience. Hunger is a potential consequence of food insecurity that results in discomfort, illness, weakness, or pain that goes beyond the usual uneasy sensation.

Food insecurity, on the other hand, occurs in a broader socioeconomic context. Food insecurity is connected to social, political, and economic factors. Land use policies at the national level may encourage producing export crops such as cotton in an attempt to increase foreign exchange

7.1 Sustainability, Overfishing, and Fish Farming

The oceans of the world have long been an important food source. Along the coast and deep in the ocean, fish stocks are protein-rich sources of food and nutrition. However, as industrial-style fishing with large boats and fish-detecting sonar now sweep the oceans, the supply of fish is endangered. **Overfishing** is now a major problem. The global increase in the distribution of severe exploitation is one reason behind the growth of **fish farming** (Photo 7.3). Some wild species, such as the bluefin tuna, are on the verge of extinction as a result of overfishing. As the fleets go into deeper water, deep-sea fish that mature and reproduce slowly (such as the orange roughy, with a life span of 149 years) are particularly vulnerable. With nonselective fishing and a weak regulatory system, deep-sea fishing will

PHOTO 7.3 Fish farming in Montenegro.

soon exhaust the fish stock. Wild fish supplies peaked in the 1980s, and their global decline will put extra pressure on farming to replace the protein provided by the world's fisheries.

In addition, the increase in overfishing has also given rise to an increase in forced labor and human trafficking in the fisheries sector. Migrant workers in particular may be deceived and coerced by brokers and recruitment agencies to work on board vessels under the threat of force or by means of debt bondage (Photo 7.4).

PHOTO 7.4 On large commercial fishing boats in the gulf of Thailand, most of the fishermen are indentured Burmese workers.

earnings, while domestic crop production declines. The global food system means that many farmers in the developing world are increasingly enmeshed within commercial networks where production for export comes at the expense of domestic food security. For example, the increase in health awareness has increased the demand for bananas in Europe and North America, with countries such as Ecuador, Costa Rica, and Guatemala scaling up banana exports. However, the production, purchase, transport, and marketing of bananas are under the control of just five big multinational trading companies, such as Chiquita, Del Monte, and Dole. Similarly, gender inequality exacerbates food

insecurity and hunger. Female farmers are responsible for growing, harvesting, preparing, and selling the majority of food in the lowest-income countries, but often lack access to finances, credit, extension, and training programs, and they are frequently underrepresented in the forums where important decisions on policy and resources are made.

Lack of access to food is also a major cause of hunger. Extreme hunger such as **famine** regularly affects millions. Amartya Sen argues that famines are less the result of lack of food and more the collapse of the system that links the supply and demand for food. He drew on the experience of the Bengal famine of 1943, in which 3 million people died. The famine did not occur because there was no food, but because wages did not keep up with growing food prices. Food price inflation, poor food distribution, hoarding, and poor government responses caused the collapse.

Three things cause famine: the redirection of food, the destruction of the productive capacity to grow food, and total neglect of the starving. Famines reflect lack of power more than lack of food. Historical geographer David Nally looked in detail at the famine in Ireland that began when the **potato blight** created multiple harvest failures beginning in 1845. More than a million people died, and 2 million left the land. He looked at the effect of the Irish Poor Law system developed prior to the famine, in 1838. Under one piece of the legislation, the Gregory Clause, tenant farmers on more than a quarter-acre holding could not receive public relief without leaving the land and entering the workhouse. Once tenants left, their dwellings were knocked down and the landholdings were consolidated. The famine provided an opportunity for the British authorities to clear "excess" population from the land and rationalize the landholding system. A colonial state used the famine to advance its agenda of agricultural rationalization and social improvement. Nally notes, "For those holding political and economic power, famine became the function of new regulatory and corrective mechanisms that unleashed the terror of the possible."

Famine occurs when the usual markets collapse, food is taken away, or the hungry are wantonly ignored. The great famine in China that occurred from 1959 to 1962 was a result of a forced collectivization of agriculture. Family farms were turned into people's communes, and impossibly high quotas were established. Grain production plummeted, but to fill

7.2 Inequality and Food Deserts

The rich world of developed economies is a place of relative food plenty. Yet even there, access to food varies. At the most extreme end are **food deserts**, parts of a city or region with very limited access to affordable, nutritious food. The US Department of Agriculture identifies food deserts at the level of census tracts, areas containing a population of approximately 5,000 (https://www.ers.usda.gov/data/fooddesert/). Almost 24 million people in the United States live more than a mile from a supermarket. Almost 2 million households live more than a mile from a supermarket and are without access to a vehicle. In total, 5.7 percent of all US households had problems getting access to the food that they wanted. Urban food deserts are most prevalent in areas with low to median incomes and the highest proportion of African Americans.

Urban food deserts raise concerns about social justice and food insecurity. For example, in Washington, DC, food deserts are concentrated in Wards 7 and 8, which not coincidentally have the lowest average incomes and are 94 percent Black (see Map 7.2). Of the city's forty-three full-service grocery stores, only three are located in Ward 8. By contrast, Ward 3, the highest-income ward, has eleven full-service stores. The opposite of a food desert is a "food swamp," which is defined as an area with an abundance of fast food, junk food outlets, convenience stores, and liquor stores that outnumber the options for affordable, nutritious food. Food swamps contribute to obesity. Wards 7 and 8 are simultaneously both food deserts and food swamps. Many studies show a correlation between food deserts and lower consumption of fruit and vegetables and higher levels of obesity.

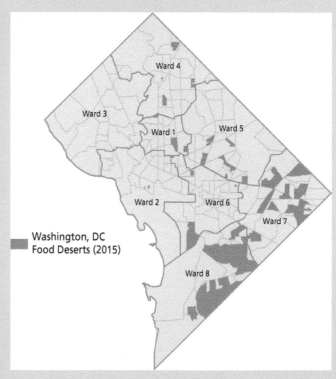

Washington, DC
Food Deserts (2015)

MAP 7.2 Food deserts in Washington, DC.

REFERENCE

Allcott, H., R. Diamond, J. P. Dubé, J. Handbury, I. Rahkovsky, and M. Schnell. 2019. "Food Deserts and the Causes of Nutritional Inequality." *The Quarterly Journal of Economics* 134:1793–1844.

their quotas, local officials continued to ship the dwindling grain supply to the cities. It is estimated that around 35 million people died in a famine that was almost entirely human-made. Heavy police control meant that people were unable to leave their villages. People were reduced to eating tree bark.

Famines continue to be present on the world stage. Their connection with deeply authoritarian regimes is exemplified in the recent famine in North Korea, when between 1 and 3 million people, of a total population of 22 million, died as the result of a collapse of the food distribution system and government incompetence. The country suffers from chronic food insecurity, with almost half of all children malnourished.

The Horn of Africa is a scene of recurrent famines. Almost 12 million people are currently at risk from famine. In 1984–1985, more than 1 million died in a famine in Ethiopia caused by conflict, drought, and economic mismanagement. Between 2010 and 2012, more than 260,000 people died in Somalia, half of them under the age of five. Persistent hunger in sub-Saharan Africa is one of the major failures of the global food system. Many causes are at work. Consider the case of Niger, which is one of the poorest countries of the world, with at least 40 percent of the population chronically undernourished. Food shortages affect more than 7 million people. Household vulnerability is made more severe by land degradation that is reducing the supply of farmland and forcing more people onto marginal lands. Low income, erratic rainfall, and the low status afforded to women are just some of the many reasons behind the recurring famine and food shortages. The growing reliance on food aid paradoxically depresses local agriculture production even further. The result is a

7.3 Sustainable Solutions: Africa's New Green Revolution

Africa has long been a continent of small farmers, half of them women. They often farm with little fertilizer, pesticide, or irrigation, on a tiny plot with a hoe. With high rates of child malnutrition, Africa must increase food production, and do so in a more sustainable way.

In 2006, the Bill and Melinda Gates Foundation and the Rockefeller Foundation joined forces to launch the Alliance for a Green Revolution in Africa (AGRA). This effort aimed to create a new Green Revolution uniquely for Africa, while avoiding some of the problems of the earlier initiatives. AGRA currently funds hundreds of projects that include research to develop and deliver better seeds, increase farm yields, improve soil fertility, upgrade storage facilities, improve market information systems, strengthen farmers' associations, expand access to credit for farmers and small suppliers, and advocate for national policies that benefit smallholder farmers. Credit programs are especially critical for poor farmers who lack collateral or creditworthiness to access traditional loans. AGRA aims to improve production of staple crops in "breadbasket" areas that have relatively good soil, adequate rainfall, and basic infrastructure and then replicate successful approaches in other areas and other countries with similar conditions.

Projects underway that focus on Africa-specific solutions including improving yields of traditional crops such as cassava, sorghum, millet, and beans, which are being bred for disease resistance and drought resistance. Cassava is especially important because it grows in poor soils and across the continent. Another crop plant, cowpea, is an indigenous African legume that grows well in semiarid regions. Cowpea is grown and marketed mostly by women and is considered an "insurance crop."

This new Green Revolution for Africa is not without its critics, including William Mosely and Timothy Wise, who point to a range of issues, including the potential social and environmental pitfalls of monocrop agriculture, the dangers of encouraging farmers to use genetically modified seeds, and the likelihood that high-cost inputs will lead to growing inequalities within African farming communities. African farmers may become more dependent on foreign imports of seed and fertilizer. In addition, as was true in the previous iterations of the Green Revolution, mechanization of farming requires land consolidation, which could result in a rural exodus.

Consider the mixed results in Rwanda, where more than 85 percent of the population farms on land that averages under two acres (the smallest farms in the region). On the one hand, general aggregate data suggest that since 2015 agricultural productivity increased and conventionally measured poverty rates have fallen. However, on the other hand, studies have shown that in many local areas household subsistence practices were disrupted, poverty was exacerbated, and land tenure security and autonomy were curtailed. All of this underscores the complexities of increasing food security.

REFERENCES

Berglus, M., and J. Buseth. 2019. "Towards a Green Modernization Development Discourse? The New, Green Revolution in Africa." *Journal of Political Ecology* 26 (1): 57–83. http://doi.org/10.2458/jpe.v26I1.

Mosely, W. 2017. "The New Green Revolution for Africa: A Political Ecology Critique." *Brown Journal of World Affairs* XXIII:177–190.

Wise, T. 2020. *Failing Africa's Farmers.* Tufts University's Global Development and Environment Institute. https://sites.tufts.edu/gdae/files/2020/07/20-01_Wise_FailureToYield.pdf.

population with precarious access to food too easily pushed over into hunger and famine by even small changes in food supply, food prices, and climate. In these circumstances, drought does not cause hunger as much as it pushes an already vulnerable population into famine.

7.4 Limits to Food Supply

The enormous gains in agricultural productivity have passed the limits imagined by Malthus. Agricultural production has increased food supply enough to dispel the original Malthusian prediction. Some see no inherent problem in feeding the world's population. And yet there are some who see a global food system precariously balanced. Whereas Charles Godfray and colleagues look to increased gains to be made, Carleton Schade and David Pimentel point to problems. A comparison of the data and arguments in the two papers makes for a compelling case study of a situation's alternative readings and is a reminder that arguments vary even when using the same facts. Godfray et al. highlight the ability of the global food system to produce even more food, but Schade and Pimentel point to systemic problems. They suggest that the rapid increase in food supply was the result of conditions unlikely to be repeated: available land, water, and energy; a benign climate; and improving crop yields. They estimate that if the population increases to 9.2 billion, there is likely to be enough food for only 6–8 billion. Food insecurity and famine are likely to afflict anywhere between 1 and 3 billion people. The problems they identify include declining land and water availability. Increasing population reduces the amount of land available for farming and degrades some of the most productive soil. They suggest a **land deficit** of between 0.1 and 0.9 billion hectares. Irrigated land is the most productive land on the planet (Photo 7.5). However, we are losing this land to **salinization** and erosion. The production of food requires large quantities of fresh water. We already have reached the ceiling of surface supply, and the supply of water from aquifers is fast running out. Climate change will also affect the global agricultural system through generating more extreme and damaging weather events, such as flooding. While the higher-latitude countries such as Canada and Russia may experience an extension of their growing season with global warming, other regions may endure harm. In general, the growth of the population combined with their enlarged human ecological footprint is having an environmental impact. We have lost or are losing forest cover, wild fish stocks in the oceans are being depleted, and global reserves of fresh water in river basins are diminishing. More than 1.4 billion people live in river basins where water sources are being depleted.

The two contrasting papers embody the different perspectives. Whereas Godfray and colleagues point to the

PHOTO 7.5 Irrigation in Midwest United States. Central pivot irrigation systems ensure high agricultural productivity in the drier parts of the United States. However, sources of fresh water are being used up, questioning the long-term sustainability of such practices.

slack in the system, Schade and Pimentel point to its limits. They suggest that we have reached, if not already passed, peak water supply, peak land availability, and peak crop yields. The future that they see is one of Malthusian limits reimposed. Adherents of either side can look to current trends to bolster their claims. On the one hand, levels of food insecurity are being reduced. On the other, there is evidence of unstable food prices (Figure 7.1). They increased rapidly in 2006, then fell as the world economic crisis hit in 2008. Since then, prices have remained historically high. Particular events can also trigger increases. The Russian invasion of Ukraine in 2022, for example, increased the price of grains by 17 percent and vegetable oils by 23 percent. Before the invasion, Ukraine was the world's leading exporter of sunflower seed oil. Only time will tell which of these trends, declining insecurity or increasing costs, is temporary and which is persistent.

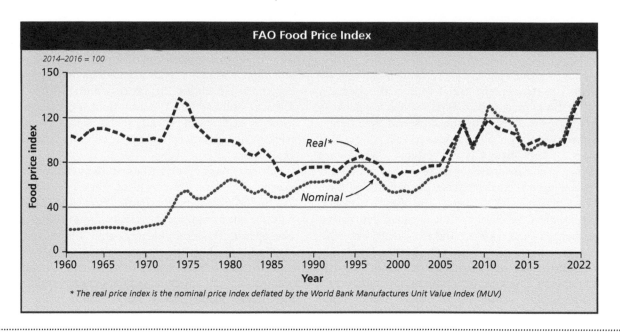

FAO Food Price Index

2014–2016 = 100

Food price index

* The real price index is the nominal price index deflated by the World Bank Manufactures Unit Value Index (MUV)

FIGURE 7.1 World food prices.

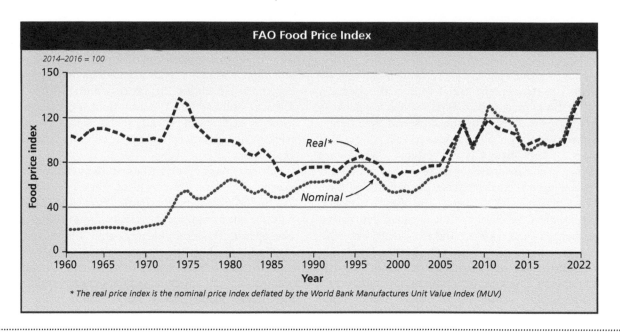

7.4 Climate Change and Food

Environmental impacts of drought or flood have exacerbated food insecurity. Drought is ever present in many parts of the Horn of Africa and can lead to widespread periodic famine in the region. Floods also affect local areas. In Kenya and Somalia, recent droughts decreased cereal output and the price of maize and sorghum rose sharply. Adverse weather has also impacted food production in Zimbabwe, where the number of people experiencing food insecurity increased from 30 percent in 2019 to 42 percent currently.

Climate change may compromise food production in countries and regions that are already highly food insecure.

Some climate experts suggest that wheat, rice, and corn yields could decrease by 10 percent for every 1 degree (Celsius) rise in temperature. In North Africa and the Middle East, climate evidence suggests there will be significant drying in the future, and therefore crop production will be stressed even more than it is today. In East and Southeast Asia, higher temperatures may stress agricultural production.

Climate change will affect food availability in the wider global food system.

7.5 Questioning the Food Production System

The Malthusian limits were bypassed because of the creation of a very efficient **industrialized agricultural system** that generated large amounts of relatively cheap food. Only seven countries produce more food than they need—the United States, Canada, Australia, Thailand, Argentina, Ukraine, and Vietnam—but a global trading system has widened the availability of food. In recent years this system has come under greater scrutiny.

The original Malthus–Boserup debate was concerned only with the quantity of food, and in this regard food supply has largely kept up with population increases. Boserup triumphs over Malthus. Gains in productivity and the emergence of a global food system have ensured large amounts of cheap food for much of the world's population. Yet there are emerging questions regarding the quality of food and the wider implications of its nutritional and environmental impacts.

Obesity is a challenge because the causes are not fully understood. Part of the cause of obesity is total caloric intake; another contributor may be physical inactivity. In many countries, high caloric intake can result from increased intake of the "wrong types of calories," such as highly processed foods (e.g., soft drinks, potato chips, and baked goods).

The World Health Organization now estimates that 1.9 billion adults are **overweight** and 1 billion are obese, as defined by the **body mass index** (BMI), which is a person's weight in kilograms divided by the square of their height in meters. Overweight people are defined as having a BMI greater than 25, and **obesity** is defined as a BMI greater than 30. It is a crude measure, because it does not distinguish between muscle and fat tissue, but until we have better metrics, it is a useful indicator. The obesity epidemic is associated with a population that is sedentary, sitting and driving rather than walking and exercising, and with diets that contain high-energy nutrients such as fat, starch, and sugar. Increasing obesity leads to cardiovascular disease and diabetes. The obesity epidemic is particularly prevalent among young children; the WHO estimates that 39 million children under the age of five are now overweight. Obesity is on the rise in nearly all countries, especially among school-age children and adults, and contributes to 4 million deaths globally.

Map 7.3 highlights the geographic distribution of obesity for women. The map for men is similar, although with a lower incidence of obesity in South America, South Africa, the Middle East, eastern Europe, and Russia.

There are high levels of obesity in affluent countries such as the United States. Obesity is also found increasingly in middle-income countries such as Mexico, where one in three adults is defined as clinically obese. Malnutrition plays a role. The bodies of malnourished young children learn to retain fat, and in later life, with more income and food choices, the children of malnourishment can become obese adults. This process, along with a decline in physical exercise and the adoption of a more sedentary lifestyle, explains the rapidly rising obesity levels of countries transitioning from low to middle income in one generation.

Obesity has complex roots. Individuals can limit their energy intake and exercise more. Yet it is more than just a case of individual choice. Jamie Pearce and Karen Witten describe environments that create higher risk of obesity as **obesogenic**. The food industry could, and in some cases is starting to, reduce the fat, sugar, and salt content of the food it produces to ensure more nutritious choices. We could make our cities friendlier to walking and cycling rather than making them convenient only for sedentary motorists.

A food movement is emerging that shifts the debate from the basic supply of food to a wider concern with nutrition. It is a debate not just about the supply of food but also about the quality of the food. In many countries, the government subsidizes the production of crops with little concern for nutritional value or environmental impact. In the United States, around 90 million acres are devoted to growing corn. A huge corn surplus is created by the combination of efficient farming and generous government subsidies. The corn surplus is turned into corn syrup that

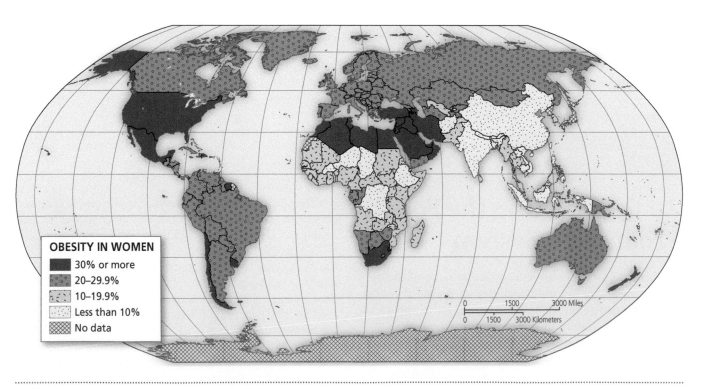

MAP 7.3 | The geographic distribution of obesity in women.

is added to the food supply, which in turn adds extra calories to the average consumer. The rise of obesity in the United States is directly linked to the corn surplus and the heavy use of corn syrup in processed foods.

Food surpluses are also the result of heavy chemical usage. The main form of pollution in much of the United States, for example, comes not from factories but from farms and agribusiness, with the runoff of pesticides and animal wastes entering streams, rivers, groundwater, and eventually the oceans. Every year between 10,000 and 20,000 farm workers suffer acute pesticide poisoning.

The rise of a **corporate industrialized food production system** resulting from the economies of scale and large investments has produced vast quantities of cheap food, but we are now beginning to question the wider costs of this system and its impact on public health and environmental quality. Another new food movement is emerging that is concerned more with health, nutrition, sustainability, and access. It is a disparate movement with a wide variety of goals and concerns, including improving the plight of agricultural workers, raising issues of animal rights in factory farming, advocating the slow food movement, promoting local food producers, and encouraging inner-city vegetable gardens. Even in the United States, large food corporations such as McDonald's and General Foods are advertising a switch to more natural and healthy ingredients. Fundamentally, this movement is concerned with the nature of the food that we produce, particularly questioning the encouragement of grain-fed meat production, assessing the wider costs of cheap fast food, highlighting the darker side of the food chain, and debating the sustainability of the entire global food system.

This system, while perhaps producing a sufficient supply of food, does little to advocate a fair or even healthy supply of food. While obesity is emerging as a major health issue, food insecurity continues to be a problem for many. Parts of the world are overfed while others are undernourished, and yet other parts are both overfed and malnourished.

We are what we eat, and for those who can afford it, there are choices to be made about what we eat and how we eat it. A nonvegetarian diet requires 2.9 times more water, 2.5 times more primary energy, 13 times more fertilizer, and 1.4 times more pesticides than a vegetarian diet. A great number of people eat a "Western diet" that is heavy in meat, dairy, and eggs. But few people calculate the resources required to produce their food. For example, it takes 4,200 gallons of water per day to produce a meat-eater's diet, while a plant-based diet uses only 300 gallons of water per day. A startling 70 percent of grain in the United States is fed to farmed animals rather than to people.

When we move beyond worrying about the supply of food, we begin to question the nutritional quality of the food we buy and eat, its environmental impact, its ethical implications, and its wider imprint on the world we live in.

What to make of the gloomy prognostication of Malthus? The constraints of food supply have been transcended by incredible gains in agricultural productivity and a global supply chain of food that links importers and exporters of food. Population has continued to increase long past the Malthusian limits. And yet if the basic limits of food supply no longer apply with the brutal force predicted by Malthus, problems remain. Of 7.9 billion people, 1 billion are hungry (undernourished), 800 million are malnourished, and a further billion are obese. It is not only the supply of food that is the problem, although there are some who see the food supply peaking and then declining as other resource inputs such as land and water become more problematic. Questions are also being raised about the distribution, quality, and price of food and the long-term sustainability of maintaining enough good-quality food for all of the world's population.

Overpopulation Re-examined

In 1968, the Stanford biologist Paul Ehrlich wrote *The Population Bomb*, predicting the imminent death of millions if population growth was not contained. **Overpopulation** was, and still is, a common assumption. Since growth rates are highest in the developing world, the debate often hinges on the belief that population controls need to be introduced into these high-growth-rate countries. But when we look at the issue in more detail, a number of important points must be made.

First, it is not simply the number of people. When we factor in the impact of different populations, the problem becomes not too many poor people, but too many people in the affluent world with their heavy ecological imprint. With only 5 percent of the world's population, the United States is responsible for 23.3 percent of the global total of carbon dioxide emissions. The average US citizen generates a much higher environmental impact and heavier ecological footprint than citizens of other countries, especially of lower-income countries. On average, Americans eat more meat and dairy than populations in low-income countries. Ehrlich saw the problem as too many people, but the real issue of food insecurity is a paradox—some people are consuming far too many calories (and are dealing with obesity), while others are dealing with chronic undernutrition.

Second, there is a politics to population control. It reflects power rather than demographic analysis. The birth control measures of the past fifty years are invariably targeted at the poor and the weak, not the rich and the strong. The World Bank used to adopt fertility targets for countries, but none for the rich countries of the world. Debates about population control reflect power, not scientific neutrality.

Third, there is an adage from farming communities: "Each mouth comes with two hands." It refers to the fact that people are a resource, not just a problem. The ultimate resource is the inventiveness of humans: the more humans, the larger the creative pool of talent.

It is not simply the number of people that is the issue; it is their relationship to the land, their use of resources, and the economic system at the heart of the people–environment relationship.

In fact, some experts argue we produce enough food to feed all 8 billion. A 2019 World Food Programme report noted if total calories from all the food produced were divided among all the people on earth, there would be 2,750 calories per person per day. Since the recommended daily minimum per person is 2,100 calories a day, there are currently enough calories to feed everyone in the world. But not everyone is getting the needed calories because food is not evenly distributed across the world. If we accept this premise, then it is not population growth that is at issue; it is the socioeconomic and political factors that influence distribution, access, and affordability.

To talk about only reducing population is to pass the blame on to the poorest and weakest, when what is really required is a fundamental reassessment of our relationship with the environment and the nature of our economic system.

Cited References

Boserup, E. 1965. *The Conditions of Agricultural Growth: The Economics of Agrarian Change under Population Pressure.* London: Allen & Unwin.

Ehrlich, P. 1968. *The Population Bomb.* New York, NY: Ballantine Books.

Godfray, H. C. J., J. R. Beddington, I. R. Crute, L. Haddad, D. Lawrence, J. F. Muir, J. Pretty, S. Robinson, S. M. Thomas, and C. Toulmin. 2010. "Food Security: The Challenge of Feeding 9 Billion People." *Science* 327:812–818.

Malthus, T. R. 1798. *An Essay on the Principle of Population.*

Nally, D. 2008. "'That Coming Storm': The Irish Poor Law, Colonial Biopolitics, and the Great Famine." *Annals of the Association of American Geographers* 98:714–741.

Pearce, J., and K. Witten, eds. 2010. *Geographies of Obesity: Environmental Understandings of the Obesity Epidemic.* Burlington, VT: Ashgate.

Schade, C., and D. Pimentel. 2010. "Population Crash: Prospects for Famine in the Twenty-First Century." *Environment, Development and Sustainability* 12:245–262.

Sen, A. 1981. *Poverty and Famines.* Oxford: Clarendon Press.

World Food Programme. 2019. *The State of Food Security and Nutrition in the World.* https://www.wfp.org/publications/2019-state-food-security-and-nutrition-world-sofi-safeguarding-against-economic.

United States Department of Agriculture. 2023. "Food Access Research Atlas." July 10, 2023. https://www.ers.usda.gov/data/fooddesert/ and https://www.ers.usda.gov/data-products/food-access-research-atlas/go-to-the-atlas/.

Selected Guide to Further Reading

Bassett, T. J., and A. Winter-Nelson. 2010. *The Atlas of World Hunger.* Chicago: University of Chicago Press.

Burgess, M. G., S. Polasky, and D. Tilman. 2013. "Predicting Overfishing and Extinction Threats in Multispecies Fisheries." *Proceedings of the National Academy of Sciences* 110:15943–15948.

Carolan, M. 2013. *Reclaiming Food Security.* New York: Routledge.

Cohen, D. A. 2013. *A Big Fat Crisis: The Hidden Forces behind the Obesity Epidemic—and How We Can End It.* New York: Nation.

De Waal, A. 2017. *Mass Starvation: The History and Future of Famine.* London: Wiley.

Guthman, J. 2013. "Too Much Food and Too Little Sidewalk? Problematizing the Obesogenic Environment Thesis." *Environment and Planning A* 45:142–158.

Hallett, L., and D. McDermott. 2011. "Quantifying the Extent and Cost of Food Deserts in Lawrence, Kansas, USA." *Applied Geography* 31:1210–1215.

Hilborn, R. 2012. *Overfishing: What Everyone Needs to Know.* Oxford: Oxford University Press.

James, P. 2013. *Population Malthus: His Life and Times.* London: Routledge.

Jinseng, J. 2012. *Tombstone: The Great Chinese Famine 1958–1962.* New York: Farrar, Straus and Giroux.

Kiple, K. F., and K. C. Ornelas, eds. 2000. *The Cambridge World History of Food.* Cambridge: Cambridge University Press.

Marlow, H. J., W. K. Hayes, S. Soret, R. L. Carter, E. R. Schwab, and J. Sabaté. 2009. "Diet and the Environment: Does What You Eat Matter?" *American Journal of Clinical Nutrition* 89:1699–1703.

McMillan, T. 2012. *The American Way of Eating.* New York: Scribner.

Morrison, N. O. 2011. "Mapping Spatial Variation in Food Consumption." *Applied Geography* 31:1262–1267.

Patel, R. 2008. *Stuffed and Starved: The Hidden Battle for the World Food System.* Brooklyn, NY: Melville House.

Pilcher, J. M., ed. 2012. *The Oxford Handbook of Food History.* New York: Oxford University Press.

Pimentel, D. 2012. "World Overpopulation." *Environment, Development and Sustainability* 14:151–152.

Pollan, M. 2008. *In Defense of Food.* New York: Penguin.

Schlosser, E. 2004. *Fast Food Nation: The Dark Side of the All-American Meal.* New York: Harper.

Toth, G., and C. Szigeti. 2016. "The Historical Ecological Footprint: From Over-population to Over-consumption." *Ecological Indicators* 60:283–291.

United Nations Children's Emergency Fund. 2021. *The State of Food Security and Nutrition in the World.* Geneva: United Nations. https://data.unicef.org/resources/sofi2021/?utm_source=newsletter&utm_medium=email&utm_campaign=SOFI%202021.

United Nations Children's Emergency Fund, World Health Organization, International Bank for Reconstruction and Development/The World Bank. 2021. *Levels and Trends in Child Malnutrition: Key Findings of the 2021 Edition of the Joint Child Malnutrition Estimates.* Geneva: World Health Organization. https://data.unicef.org/resources/jme-report-2021/=.

United Nations Food and Agriculture Organization. 2014. *The Water–Energy Nexus: A New Approach in Support of Food Security and Sustainable Development.* Rome: United Nations. http://www.fao.org/3/bl496e/bl496e.pdf.

United Nations Food and Agriculture Organization. 2018. *Transforming Food and Agriculture to Achieve the SDGs.* Rome: Food and Agriculture Organization of the United Nations. https://www.fao.org/policy-support/tools-and-publications/resources-details/en/c/1140709/.

Woodham-Smith, C. 1962. *The Great Hunger, Ireland 1845–9.* London: Hamish Hamilton.

The Economic Organization of Space

The economic organization of space is the subject of Part 3. In Chapter 8, we examine the economic geography of uneven development at the global and regional scales. In Chapter 9, we identify the economic geography of the primary, secondary, and tertiary sectors and discuss recent trends. In Chapter 10, we turn our attention to understanding the global economy.

OUTLINE

8

Uneven Development

In this chapter, we will look at the economic organization of space through a discussion of uneven development at different scales.

8.1 Global Differences

Across the globe, uneven development of economic activity has caused marked disparities in living standards and quality of life. There are places of wealth and places of poverty. There are places where people enjoy access to healthcare, jobs, and social services; there are also places of deep inequality, discrimination, and poor health. According to the United Nations Development Programme, the wealthiest 20 percent of the global population earns 82 percent of the total global income. The richest 1 percent has 50 percent of the world's wealth. Table 8.1 highlights data from three countries: Brazil, Cameroon, and the United States. There are not only significant differences in income, but also impacts of these differences. In Cameroon, many people do not live past their mid-fifties, while in the United States people regularly live into their seventies and beyond.

TABLE 8.1 Global Differences			
	CAMEROON	BRAZIL	UNITED STATES
Gross domestic product per capita (USD)	3,600	14,100	60,200
% Labor force in agriculture	70	9	0.7
% Labor force in manufacturing	13	32	17
% Labor force in services	17	58	79
Life expectancy at birth (years)	63	75	80

Source: CIA World Factbook. 2022.

Uneven development is both a material condition and a theory of how this material condition came to be. To explain uneven development and its consequences in the contemporary world is a difficult and complex task, particularly because many of the causes have been unfolding over centuries. To explain these differences, some, such as Jared Diamond, take a very long-term view—13,000 years, to be precise—and focus on how initial geographical advantages were turned into economic advantages. Geography plays a part in shaping the global economic order of marked disparities, but it is rarely pivotal. Japan, with few natural resources, has reached a high level of economic development and good living standards, while much of tropical Africa, despite a bounty of resources and minerals, remains poor.

The patterns not only vary over space, but also change over time. In the fifteenth century, China was far ahead of Europe and had enough scientific and technical knowledge to inaugurate its own industrial revolution. This did not happen, and China focused inward. In contrast, technologically backward Europe became the center of massive economic change as Britain's Industrial Revolution transformed the old patterns into a new global order in which manufacturing and economic vitality became the monopoly of the West. In recent years, even this order has changed with the global shift of manufacturing and the rise of new industrial powers such as Brazil, China, and India. The global economy is dynamic and ever-changing.

Global inequalities raise the question of why some countries are rich and some are so poor. The answer, according to some, lies in the colonial legacy of the Global North getting rich off the exploitation of the Global South. Others point to the differential ability to adopt economic opportunity and harness technological possibilities. Cultures that promote innovation and openness tend to do better than others. Small differences over time became sources of major divergences. For instance, when the Ottoman sultan forbade the establishment of a printing press in 1485, it reduced the opportunities for a spread of learning and an increase in technical proficiency. Institutions, especially those that secure property rights and allow broad economic and political participation, play a significant role in economic growth.

The Industrial Revolution in Britain was the result of many factors, including the ability to import cheap raw materials from its colonies that also provided a captive market for its manufactured goods. There were also important institutional shifts, including the creation of an efficient banking system to connect savings to investment and the establishment of a mass education system to train people for work and industrial employment.

The World Bank employs a classification of counties based on their gross national income (see Table 8.2). Some countries shift categories as their economies grow or decline. Even in just the past five years, Haiti and Tajikistan moved from low income to lower middle, while Moldova climbed from lower middle to upper middle. In the period 2011 to 2016, only poor, war-wracked South Sudan shifted downward, from lower middle to low. However, in the five-year period from 2016 to 2022, a number of countries, in large part because of economic contraction caused by COVID-19, fell into a lower income category. Belize, Indonesia, Iran, and Samoa dropped from upper middle to lower middle, while Mauritius, Panama, and Romania tumbled from high to upper middle. The pandemic led to an increase in poverty in many countries around the world.

There are substantial variations across the world, with the average gross national income of high-income Switzerland 134 times that of low-income countries. We live in an unequal world with national concentrations of wealth and vast areas of poverty. These income figures are more than just abstract statistics; they are a measure of the quality of life, because higher national incomes translate into better jobs, housing, and healthcare. There is a chance element to the lives we lead. If you are born into poverty in a poor country, your life chances are more limited and circumscribed than if you are born the son or daughter of affluent Swiss parents where private affluence is reinforced by public bounty. While we may imagine that the character of our lives is self-constructed, it is in large part based on where we were born. Geography in this particular case is destiny. The Economist Intelligence Unit estimates an index for eighty countries based on the number and quality of opportunities for citizens to lead a healthy and prosperous life. The top five countries are Switzerland, Australia, Norway, Sweden, and Denmark. The United States ranks

TABLE 8.2 Income Categories		
INCOME CATEGORY	**COUNTRY EXAMPLES**	**GROSS NATIONAL INCOME PER CAPITA**
Low	Afghanistan, Mali, Syria	Below $1,046
Lower middle	India, Morocco, Vietnam	$1,046–$4,095
Upper middle	China, Mexico, Peru	$4,096–$12,695
High	Switzerland, Norway, United States	Above $12,695

Source: Adapted from World Bank data, 2023.

sixteenth and the United Kingdom twenty-seventh. The bottom five are Angola, Bangladesh, Ukraine, Kenya, and Nigeria. You are luckier if you are born in Australia or Switzerland than in Angola or Nigeria.

Assessing and comparing national wealth is an exercise fraught with many dangers. A UN report took up this difficult task. The report's authors measured three kinds of assets: physical capital (buildings, infrastructure, etc.), human capital (including the educational skill levels of the population), and natural capital (land, forest resources, etc.). The United States was the richest country, with over $117.8 trillion of wealth, compared to Japan ($55.1 trillion), China ($20.0 trillion), and Germany ($19.5 trillion). When wealth per person was measured, the rankings changed to Japan, the United States, Canada, Norway, and Australia. These are rough and ready measures, but they begin the process of identifying national wealth and widening the definition of national assets.

The Core–Periphery Model

In the **core-periphery model** (see Figure 8.3) the global economy has winners and losers, not just as a historical accident but as part of the very process of economic development. Some countries are rich because other countries are poor. The initial advantage of some countries was reinforced by colonialism and economic imperialism, which led to the differential distribution of wealth and poverty. Instead of being located at different stages in time, countries are part of a spatially connected order, a world system comprising core, periphery, and semiperiphery. For 200 years the core was in western Europe and later in

---

8.1 The Kuznets Curve

The bell-shaped **Kuznets curve** (see Figure 8.1) depicts inequality increasing with rapid urbanization and industrialization. Growing inequality in China, for example, is to some extent a measure of recent industrialization and the growing economic differences between the interior rural areas and the faster-growing, industrial, coastal urban regions. The economist James K. Galbraith has produced an amended Kuznets curve to suggest that in some higher-income countries, rising inequality is created when high-tech and finance become larger parts of the economy. The large wealth amassed by owners and top earners in these sectors skews the national figure toward increased inequality. The smooth bell shape of the original Kuznets curve then starts to trend upward. Notice how in the modified Kuznets curve (Figure 8.2), the United States, a country that has long since moved from a mainly

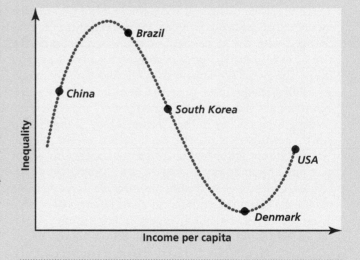

FIGURE 8.2 Modified Kutznets curve.

agricultural to an industrial and service economy, once more has a relatively high level of inequality. The United States is situated after the inflexion in the curve. Thomas Piketty argues that since the returns to capital are higher than the rate of economic growth and so more wealth is appropriated by the richest, then inequality will continue to persist unless there is a marked redistribution of wealth.

REFERENCE

Piketty, T. 2013. *Capital in The Twenty-First Century*. Cambridge, MA: Belknap Press.

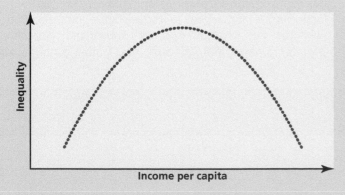

FIGURE 8.1 Kuznets curve.

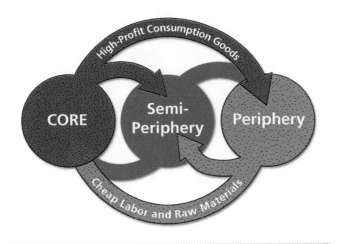

FIGURE 8.3 The core–periphery model.

North America. Economic transactions were marked by an unequal exchange of cheap raw materials from the periphery to the core and the export of manufactured goods from the core to the periphery. This essential dynamic explains the colonial expansion of core countries as they sought to establish monopoly control over regions that were then transformed into colonial peripheries used to feed and sustain the core economies by supplying cheap raw materials and purchasing manufactured goods. Britain was a core country that incorporated much of India into commercial domination and political control. Raw cotton grown in India was shipped to Britain, processed into cloth in Lancashire cotton mills, and then exported abroad. The indigenous cotton manufacturing industry in India was destroyed, while the domestic cotton industry in Britain was a leading force in the Industrial Revolution, which propelled national economic growth.

This core–periphery model allows us to globalize such issues as the Industrial Revolution, presenting them less as unique national experiences and more as part of a global drama. The model is dynamic. The category of semiperiphery is something of a transitional condition situated somewhere between the dominance of the core and the relative weakness of the periphery. Some countries can move from the periphery or semiperiphery toward the core. In the past fifty years, China, South Korea, and others have undergone marked industrialization to place them closer to the core. And even some strategic primary producers of raw materials, such as the oil-producing countries of OPEC or mineral-rich Australia, can use their collective interests and

8.2 Measuring Poverty

Although the UN's Human Development Index remains widely used to measure "development," its use of only three indicators makes it susceptible to bias. The Global Multidimensional Poverty Index (MPI) was developed in 2010 by the Oxford Poverty and Human Development Initiative to identify the most vulnerable people—the poorest among the poor. The MPI assesses poverty at the individual level. If someone is deprived in a third or more of ten (weighted) indicators, the global index identifies them as "MPI poor," and the extent—or intensity—of their poverty is measured by the number of deprivations they are experiencing. The MPI can be used to create a more comprehensive picture of people living in poverty and allows comparisons across countries and regions, as well as within countries by ethnic group, urban/rural location, and other community characteristics.

The poor lack not only income, but also education, health, justice, credit, and other productive resources and opportunities. Many economists and development experts prefer to use the MPI because it better captures the true reality of poverty. The MPI accounts for severe deprivations in traditional areas like education, health, and living standards, but also includes other indicators (e.g., child mortality) that have a daily impact on a person's poverty status. As Sabina Alkire and Selim Jahan explain, the MPI is valuable for highlighting the different ways that people are poor and has become an important way to understand what poverty really looks like for many of the most vulnerable people.

The most recent MPI report measures acute poverty for 105 countries, covering 5.7 billion people, approximately 77 percent of the global population, A key finding was that 1.3 billion people live in multidimensional poverty in 105 countries, nearly all located in the Global South. Half of all poor people are children under the age of eighteen years; over 665 million children (one of every three children) are spending their childhood in multidimensional poverty.

REFERENCES

Alkire, S., and S. Jahan. 2018. "The Global MPI 2018: Aligning with the Sustainable Development Goals." UNDP Human Development Report Office Occasional Paper. New York, NY: United Nations. https://hdr.undp.org/content/new-global-mpi-2018-aligning-sustainable-development-goals.

Oxford Poverty and Human Development Index. 2023. "Global Multidimensional Poverty Index: Global MPI 2023," https://ophi.org.uk/global-mpi-2023/.

combined power to increase prices and force redistribution in global wealth. The core–periphery model has the enormous advantage of globalizing historical events and adding a much-needed dose of political economy to standard historical narratives while being flexible enough to explain the moves of some national economies from periphery to semiperiphery, with the possibility of core membership changing and shuffling. From their initial advantages, the core countries became richer.

A Political Economy Framework Matures

Scholars have continued to build and expand on the core–periphery model. Geographers such as Richard Peet, Elaine Hartwick, and David Harvey have helped advance a political economy perspective. Political economists argue that global inequalities and uneven development were an *inevitable* consequence of the accumulation of capital in a capitalist system. Capitalism created uneven development.

A political economy framework provides an explanation for uneven development and global inequality. Jason Hickel, in his book *The Divide: A Brief Guide to Global Inequality and its Solutions*, asserts that poor countries are poor because they have been integrated into the global economic system on unequal terms. Colonial and imperial histories are vitally important in understanding the context for levels of development and the "gap" between the richest and the poorest. These legacies still impact economic, political, and social structures and institutions, even half a century after decolonialization. Political economy scholars argue these structures have caused economic development that favored some countries more than others and that some countries are rich not because they are further along some development trajectory, but because they had an initial advantage that was then reinforced by colonialism and economic imperialism.

The term "political economy" has expanded in meaning beyond uneven economies, to include uneven resource distribution and use, uneven sociocultural indicators (gender, class, race, ethnic group, age), and more. The term also implies uneven geographical development, which involves a number of metrics (employment rates, income levels, rates of economic growth, and so on), and it has been described at all geographical scales—from intraurban disparities all the way through subnational regional differences to uneven international development.

8.2 Regional Differences

Economic disparities are also evident at the regional level. Consider the case of China. China's economy, like most national economies, is a patchwork of very different regional economies, some growing faster than others, some more industrialized than others. There is a marked regional variation in income levels. The fastest-growing coastal regions, where the bulk of export-oriented manufacturing takes place, are wealthier than the more distant inland regions with much less manufacturing activity. These differences are the principal reason behind China's large-scale internal migration from inland provinces to the coastal regions.

We can also map the flip side of income. Map 8.1 maps the poverty rate for people under age eighteen across the United States. Note the high figure along the Mississippi valley and in Appalachia, Piedmont, and rural areas of the West. Lower poverty rates are found in the more dynamic metropolitan regions of Megalopolis, Chicago, and along the West Coast.

Models of Regional Economic Growth

Regional disparities are best understood with reference to models of regional economic growth. The process of **cumulative causation** occurs when growth feeds on itself. Growth in a leading region spreads out into other lagging regions through diffusion of innovations and access to markets (Figure 8.4). There are also backwash effects that include the flow of capital and labor from the lagging to the leading regions. In the United Kingdom, the backwash effects are clearly more pronounced than the spread effects, with growth concentrated in London and the southeast part of the country, which attracts more labor and capital, that in turn, through cumulative causation, generate even more growth. The flow of capital and skilled labor from the peripheral regions, such as Scotland, limits economic growth and ultimately leads to pronounced regional disparities. As Figure 8.5 shows, Britain has the widest regional disparities among the larger economies, while Italy has the lowest.

Uneven regional development is not a temporary condition that will be resolved through time, but rather is absolutely essential to a capitalist mode of production. Waves of investment are uneven across space, making some regions more developed than others. These investments are spatially fixed and form the basis for subsequent rounds of capital investment. Some regions are then bypassed by subsequent waves of investment as capital searches for areas of greatest profit. The result is the constant production of uneven development.

We can also consider regional economic dynamics from the perspective of stages of product and **profit cycles**. A **product cycle** is the trajectory of a new product. At early stages of a product cycle, when new products are just being created, it is important to be close to research and development. Sales, shown in red in Figure 8.6, go through a steep rise during the growth and maturity phases and then drop off. Profits, shown in the shallower curve, are negative until

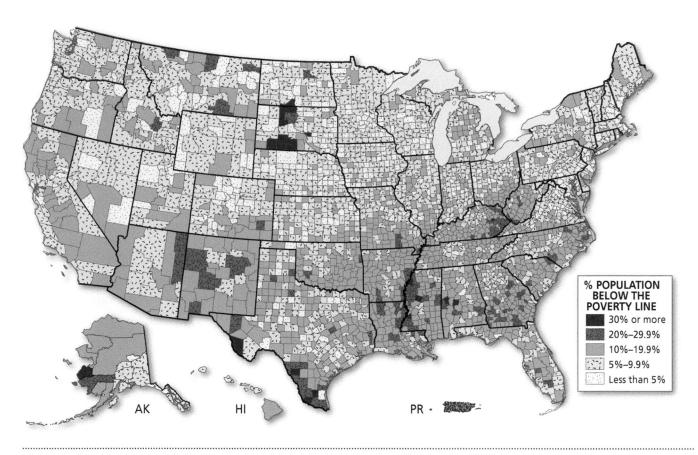

MAP 8.1 Regional differences in poverty in the United States.

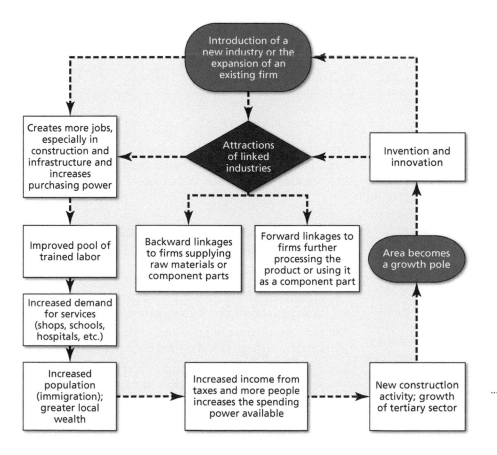

FIGURE 8.4 Cumulative causation. Diagram courtesy of Barcelona Field Studies Centre.

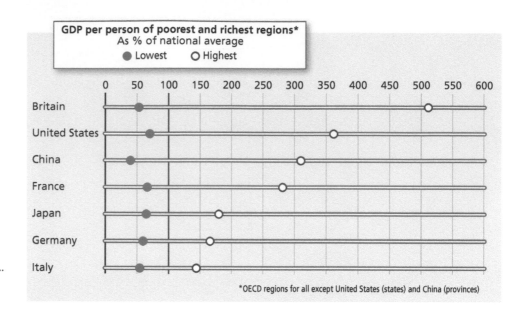

FIGURE 8.5 Regional disparities in GDP per person.

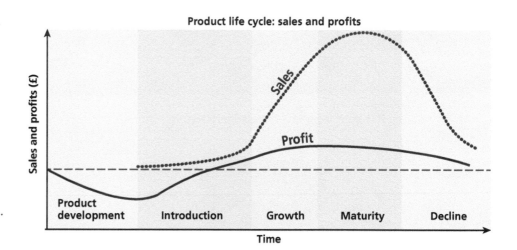

FIGURE 8.6 Product and profit cycles.

the growth period. Innovation occurs in the more specialized regions, but as the products became more standardized, the crucial locational component is less closeness to innovation and more reducing labor costs. Consider the case of typewriters, which were introduced in the early twentieth century but then declined in use by the end of the century as computers replaced them. Typewriter manufacturing shifted from the early specialized centers where labor was expensive, such as Syracuse, New York, to closer but cheaper labor areas such as Cortland, New York, to extremely cheap labor areas in Mexico before finally becoming obsolete. Similarly, today, while new computer technologies are generated in Silicon Valley, the production of actual computer chips can occur in newly industrializing regions with good infrastructure and cheap labor, such as Thailand. In the early stages of product development, when super profits can be made, new firms are attracted to the location of the initial innovation, but as more firms enter the markets, new spatial forces are at work and firms may relocate closer to markets or away from higher-wage areas to maintain and increase profitability.

Regional Economic Clusters

There is a distinct clustering of similar economic activities at the regional scale. Four factors are important:

- locational pull of resources, such as proximity to transportation links or energy sources;
- the concentration of firms and workers, which provides a pool of specialized labor that benefits both labor and companies;
- subsidiary trades and services that are attracted to the core industries; and
- the spillover of innovative ideas from firm to firm.

Regional economic complexes emerge because of the locational pull of reduced transaction costs. However, these are maintained by the tight social networks between firms that require interpersonal trust, proximity, and regular face-to-face contact. These social networks are particularly important in highly specialized sectors. Even in the information age, the spatial clustering of activities, firms, and workers provides the efficient means of communication that help solve problems and facilitate mutual cooperation.

Regional economic clusters provide a **thick market** (a large pool of labor and specialized firms), market access, and savings in public goods provision. In effect, economic **agglomerations**, because they concentrate things in the same place, make markets more efficient. There are efficiency gains from the scale effects. And, in

turn, this spatial ensemble of benefits, through cumulative causation, creates path-dependent development. Once the movie industry was established in the Los Angeles area, for example, it continued to attract capital and skilled labor, which made it even more attractive to make movies in that location.

There are also countervailing forces at work. The more firms cluster in one specific region, the more land and labor costs go up and congestion costs increase. These **centrifugal forces** can disrupt the clustering of routine production but are less compelling for specialized sectors such as entertainment, financial services, or cultural-creative industries. The high-tech sector is still concentrated in Silicon Valley despite the high cost of land and labor. Industrial clustering is also found in the small-craft-based

8.3 The Changing Concerns of Economic Geography

The subdiscipline of economic geography has undergone important shifts as its practitioners look at the economy in different ways. Early work, produced at a time of industrial growth, focused on theories of industrial location. One of the earliest studies was Alfred Marshall's *Principle of Economics*, which first appeared in 1891. A strong German contribution is embodied in Albert Weber's *Theory of the Location of Industries* (1909) and August Losch's *The Economics of Location* (1939). By the middle of the twentieth century, geographers sought to explain the location of industry through looking at the various factors of production from the standpoint of the individual enterprises. Estall and Buchanan (1961) discuss the role of various factors, including materials, markets and transfer costs, energy sources, labor and capital, technological change, and the role of government. In their case study of iron and steel, for example, they point to the geographical pull of coal, iron ore, material assembly, market, capital, and labor, as well as the importance of inertia.

Critics of standard industrial location theory pointed to the assumption of perfect economic rationality. Decision makers, in the models at least, had all the necessary information and were blessed with perfect rationality. Scholars pointed to the need to incorporate suboptimal decision-making and satisfying rather than maximizing behavior. Critics also pointed to the emphasis on single-plant, single-product establishments and raised the need for a fuller understanding of agglomerations, corporate context, and multinational transactions. An institutional approach emerged that stressed the interaction between firms rather than the behaviors of individual firms and showed that location was the result of complex negotiation and bargaining with a variety of other agents in a wider social and political context.

The concerns of economic geographers have also widened. Some economic geographers have utilized evolutionary economics to refine location and relocation theories. A **Marxist political economy** approach explains how uneven development is an essential part of capitalist economies. A new economic geography now locates economic processes in deeper social, cultural, and political contexts in the exploration of themes such as commodity chains, technology and agglomeration, commodification of the environment, and the role of knowledge-based economies and creativity in regional and city development.

There has also been a fruitful infusion of feminist perspectives into economic geography. Gender relations intertwine with uneven development, globalization, locational strategies, and employment practices. Issues such as the feminization of poverty, the dominance of young women in the factories of coastal China, the feminization of domestic service, and the double bind that affects many women as they work both outside and inside the home all highlight the close connections between gender, work, and the economy. The workplace is structured with distinct sexualized and gendered scripts. Gender relations are created and contested in the workplace.

REFERENCES

Clark, G. L., M. Feldman, M. Gertler, and D. Wojick, eds. 2018. *The New Oxford Handbook of Economic Geography*. Oxford: Oxford University Press.

Coe, N. M., P. F. Kelly, and H. W. Yeung. 2020. *Economic Geography: A Contemporary Introduction*. 3rd ed. Oxford: Wiley–Blackwell.

Estall, R. C., and R. O. Buchanan. 1961. *Industrial Activity and Economic Geography*. London: Hutchinson.

industries of northern Italy that specialize in clothing, design, and furniture.

Manuel Castells and Peter Hall identify contemporary industrial complexes, which they term **technopoles**, established around high-technology industry. They include semi-planned territorial complexes such as Silicon Valley in the United States, science cities such as Daejeon in Korea, technology parks such as Sophia Antipolis in France, twenty-six technopoles spread through Japan, and large metro areas around the globe, including more established ones such as London and Tokyo and newer ones such as Seville, Spain, and Adelaide, Australia. They identify four foundational elements: a determined administration, liminal individuals who straddle the business and academic worlds, an environment of innovation, and easy access to venture capital.

Regional Economic Clusters in the United States

As an example of regional differences, we can consider the new geography of employment in the United States. Enrico Moretti identifies a divergence from around the mid-1980s, when selected cities with propulsive economic sectors kept attracting more growth and generating higher-wage jobs. Meanwhile, certain urban regions failed to attract new growth and were unable to replace the jobs lost by deindustrialization. In these declining urban regions, wage levels dropped and the more mobile and higher-skilled workers left. For urban regional economies to be successful, they need to constantly push up the steep slope of innovation adoption, which requires expertise, education, and knowledge. The more the national economy shifts toward a knowledge-based economy, the more marked the differences become between those regions that attract and those that fail to attract these sectors. There are now three types of economic regions in the United States:

- the brain hubs of a well-educated labor force, higher wages, and innovation; examples include San Francisco and Palo Alto;
- declining manufacturing hubs of deindustrializing cities, such as cities in the **Rust Belt** (Map 8.2); and
- city regions in the middle that could go either way.

8.3 The Role of the State

Economic globalization is creating a more integrated global economy, but even in this new world of interconnections and transactions that span the globe, the state remains important. Many firms operate within national

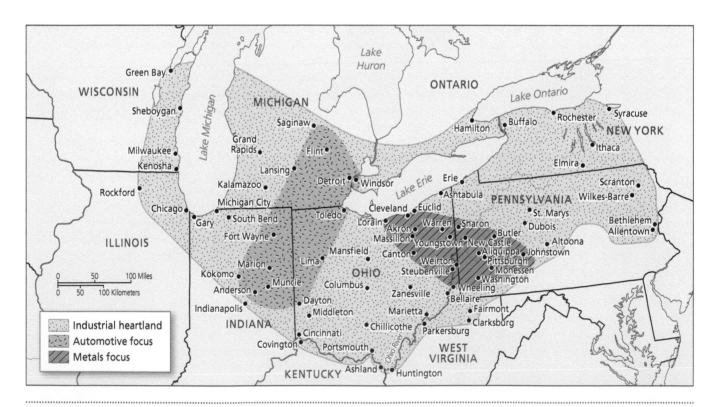

MAP 8.2 Rust Belt. Since the 1980s, cities in this region have experienced deindustrialization, declining economic growth, and population loss as people seek jobs elsewhere.

markets, and national regulations play an important role in shaping the economic decision-making context.

The state is not a singular organization, but a collection of different institutions and agencies. Sometimes their interests coincide; other times they collide. The US Department of Defense, for example, wants to maintain a large and expensive military presence funded by the government. The US Treasury, in contrast, is concerned with controlling government expenditure. The clashes are resolved through the political process.

Through its multiple and varied roles, including employer, investor, policy maker, and regulator, the state shapes economic geographies. Within the nation's borders, the state has fiscal powers of taxation and spending, currency control, fixing exchange rates, and setting interest rates. Raising interest rates, for example, slows down industries such as house building that rely on long lines of credit. Lowering interest rates in turn can stimulate such industries.

The state sets the regulatory framework, which affects economic activity. The economic geographer Jamie Peck examines the political economy of **workfare**, the welfare-to-work initiatives that have developed since the 1990s. He traces the development of workfare policies in Canada, the United States, and the United Kingdom and shows that workfare is not so much concerned with creating jobs as with deterring welfare claims and facilitating the acceptance of low-paying jobs.

The State and the Global Economy

The state also has a role in shaping the connection between national economies and the global economy. At one extreme are the closed economies that limit trade and economic connection with the outside world; examples include Cambodia under the Khmer Rouge and present-day North Korea. Such instances are rare, because global trade is the lifeblood of most national economies. At the other extreme are more loosely regulated economies that facilitate penetration by foreign capital. Most countries fall somewhere in the middle.

The governments of developing countries, seeking to support the industrialization of the national economy, follow either an **import substitution** strategy, in which they protect local industry, or an export-oriented form of industrialization. This was followed in Latin America up until the 1980s. In contrast to **protectionism**, there is the ideology of free trade. One dominant view is that countries develop through the adoption of free-trade policies. In a provocative argument, Ha-Joon Chang takes issue with this idea and argues for protectionism rather than free trade. He claims that promotion of a free-trade strategy works against the development interests of national economies. Protectionism allows local firms to grow and prosper by borrowing ideas and techniques from more efficient producers. Since the rise of Britain, most countries, even

successful ones such as the United States, have competed initially with forms of protectionism for local industries. More recently, the spectacular success of countries such as South Korea has been based on protecting local industries before they became internationally competitive.

The empirical evidence suggests that more open trade leads to rising per capita income. However, per capita income is a crude measure that does not measure the distribution of income, but merely the average. The evidence on income distribution is inconclusive. There are examples of countries where income inequality actually increased in the wake of trade liberalization: Argentina, Chile, Colombia, Costa Rica, and Uruguay all saw increasing inequality. In the United States, the declining incomes of the middle class are commonly linked to open trade policies.

The Development State

The **development state** is the name given to countries that have successfully crafted their economic policies to move from the periphery of the global economy closer to the core. The term implies a state that is focused on economic development. Examples include Japan, South Korea, and China. While they adopted a variety of measures, they all aid specific industries, protect domestic industries in the early years, enhance infrastructure, and spend on education. There is a strong commitment to export-led growth.

The most dramatic and recent example is China. The China model, which consists of directing public spending to foster infrastructure improvement and educational attainment, while lacking democratic accountability, was spectacularly successful in raising 600 million people from chronic poverty in less than two decades.

The development state was primarily an East Asian phenomenon, and that raises the question of whether it is transferable. Francis Fukuyama argues that there are limits to this transferability, because the model is culturally specific to Confucian societies with a meritocracy in public service, a heavy emphasis on education, and a deference to authority. His argument would seem to be undermined by the relative success of countries like Brazil that have different traditions from East Asian societies but are still achieving economic growth.

Does the development state have to be an authoritarian government? In the early years of industrialization, this tends to be the case, and modernization does not necessarily lead to democratization but may make democratization more possible. The more developed and richer a country becomes, with wealth widely spread, the more it tends to move further to the democratic end of the political spectrum.

The development state can achieve spectacular success as public and private interests are focused on the same goal of achieving rapid growth. However, the reliance on export-led growth makes the economies vulnerable to global shocks.

8.4 Addressing Poverty

To effectively reduce poverty, a government must adopt public policy interventions that help to modify the social, cultural, and economic conditions that created poverty in the first place. While economic growth is an essential factor for success in the fight against poverty, it is often insufficient. There is the need for growth to be augmented by policy interventions such as social protection to mitigate vulnerability. A 2015 United Nations Food and Agricultural Organization report said that between 1990 and 2015, globally social protection intervention has helped lift about 216 million people out of hunger and vulnerability.

Social protection refers to policies or actions directed at reducing weakness and shocks through alleviation of poverty perils by promoting effective labor markets, minimizing individuals' vulnerability to risks, and building their capabilities to coordinate economic and social disturbances, including old age and ill health, disability, unemployment, and financial exclusion. Some of the more effective social protection measures are public or government cash transfers, which are a form of income redistribution from the rich to the poor. For example, in agricultural communities, social protection could include subsidizing the price of fertilizers sold to farmers and leasing of land to farmers at lower rates, among others. This can be a critical intervention because in many instances agriculture is associated with various shocks and risk, such as floods, pests, and disease, which negatively affect crops and therefore livelihoods.

In the United States, social protection includes Social Security, unemployment benefits, and the Supplemental Nutrition Assistance Program, sometimes called "food stamps," that has reduced child hunger. A 2019 study by the National Academies found that social protection programs that alleviate poverty directly (such as income transfer) or indirectly (by providing food, housing, or medical care) can substantially improve child well-being.

Social protection systems are fundamental not only to lifting people out of poverty, but also in preventing them from falling back into poverty. A study by Oluwatoyin Matthew and her colleagues found that social protection programs in Nigeria, when properly earmarked and structured, can bridge the income shortcomings of poor families and protect health.

REFERENCES

Matthew, O., R. Osabohien, F. Fagbeminiyi, and A. Fasina. 2018. "Greenhouse Gas Emissions and Health Outcomes in Nigeria: Empirical Insight from ARDL Technique." *International Journal of Energy Economics and Policy* 8:43–50.

National Academies of Sciences, Engineering and Medicine. 2019. *A Roadmap to Reducing Child Poverty.* Washington, DC: National Academies Press.

United Nations Food and Agricultural Organization. 2015. *The State of Food and Agriculture 2015: Social Protection and Agriculture: Breaking The Cycle of Rural Poverty.* https://www.fao.org/3/i4910e/i4910e.pdf.

8.4 Capital and Labor

An important element of economic geography is the changing relationship between **capital** and **labor**. The capital–labor relationship is crucial to the modern economy. Capital seeks to keep the costs of labor down and uses its influence over the state to ensure legislation that minimizes the costs and power of labor. Labor in turn seeks to organize in order to have bargaining power to raise wages and improve working conditions. Changes in economic geography can be partly understood in relation to capital–labor relations and their differential mobility. Capital has generally become more mobile than labor, which strengthens the hand of capital in relation to labor.

The differential in mobility has widened. The suburbanization of industry was a response to the growing power of labor within cities. The shift of car manufacturers from Michigan to the southern states was a move by capital to reduce the costs and power of labor. More recently, the global shift in manufacturing is partly a move to cheaper regions of the world with less organized labor, such as China.

Capital

"Capital" has a number of meanings. In the standard literature, it is used to refer to real assets, such as factories, property, and machinery, or financial assets, such as bonds, securities, and stocks. But this is only the physical expression; capital is also a social relationship.

The mobility of capital has been reinforced by changes in production. In the system known as **Fordism**, plants were huge, fixed capital investments. Bargaining between capital and labor thus took place against a fixed location. Recently, however, a more flexible form of production has been introduced. Nike, for example, has no shoe factories and no Nike workers. Nike's shoes are made under contract by a range of shoe manufacturers. Factories compete to obtain Nike orders and are then licensed by Nike if they are capable of making shoes to cost and design specifications. The old model of manufacturers making things and retailers simply selling them has been replaced by the power of retailers and brands. Now retailers tell the manufacturers what to produce. Contracts are for short-run lines rather than long-run batches. This just-in-time production system ensures that goods are made to meet demand. Just-in-time, flexible production allows low prices and high turnover. It also means a marked change in capital–labor relations. Capital is now hypermobile. Workers in one factory cannot bargain in the same effective way that the workers of the old Fordist system could. Capital is no longer fixed in place. Retailers can move their production contracts to another factory in another country. While capital can roam the world, labor is more fixed in place. The result is an uneven bargaining arrangement that has tended to favor management over labor.

Capitalism is a dynamic, crisis-laden economic system with tremendous vitality. Perhaps the most sustained geographical analysis of capital as a social relationship is the work of David Harvey, who since 1973 has been developing a historical geography of capitalism. In his work he outlines the recurring crises of capitalism. For the system to continue, money needs to be mobilized and invested, technology employed, nature exploited, labor assembled, and credit made available. At any of these stages, problems may occur; there may be a lack of money to make suitable investments, technological innovation may undermine profitability, labor may become too powerful and demand too much in wages, environmental costs may restrict growth, or access to credit may dry up. A profit squeeze or credit crisis can cause blockages in the smooth functioning of the system. In fact, smooth functioning is rare. Capitalism is capable of amazing growth and innovation, but it is subject to recurring crises and continual turmoil as the crisis solved in one area becomes a crisis in another area. The globalization of capitalism has kept the system going but has widened the opportunity for recurring crises worldwide.

Labor

There is a geography of labor that includes the importance of the geography of labor legislation, the political geography of union organizing, and the power of place. Geographer Andy Herod, for example, has shown how organized labor impacts and influences the geography of capitalism with reference to selected case studies of garment workers, steel workers, and dockworkers.

The overall nature of the capital–labor relationship has shifted. From the 1930s to the 1980s, organized labor in Europe and North America was powerful and played an important role in creating public policies that guaranteed relatively high levels of social welfare. Skilled labor was in short supply and so could use its bargaining power to increase both wages and what is termed the social wage, the range of social benefits provided by the state. There were national variations in the affluent West, with social welfare levels significantly higher in the Scandinavian democracies than in the United States. Since the 1980s, the power of labor has been weakened by global shift and mechanization. The deindustrialization of the economies has led to a weakening of organized labor and a resultant shift in capital–labor relations, with wages stagnating (Figure 8.7 and Photo 8.1).

The basic geographic difference that strengthens capital and weakens labor is differential mobility. Capital is freer to roam the globe in the search for profitable investments. Labor, in contrast, is more fixed in place, unable to freely move across national borders. Mobile capital has more power in a globalizing economy than a labor force more fixed in space.

And here we come to a paradox. Even as firms and companies seek to minimize labor costs, the high-mass-consumption economies need consumers to have enough disposable income to buy goods and services. This is the Walmart economy, where companies pay low wages but then have to provide cheaper goods. The goods are cheap because the wages of the producers are low.

Labor is not just an economic entity. It is embodied in people. An important element of the new economic geography considers how social differences are reinforced, undermined, and contested in the workplace. Who works in what jobs and how much they are paid is not a simple labor market process but a wider economic and social

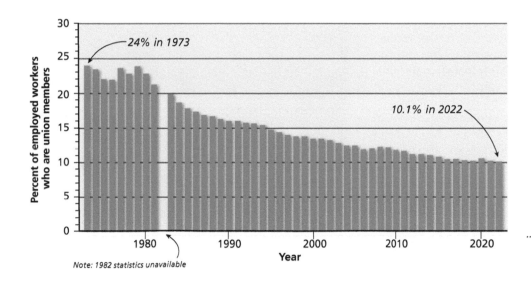

FIGURE 8.7 Declining union numbers.

PHOTO 8.1 Union Hall in Baltimore.

differentiator. There is considerable work on ethnicity, for example, that examines ethnic economies (where workers in one sector or trade share common ethnicity), the geography of ethnic networks, and the connection between ethnic networks and accelerating economic globalization. Geraldine Pratt explores the geography of Filipina domestic workers in Vancouver, Canada. She found that women from a variety of occupations were restricted through immigration laws, which did not recognize their national qualifications, to the domestic service sector.

Households get by through using the different sectors of the economy. The available getting-by strategies will depend on the resources. A household with a large income from the formal economy will not have to sell cigarettes on the street corner. A household with few formal opportunities from the market or state, in contrast, must look at alternative sources and rely on the informal economy.

At the very bottom are the poor, who must be creative in their survival strategies (see Photos 8.2 and 8.3). Around 40 percent of the world's population lives on less than $2 a day. One billion people live on less than $1.25 per day. Lacking income, they are trapped in extreme poverty. One proposed solution is the use of microcredit—lending small amounts to people to provide them the financial means to become more self-reliant. Muhammad Yunus pioneered the use of microloans in Bangladesh. One study looked at the detailed diaries of 250 households in Bangladesh, India, and South Africa.

It was clear that the problem is not only the meager income but also the unpredictable and irregular nature of that income, which makes saving and planning very difficult. One proposal is to shift emphasis from the encouragement of microfinance toward easier access to savings opportunities.

8.5 Different Economies

We can distinguish between different types of economies. Employment in the **formal economy** involves the sale of labor in the marketplace, formally recorded in government and official statistics. The **informal economy** is the unrecorded sector, where few, if any, taxes are paid. The informal sector is a response to the lack of formal employment opportunities. This sector is just as complex as the formal sector. Based on his work in Latin American cities, Ray Bromley identifies different sectors, including retail distribution, small-scale transport, personal services, security services, gambling services, and recycling enterprises. Work conditions vary from short-term wage work to precarious self-employment. The workers in this sector may be disguised as wage workers (they are paid a wage but it is not recorded), workers on short-term

PHOTO 8.2 Informal employment in Panama.

PHOTO 8.3 Informal employment in Kenya.

wage work, dependent workers who may lease a personal transport vehicle from an employer, and the self-employed. In his research on the city of Cali, Colombia, Bromley found that 40 to 45 percent of people who worked in the street were in precarious self-employment, 39 to 43 percent were disguised wage workers, 12 to 15 percent were dependent workers, and only 3 percent were short-term wage workers. Subsequent work on Cali done by Lina Martinez and her colleagues revealed the granularity of street vending, with a wide variety in conditions, characteristics, stability, and incomes of street vendors that range from the relatively affluent through to the lower income and more marginal and from the long established to the recently arrived.

The divisions between the formal and informal sectors fluctuate according to circumstance. When and where the formal economy provides many employment opportunities, the informal sector becomes less significant. However, if the formal sector provides few jobs or only jobs with low wages, then the informal sector is a sophisticated coping strategy as people use a variety of tactics to make a living.

The **illegal economy** is the shadowy world of prostitution, illicit drug sales, racketeering, and the like. Bromley noted the importance of prostitution, begging, and property crimes involving illegal appropriation through stealth (theft), the threat or use of violence (robbery), or deception (conning). It is difficult to estimate the sector's full extent. Much of the global illegal sector consists of the production and distribution of goods and services for which there is a demand even though they are illegal. The demarcation of legal and illegal varies: alcohol is not illegal in the United States, but cocaine is; prostitution is legal and regulated in both Germany and the Netherlands, but is illegal in most of the United States; one can buy marijuana in Amsterdam without breaking the law, but the same transaction in Charleston, South Carolina would be considered criminal. The line between legal and illegal is constantly being crossed: police who take bribes, chemical companies that illegally dump waste, corporations that form illegal cartels to charge high prices to the federal government. The distinction between the three sectors, formal, informal, and illegal, is at times fuzzy. Rather than look at the differences, a more rewarding strategy may be to look at how transactions transgress the lines.

A **social economy** consists of not-for-profit activities; it is also sometimes referred to as the "third sector." It includes cooperatives, charities, voluntary groups, trusts, and religious organizations. This sphere of economic activity, somewhere between the market and the state, plays an important role in certain parts of the economy as both funding source and employer.

The **communal economy** involves the cashless exchange of goods and services. It is common in neighborhoods and extended family systems. If someone wants a babysitter for an evening, they have a number of possibilities: they can hire someone and record the transaction and pay taxes (the formal economy); they can hire a teenager and pay them an agreed sum without informing the authorities (the informal sector); or they can ask a neighbor, with the often unspoken understanding that they will return the favor. The latter, and often the most common, response is an example of the communal economy in action. It can range from reciprocal favors, such as babysitting, grass cutting, and garbage removal, to a host of household chores, from building maintenance to carpooling.

The **domestic economy** is the amount, type, and division of labor within the home. A hundred years ago, there was much greater use of paid domestic labor for middle- and upper-income groups. The increasing cost of labor has meant the decline of mass domestic labor, though it remains very popular in East and Southeast Asia and persists in the West among the wealthy. Foreign, vulnerable, and illegal workers do a high proportion of such work.

The division of labor within households is a source of change, conflict, and negotiation. As more women have joined the formal labor force, they feel the double demand of work and home because they carry much of the domestic load. The division of domestic labor is not simply an apportionment of necessary work; it involves questions and representations of femininity, masculinity, and the family.

Recent years have also witnessed the rise of the **sharing economy**, in which owners of assets such as cars or rooms rent them out in peer-to-peer exchanges. Examples include Uber and Airbnb. The sharing economy has grown with the widespread adoption and use of social media and specific apps that link providers with consumers. The sharing economy disrupts traditional markets: the introduction of Uber, for example, is seen as a threat to many cities' traditional taxi systems. The consequences of the widespread diffusion of the sharing economy on traditional economies are only just unfolding.

8.6 The Rise of Mass Consumption

We can make a distinction between different types of capitalist economy. In the nineteenth and early twentieth centuries, capitalist economies were based on the production of commodities and most people were relatively poor. Beginning in the United States in the 1920s, a new economy emerged based on the mass production and mass consumption of goods. In societies of high mass consumption, the consumer emerges as an important element in the making of economic geographies. Consumer spending in the United States is now responsible for two-thirds of all economic activity. The enthronement of the consumer has many implications; here we will discuss only five.

First, as consumer spending shifts from meeting basic requirements to discretionary spending, desire rather than need becomes a more important driver of consumption. Consumers have to be persuaded to buy things, and this basic fact is the reason behind the growth and emergence of advertising and promotion of goods. While we may want clothes, we have to be persuaded to buy particular brands. Consumption is now tied to the creation of identities and the pursuit of lifestyles. To buy a pair of sunglasses is to shield your eyes from the Sun, but to buy a pair of Ray-Bans or Guccis is to evoke other images about your status. Consumption is now tied to aspirations, images, and desires as much as to basic needs and requirements.

Second, forms of consumption are markers of social status and social difference. People are distinguished not only by how much they have, but also by what they have and how they use it. Forms of consumption mark different levels of society. While all middle-income groups can have a beach holiday, only the richer can afford a Caribbean vacation, and only the very rich can rent a villa in St. Bart's. Luxury fever can filter down the income scale as the consumption trends of the very rich become embedded in popular consciousness.

Third, the full flowering of a society of high mass consumption is predicated on access to credit so that consumers can buy big-ticket items. The rise of credit to lubricate mass consumption has its upside, because it enables purchase before the full price is saved. It also has its downside, because it makes the economy vulnerable to credit squeezes. That is what happened after the 2008 meltdown as many housing consumers in the United States had extended credit on homes that were depreciating in value. With consumers unable to gain access to any more credit, the economy sputtered and declined as demand fell.

The more consumer spending is reliant on credit, the more the economy is vulnerable to credit crises.

Fourth, the economic geography of places is shaped by the places of consumption. As shopping becomes an increasingly important social and economic activity, retail sites become bigger. The typical mall has increased in size and scope. The mall is the cathedral of consumerism, an alternative retail environment where consumers are enticed, entertained, and encased in a totalizing retail space.

All of this changed with COVID-19. The pandemic transformed the retail landscape dramatically and quickly, with many stores closing and more people using online and door delivery services. As people stayed home, there was an increase in having things delivered. Although some stores and shops have started to reopen, the pandemic probably caused an irreversible change in retail shopping and greater use of home delivery services. Retail establishments are clustered together. We can understand this phenomenon with reference to **Hotelling's law**, named after economist Harold Hotelling, and his classic paper, first published in 1929. The law states that it is rational for producers to make products similar to their competitors'. If we substitute retailers for producers, then the law explains the clustering of retail distributions. Assume a straight street. If there is only one coffee shop, the owner is indifferent to location anywhere along the street. But if there are two or more coffee shops, it is more rational for them to be together at the center of the street, drawing on 50 percent of the street's coffee drinkers. Retail establishments cluster together because it makes rational economic sense even for providers of the same product—otherwise, they might lose customers (see Photo 8.4). That is why a Burger King and a McDonald's are

PHOTO 8.4 Retail clustering in a market in Seoul.

often close together when you turn off the highway. Even retailers with different merchandise cluster, because they get the extra advantage of possible market capture of consumers of other producers. That is why you often find a Starbucks clustered with the two hamburger sellers.

Fifth, and finally, there is the growing dominance of retailers over producers. Traditionally, manufacturers made things and then sold them to retailers, who sold them to customers. Retailers were simply in the middle, buying goods from producers and then selling them to customers. But with the rise of the big-box retailer and a more flexible production system, with many small companies producing small runs of goods relatively quickly, power shifts from the large number of small manufacturers to the small

number of larger retailers. The very large retailers have such enormous power that they can force producers to compete with each other to meet strict cost limits. Manufacturers are willing to do this because access to a big retailer ensures a large customer base. The lower price per item is offset by the higher turnover of items. Take the case of Walmart, which in 2022 had 10,593 stores in twenty-four countries. Walmart uses its enormous leverage to drive down the prices of the manufacturers who supply these stores. Walmart reduces the price of goods by relying on producers from low-wage countries, thus pushing North American producers to reduce their costs. A number of North American producers have offshored their production to reduce costs in order to sell to Walmart. Walmart

8.5 The Costs of Climate Change

Climate change will have significant economic costs. There are the obvious costs—such as the costs of rebuilding after wildfires or floods. Global grain yields may decrease, causing prices for food staples to increase. But there are also the less obvious, "social" costs. A collaboration of climate scientists, economists, computational experts, and analysts at multiple universities used data from forty countries to calculate both costs and benefits of climate change and adaptation. The results allow researchers to monetize those costs to society, referred to as the social cost of carbon. They showed that for every ton of carbon emitted, there are about $51 in "social" damages, such as increased mortality rates. Predicted higher temperatures and pollution from fossil fuels can cause heat stroke and cardiovascular disease. Not surprisingly, many of those who are most vulnerable to the social costs of climate change will be poor.

A 2022 study by the Asian Development Bank on climate risk and poverty listed a combination of actions as critical for promoting resilient livelihoods for the urban poor. First is to strengthen coping mechanisms, such as stockpiling food for flood seasons. Second is to implement incremental adaptation to accommodate changes from climate change, such as building higher dikes to protect slums from increased floods. Third, and more challenging, is to undertake transformational adaptation that introduces fundamental systemic change by addressing the root causes of vulnerability to climate change, such as land use changes that introduce nature-based solutions to manage flooding and the involvement of local women in protecting natural resources.

The report suggested governments build resilience among the poor by increasing social protection measures, adapting public health systems, and creating policies that ensure safe housing and robust community infrastructure. For example, pro-poor policies that enhance social and economic safety nets, improve education, and teach job skills can also help strengthen climate

resilience. In terms of public health systems, building resilience will require health policies and plans that recognize, predict, and provide services in response to the likely health impacts of climate change (such as increased heat stress). Healthcare systems could introduce new heat stress–related programs that support those who work outdoors. In addition, safe housing will be critical if disasters and weather impacts damage the housing of poor households that have used substandard materials or are more exposed to hazards.

The report recommends governments adopt "no regret" or "low regret" solutions: in other words, see the value of poverty reduction strategies that generate social and economic benefits irrespective of how the future climate pans out.

The 2022 United Nations climate change conference, commonly referred to as COP 27, involved establishing a dedicated fund for countries hard hit by climate disasters, a commitment to keep global warming to below 1.5 degrees centigrade, holding businesses and institutions to account, mobilizing more financial support for developing countries, and moving from pledges to concrete action.

REFERENCES

Asian Development Bank. 2022. *Building Resilience of the Urban Poor in Indonesia*. https://www.adb.org/sites/default/files/publication/763146/building-resilience-urban-poor-indonesia.pdf.

Resources for the Future. Social Cost of Carbon Initiative. Various publications. July, 2022. https://www.rff.org/topics/scc/social-cost-carbon-initiative/

United Nations. 2022. "Climate Action." https://www.un.org/en/climatechange/cop27.

US Congress, House Committee on Oversight and Reform, Subcommittee on Environment. 2019. *Hearing on "Economics of Climate Change."* 116th Cong., first session.

and stores like it are a major cause for the shift in manufacturing to lower-wage regions.

Walmart uses its market power to reduce costs, leading to more offshoring and more imported goods from cheap-labor regions, which in turn leads to fewer jobs and lower incomes in the formerly richer regions and consequently lower wages, forcing more people to seek deals at Walmart. Although big-box retailers reduce costs and increase productivity, there is also the associated downward spiral of job loss and wage reduction.

Cited References

Bromley, R. 1997. "Working in the Streets of Cali, Colombia: Survival Strategy, Necessity or Unavoidable Evil?" in J. Gugler, ed. *Cities in the Developing World: Issues, Theory and Policy*. Oxford: Oxford University Press. 124–138.

Chang, H.-J. 2007. *Bad Samaritans: The Myth of Free Trade and the Secret History of Capitalism*. London: Bloomsbury.

Central Intelligence Agency. 2022. "The World Factbook," https://www.cia.gov/the-world-factbook/.

Diamond, J. M. 1998. *Guns, Germs and Steel: A Short History of Everybody for the Last 13,000 Years*. New York: Vintage.

Fukuyama, F. 2012. "The Future of History." *Foreign Affairs*, January–February, 53–61.

Hall, P., and M. Castells. 1994. *Technopoles of the World: The Making of 21st Century Industrial Complexes*. New York: Routledge.

Harvey, D. 2007. *A Brief History of Neoliberalism*. Oxford: Oxford University Press.

Harvey, D. 2010. *The Enigma of Capital and the Crises of Capitalism*. London: Profile.

Herod, A. 1998. *Labor Geographies: Workers and the Landscapes of Capitalism*. New York: Guilford Press.

Hickel, J. 2017. *The Divide: A Brief Guide to Global Inequality and Its Solutions*. London: Heinemann.

Hotelling, H. 1929. "Stability in Competition." *Economic Journal* 39:41–57.

Martinez, L., J. R. Short, and D. Estrada. 2018. "A Case Study of Street Vending in Cali." *Cities* 79:18–25.

Moretti, E. 2012. *The New Geography of Jobs*. Boston: Houghton Mifflin.

Oxford Poverty and Human Development Index. 2023. "Global Multidimensional Poverty Index: Global MPI 2023," https://ophi.org.uk/global-mpi-2023/.

Peck, J. 2001. *Workfare States*. New York: Guilford.

Pratt, G. 1999. "From Registered Nurse to Registered Nanny: Discursive Geographies of Filipina Domestic Workers in Vancouver, BC." *Economic Geography* 75:215–236.

World Bank Group. 2023. "The World By Income and Region," https://datatopics.worldbank.org/world-development-indicators/the-world-by-income-and-region.html.

Select Guide to Further Reading

Acemoglu, D., and J. Robinson. 2012. *Why Nations Fail: The Origins of Power, Prosperity, and Poverty*. New York: Crown.

Allen, R. C. 2011. *Global Economic History: A Very Short Introduction*. Oxford: Oxford University Press.

Chant, S., ed. 2010. *The International Handbook of Gender and Poverty: Concepts, Research, Policy*. Cheltenham, UK: Elgar.

Coe, N. M., P. F. Kelly, and H. W. Yeung. 2020. *Economic Geography: A Contemporary Introduction*. 3rd ed. Oxford: Wiley–Blackwell.

Escobar, A. 2011. *Encountering Development: The Making and Unmaking of the Third World*. Princeton, NJ: Princeton University Press.

Frank, A. G. 1966. *The Development of Underdevelopment*. Cambridge, MA: Free Press.

Hudson, R. 2020. "The Illegal, the Illicit and New Geographies of Uneven Development." *Territory Politics Governance* 8 (2): 161–176. https://doi.org/10.1080/21622671.2018.1535998.

Inglehart, R., and C. Welzel. 2005. *Modernization, Cultural Change, and Democracy: The Human Development Sequence*. Cambridge: Cambridge University Press.

Landes, D. S. 1999. *The Wealth and Poverty of Nations: Why Some Are So Rich and Some So Poor*. New York: Norton.

Smith, N. 2008. *Uneven Development: Nature, Capital and the Production of Space*. 3rd ed. Athens, GA: University of Georgia Press.

ON ECONOMIC AND HISTORICAL DISCUSSIONS OF THE CORE–PERIPHERY MODEL, SEE:

Krugman, P. 1991. "Increasing Returns and Economic Geography." *Journal of Political Economy* 99:483–499.

Wallerstein, I. 1974. *The Modern World System*. Vol. 1. New York: Academic Press.

Wallerstein, I. 1980. *The Modern World System*. Vol. 2. New York: Academic Press.

Wallerstein, I. 1989. *The Modern World System*. Vol. 3. San Diego, CA: Academic Press.

9

Economic Geographies

Economic geography looks at the changing nature and spatial distribution of different sectors of the economy. The mix of sectors varies over time and space. Examining the changing configurations of economic growth at the regional and global scales provides an intriguing insight into economic change and its consequences.

Three distinct sectors of economic activity can be identified: primary, secondary, and tertiary. These are outlined in Figure 9.1. Some care should be taken with this division, because it suggests strong demarcation lines between the sectors, whereas in reality there is considerable interaction. Consider the case of a financial institution (tertiary sector) that lends money to a farmer to buy machinery to grow potatoes (primary) that are then transformed in a factory into potato chips (secondary). All three sectors are connected in complex arrangements, networks, and flows. In the rest of this chapter, we will explore some of the socio-spatial dimensions of each of these sectors separately but remain mindful of the flows and connections that link them together.

As economies grow and mature, the secondary and tertiary sectors tend to become more important in generating economic growth. Table 9.1 highlights the changes in the sector mix in China. As the economy developed, the **tertiary sector**

LEARNING OBJECTIVES

9.1 Identify the three sectors of economic activity and **explain** how they relate to von Thünen's model of land use.

9.2 Describe the agrarian transition, and **explain** its changes and challenges.

9.3 Explain the commercialization of agriculture and the impacts of the Green Revolution.

9.4 Explain the connection between the secondary sector of economic activity and the four Kondratieff cycles of historical periods of manufacturing innovation.

9.5 Describe the rise and location of the tertiary and quaternary sectors and **explain** their impact.

9.6 Compare the major trends in the three sectors of economic activity in the Global North and the Global South.

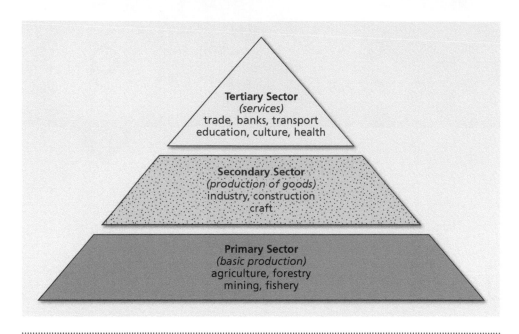

FIGURE 9.1 Three sectors of the economy.

became an increasingly important element.

The relative decline of the primary sector and rise of the secondary and tertiary sectors in a national economy is linked to profound socio-spatial changes, including the commercialization of agriculture, high levels of rural-to-urban migration, and increasing levels of industrialization and urbanization.

TABLE 9.1 Changing Sector Composition in China, 1978–2019

YEAR	1978	1996	2019
Gross domestic product	100	100	100
Primary	28	20	3
Secondary	48	49	36
Tertiary	23	30	59

Source: China Statistical Yearbooks, 2020.

9.1 Agriculture

The **primary sector** includes agriculture, forestry, mining, and fishing (Photo 9.1). Here we will focus on agriculture.

Agriculture is very much influenced by location. The type of agriculture possible depends largely on climate and weather, soils, and water availability. Natural rubber production, for example, takes place in tropical areas, especially in Thailand, Indonesia, Malaysia, India, and Vietnam. Wheat growing, in contrast, is concentrated in the temperate grasslands of the world. The **wheat belt** countries include Canada, Russia, and the United States in the Northern Hemisphere and Argentina and Australia in the Southern Hemisphere.

Geography obviously plays an important role in the distribution of primary activities. But it would be incorrect to see the primary sector only as the outcome of locational opportunities and constraints. Take the case of sugar. Traditionally, it was a tropical plant that could be grown only in the hot, humid areas of the world. By the seventeenth century, a market for sugar had developed in Europe to such an extent that the search for sugar-growing areas constituted an important element in the European colonization and annexation of Caribbean islands. The work was brutally hard, and diseases from the Old World soon killed off the Indigenous people. The need for labor fueled the slave trade. European countries without tropical colonies were at a huge disadvantage, having to pay others for an expensive good. It is not incidental that sugar beet was developed in Germany, a country with no tropical possessions. By the early 1800s, the first beet sugar factory opened in Germany. Sugar beet is a plant with a tuber that contains a high concentration of sucrose. The plant can be grown in colder climates, and so the tropical source became less important. As beet sugar production increased in Europe and North America, the sugar plantation economies of the Caribbean dramatically declined.

The von Thünen Model

Markets play an important part in affecting agricultural production. In 1826, wealthy German landowner Johann Heinrich von Thünen (1783–1850) proposed a model of land use around a city (Figure 9.2). His model was highly idealized in order to identify the independent role of location. The "isolated" city of his model was situated on a flat plain with homogenous fertility and unvarying transportation costs.

PHOTO 9.1 The primary sector includes a range of resource-extractive activities such as mining, seen in this quarry in Pennsylvania.

The city was the main market for farmers. Since farmers closer to the city paid less for transport, they could bid more for land, which resulted in higher land costs closer to the city. Only farmers growing the more intensive crops, with high returns, could thus afford the land closer to the city. The net result, according to von Thünen, was a concentric ring pattern with more expensive land and thus more intensive agriculture closer to the city. At the time he was writing, market gardening and firewood production were intensive land uses, while livestock farming was more extensive.

While the **von Thünen model** operates on the assumption of a flat plain with unvarying fertility, conditions that occur rarely in the real world, it has been useful as a platform for subsequent studies. The von Thünen model is useful to explain historical developments. Environmental historian William Cronon, in his 1991 book *Nature's Metropolis*, examines the relationship between Chicago and its hinterland from 1850 to 1890. He shows how the physical world was turned into a commodified human landscape as grain, lumber, and meat production transformed prairies and woodlands into the physical basis for the city's growth and development. More detailed analyses of contemporary conditions look at the rural–urban fringe of cities. In capitalist land markets, the ring of land closest to the city is often underutilized, in contrast to von Thünen's predictions, because of the heavy shadow of possible future land use conversion. With rural-to-urban conversions a very real possibility, landowners make few investments in agricultural production. The land use planning system that directs which areas are available for rural-to-urban conversion will clearly affect the simple linear-distance model proposed by von Thünen.

In the case of Megalopolis, as the urbanized seaboard of the United States from Boston to Washington is sometimes called, there is empirical evidence of the validity of the von Thünen model. Land closer to the cities is much more expensive than land on the periphery of Megalopolis. If farmers do locate in areas closer to the city, they need to engage in intensive high-yield farming, such as market gardening of fruit and vegetables. In some cases, it is not so much distance as time of transportation that is important. Large cities require quick and immediate supplies of fruit and vegetables and other

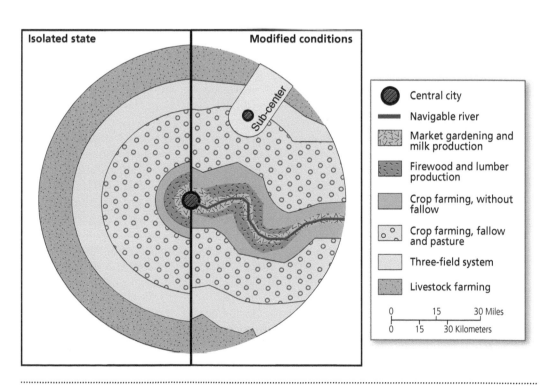

FIGURE 9.2 The von Thünen model.

specialized agricultural products. Restaurants, for example, require daily supplies for their diners. Proximity to the city provides swift access, but at the price of high land values, which in turn means that only intensive farming with high yields makes economic sense. Organic fruits and vegetables is another emerging market that yields a high dollar return per acre. Affluent consumers are willing to bear the greater costs associated with these products. High-yield counties are located around New York City, Philadelphia, and Boston. There are other areas of relatively high yield, such as New Jersey (giving credence to the nickname Garden State), Rhode Island, and southern Connecticut. Agricultural land is being lost as the suburban spread flows across the landscape, but the remaining farmland that is safe from urban development, even temporarily, and close to the city has become so expensive that very intensive forms of farming are required, including nurseries, greenhouse crops, and higher-value specialty products.

There are now many case studies of land use change that draw on the von Thünen model, including tropical forest degradation close to Dar es Salaam in Tanzania and crop production outside of Yaoundé in Cameroon. In their study of land use changes in forest areas of Indonesia, Miet Maertens and her colleagues used satellite imagery, GIS data, and survey techniques. The von Thünen model, although useful, did not explain all the variation in this Indonesian forest frontier area. Not just distance to markets was important; so were population levels, access to technology, and individual household characteristics.

The Agrarian Transition

The **agrarian transition** occurs as agriculture shifts from subsistence to commercial and moves from meeting local market demands to provisioning national and global **food supply chains**. It involves changes to agriculture, including intensification and expansion, market integration, and environmental change.

The extension and intensification of agriculture are pronounced (see Photo 9.2). There is a global expansion of arable land, an expansion of irrigated surfaces, and the development of intensive mixed farming systems in dense population areas. Higher-income countries have adopted large-scale mechanization, biological selection, and increasing use of chemicals to boost yields. Even in some lower-income countries, there is a marked commercialization of agriculture involving the marketization of peasant economies, land grabs for commercial farming, and, overall, a greater integration into global food supply chains.

Market integration takes two main forms. On the one hand, there is the commercialization of the peasantry, as subsistence smallholders become commercial farmers. Geographer Sarah Turner looked at the contemporary process among Hmong farmers in the upland areas of northern Vietnam. She conducted more than 200 interviews and

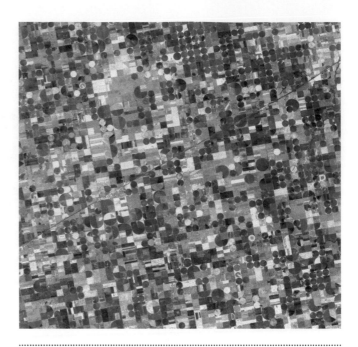

PHOTO 9.2 Intensive center-pivot irrigated agriculture in Kansas.

measured the expansion of rice paddies as a replacement for opium cultivation, which was banned by the government.

There are enormous barriers to this process, as smallholders and subsistence farmers often lack access to capital and technology. There are many cases, however, where traditional peasant societies respond creatively to the new commercial realities despite the barriers. Small-scale farmers can, for example, market their ecological sensitivity to sophisticated consumers, as in the case of smallholding organic coffee growers in Costa Rica. The growth of fair-trade initiatives, for example, often links ethical consumers with small-scale producers.

On the other hand, there is also evidence of a **land grab** as peasant farmers are displaced in the wake of the forced commodification of land with the resultant commercialization of agriculture. This is the second and very brutal form of market integration, and it has a long history. In England, for example, the **enclosure movement** privatized the common land of fields, marshes, heaths, and woodlands. Between 1760 and 1820, almost 20 percent of England's total acreage was enclosed, and the English peasantry was destroyed. A system of traditional rights and obligations was replaced with the cash nexus of agrarian capitalism.

Similar processes are occurring now throughout the world, reinforced by a growing commercialism of agriculture and demand for goods such as sugar and soya, and turbocharged by the demand for biofuels such as ethanol. Land grabbing involves the buying, leasing, or simple appropriation of large parts of land illegally, unfairly, or in an underhanded manner. It is especially evident in the Global South,

9.1 Food Supply Chains

Supply and demand for food are linked through food supply chains, which have become longer and more complex for three interrelated reasons. First, with urbanization, more people live in towns and cities. This growing urban population needs to be supplied with food produced elsewhere. Long and complex transportation links connect urban consumers with rural providers. Second, there is also dietary transition. As household incomes grow, so does the consumption of dairy, meat, and fish. **Bennett's law** states that the consumption of starch staples declines as household income increases. In China, for example, economic growth and the development of a middle class has led to increased consumption of meat, particularly chicken. Third, there is the increased globalization of trade in foodstuffs. Food supply chains are increasingly multinational and global. The longest chains are often associated with very large retailers, such as supermarkets that are large enough to have substantial economies of scale. A study by Miet Maertens and colleagues looked at exports of horticultural products from sub-Saharan Africa to Europe and found important positive welfare effects as rural providers were able to make more money and have a more dependable source of income.

There are least four types of food supply chains. In **buyer-driven chains**, a small number of retailers consolidate their supply around a few suppliers, although there are opportunities for smallholders where crops need intensive care, as in Peru's asparagus sector. In **producer-driven chains**, such as coffee, cocoa, and tomatoes, producers can establish cartels. In **bilateral oligopolies**, such as bananas and pineapples, large-scale producers combine with large-scale retailers. In **traditional markets**, smallholders cater to domestic and local markets.

Consumers, especially more affluent food consumers, can access different types of food chains, shopping at supermarkets as well as at local organic markets and buying food from distant international food supply chains as well as from local producers.

REFERENCES

Lee, J., G. Gereffi, and J. Beauvais. 2012. "Global Value Chains and Agrifood Standards: Challenges and Possibilities for Small Holders in Developing Countries." *Proceedings of the National Academy of Sciences* 109:12326–12331.

Lezoche, M., J. Hernandez, M. Alemany Díaz, H. Panetto, and J. Kacprzyk. 2020. "Agri-food 4.0: A Survey of the Supply Chains and Technologies for the Future Agriculture." *Computers in Industry* 117. https://doi.org/10.1016/j.compind.2020.103187.

Maertens, M., B. Minten, and J. Swinnen. 2012. "Modern Food Supply Chains and Development: Evidence from Horticulture Export Sectors in Sub-Saharan Africa." *Development Policy Review* 30:473–497.

---

where Indigenous peoples and peasant societies often do not have legal documentation of their land claims, and if they do, they are easily pushed aside by powerful foreign investors in alliance with corrupt government officials. Estimates suggest at least 81 million acres in the past decade were purchased by foreign investors that led to the subsequent eviction of local people. In just one example among many, 769 families were forced off the land in Guatemala's Polochic valley in 2011 to make way for sugar cane plantations.

Land grabs worldwide are fueled by the commercialization of agriculture as food production in many regions becomes more linked to global markets and the growing demand for biofuels. When agricultural land becomes more valuable as a commodity, rich and powerful interests seek to possess the land. This often involves the dispossession of local communities with long-established links to particular parcels of land.

The land grab is particularly pronounced when the peasantry is of a minority ethnicity compared to the ruling elites. Often marginalized by political structures and denied full access to political power, peasant societies struggle to maintain their land rights against well-connected corporations and powerful state bureaucracies. For example, the Mapuche tribal group was restricted to land reserves by the Chilean government in 1883. Even this "reserved" land is now subject to state and private appropriation as hydroelectric dams are built, logging companies move in, and plantations are created to grow sugar and soya.

In many countries, legal land rights are uncertain. The traditional claims to land often do not hold up in modern legal frameworks more concerned with adjudicating private property rights than communal rights. In the Philippines, for example, only 3.5 million of a total of 12 million hectares are private property. The rest of the land has uncertain formal legal status, making it more susceptible to a land grab.

There are also instances where peasant communities resist the land grab. The Bajo Aguán region in Honduras has been a scene of intense conflict over the past two decades. In the 1990s, large landowners purchased land from farmer cooperatives to harvest palm oil for export. Local activists complained that the deals were unfair and people were not made aware that they were signing away land rights. One company owned by a member of the very wealthy Honduran elite amassed one-fifth of all the land in the region. In 2009, peasants invaded and occupied company land. Troops were sent in, people were evicted, and more than forty people were killed. The peasants resisted, and the bad publicity led one German bank to withdraw loans to some of the large companies. In 2011, the Honduran government passed a law that allowed peasant farmers in Bajo Aguán to purchase 4,000 hectares at favorable interest rates. This particular story has all the elements of the global land grab: marginalized peasants, rich landowners, and a growing market for food exports and biofuels. In this case, however, the peasants resisted and forced a change in the government attitude.

9.2 The Commercialization of Agriculture

The commercialization of agriculture, often referred to as the Second Agricultural Revolution, involves greater capital investment in machinery and biotechnology, which leads to greater productivity. To take just one example, the average American farm produced 40 bushels of grain per acre in 1900. A hundred years later, the same acre was producing 100 bushels (see Table 9.2). This enormous increase in productivity is the result of increased use of pesticides, genetically engineered crops, and heavy doses of fertilizers. All of these come at a price.

Pesticides, for example, increase productivity by reducing the harm of insects on crop yields. In her 1962 book *Silent Spring*, Rachel Carson highlighted the negative environmental consequences of heavy pesticide use. The title of the book describes a spring devoid of birdsong after pesticides have decimated songbirds. Almost 3 billion kilograms of pesticides are used around the world each year, 500 million kilograms in the United States alone. One study looked at the impact on public health, the development of pesticide resistance in pests, honeybee and bird losses, and groundwater contamination, estimating the total costs in the United States in 2002 at $10 billion. The costs are probably much higher in countries with less stringent environmental protection legislation and enforcement.

Increased use of fertilizers, especially those that restore soil nitrogen, also has deleterious effects on groundwater contamination and air quality through emission of nitrous oxides. Runoff from fertilized fields finds its way into streams, rivers, and seas, creating environments superenriched with nutrients, where algae and plant overgrowth can reduce the oxygen so much that dead zones occur. **Eutrophication** is the name given to the process of enriching ecosystems with chemical nutrients, which creates algal blooms that result in dead zones. Nutrient enrichment also causes new patterns of disease for humans and wildlife. In Caribbean wetlands, for example, increased nutrient runoff creates lusher vegetation that provides a home for malarial mosquito larvae.

The issue of using **genetically modified organisms (GMOs)** to grow crops is controversial. Plants are genetically altered to make them more resistant to disease and predators and to produce higher yields. While the proponents of GMO highlight the advantages of cheaper food, critics point to the risks of unintended health effects on consumers and of contamination between GMO and non-GMO crop varieties with unknown consequences. The risks and full costs of GMO agriculture have yet to be fully assessed.

The most sustained commercialization of agriculture in the developing world that involved large inputs of biotechnology was the Green Revolution, the name given to a series of innovations and transfer, especially marked from the late 1940s to the 1970s, that increased agriculture production through new high-yielding varieties of crops, irrigation, and widespread use of pesticides and fertilizers. The Green Revolution integrated all three strategies: using hybrid varieties of maize, rice, and wheat (genetically engineered crops), which produced substantially higher yields; increasing use of synthetic pesticides and fertilizers; and increased use of mechanization of equipment (tractors, harvesters, etc.), which required higher inputs of capital.

During the 1970s and 1980s, varieties of rice were widely planted in the paddies of Southeast Asia and China and new varieties of wheat in the drier regions of Asia and Latin America. In Latin America, more than 80 percent of the wheat land now grows Green Revolution varieties.

TABLE 9.2	Corn Production in the United States		
YEAR	**YIELD (BUSHELS PER ACRE)**	**HOURS REQUIRED FOR 100 BUSHELS**	**EQUIPMENT USED**
1850	40	75–90	Plow
1900	40	35–40	Gangplow, disk, harrow, two-row planter
1950	50	10–14	Tractor, three-bottom plow, disk, harrow, four-row planter, two-row picker
2000	100	2.5	Tractor, five-bottom plow, 25-ft tandem disk, planter, 25-ft herbicide applicator, 15-ft self-propelled combine, trucks

Source: American Farm Bureau.

Agroecology: Sustainable Agriculture?

Agroecology draws on a rich history of traditional farming systems in which local farmers have developed long-standing livelihood strategies and adaptive responses to environmental change. Agroecology entails the use of ecological principles such as crop rotation, mixed cropping, livestock integration, agroforestry, and composting to design resilient, sustainable, low-input farming systems that all draw on local farmers' agricultural knowledge and complex understanding of environmental conditions. The UN's Food and Agriculture Organization reports there is evidence that restoring and scaling traditional agroecological practices through farmer-led education could help improve food security and build social resilience among poor populations.

However, feminist political ecologists are cautious about adopting agroecology without the transformation of social and political roles and systems. Ruth Nyambura, a Kenyan ecofeminist and researcher, writes that using alternatives like agroecology will be useless if questions of power, by class–gendered–racial, rural–urban, young–old are not constructively addressed. She concludes that agroecology and family farming are limited in their revolutionary potential if women continue to face the violence of patriarchy in their immediate surroundings and especially with relation to access and control over the ecological resources and the exploitation of their reproductive and productive labor.

REFERENCES

Nyambura, R. 2017. "Agrarian Transformation(s) in Africa: What's in It for Women in Rural Africa?" *Development* 58:306–313, https://link.springer.com/article/10.1057/s41301-016-0034-0.

United Nations Food and Agriculture Organization. 2018. *Transforming Food and Agriculture to Achieve the SDGs*, http://www.fao.org/3/I9900EN/i9900en.pdf.

The Green Revolution did improve yields, but at the cost of environmental damage and harming the health of local communities. Studies have shown a reduction in soil fertility and increased erosion resulting from the application of pesticides and fertilizers. Expanding irrigation to meet increased water needs of new hybrid varieties has led to groundwater depletion and water shortages in some areas. The social impacts were also pronounced. The increased mechanization of farming resulted in the consolidation of small farms and fields that displaced many tenant farmers and intensified rural poverty. The Green Revolution is now seen as a deeply flawed experiment with some positive consequences, such as increased crop production and cheaper food for consumers and some negative consequences. We need a real green revolution, one that is sensitive to environmental impacts, community concerns, and translating sustainable and safe practices into mainstream agriculture. Agroecology may be one answer to a more sustainable form of agriculture.

9.3 Manufacturing

The manufacturing of goods constitutes the bulk of the **secondary sector**. Although craft industries have existed for centuries—glass making in Venice, for example, dates from the fourteenth century—the large-scale manufacturing of goods as a major component of an economy is a comparatively recent phenomenon associated with the Industrial Revolution.

The Industrial Revolution is one of several long waves of production based on the clustering of innovations. The Soviet economist Nikolai Kondratieff first identified these long, fifty-year cycles in the 1920s. Kondratieff made a distinction between inventions and innovations. Inventions are new ways of doing things; they occur almost randomly, but tend to be adopted into production techniques in waves of innovation. The basic nature of these **Kondratieff cycles** is noted in Table 9.3. Four cycles are commonly noted: textiles, iron and steel, mass production, and high tech. Each Kondratieff cycle is associated with key innovations that structure society and space. The first two, from 1785 to 1895, are associated with the development of factories, the emergence of industrial districts, and the growth of towns and cities.

The First Cycle

The first Kondratieff cycle, from 1787 to 1845, is associated with textile manufacturing and saw its full flowering in Manchester, England. In 1750, Manchester was just one of hundreds of textile towns located all over Europe. What made it the first industrial city were developments in technology, new sources of supply and demand, and social networks that fostered innovation and risk taking. In the 1760s, new steam-powered machines increased production levels. In 1774, the population of Manchester was only 41,032; by 1831 it was 270,901. Mills were working day and night. Other ancillary industries also developed. Railway links were built to transport its goods, and by 1840 the city was served by six railway lines and was a center of locomotive construction.

9.3 The Industrial Revolution

The Industrial Revolution occurred around 1800 and was centered in Britain. At its core, the Industrial Revolution was a new way of making things. Production capacities were increased by mechanization. Steam power released capacities beyond the limits of even the most strenuous human exertion. Textile production was increased tenfold by mechanization. Once started, the Industrial Revolution took off in an upward cycle of increased growth and expansion as inventions were turned into innovations that improved, increased, and streamlined the making of things. The Industrial Revolution was also an institutional revolution, as economic transactions shifted from one based on patronage and personal contact to a greater reliance on anonymous markets, performance measurements, and impersonal exchange.

Britain was the first industrial nation. A number of reasons lay behind its transformation. Britain was a relatively small, politically stable, densely populated country. Colonial expansion ensured cheap imports and secured export markets. Because of high wages, British companies were forced to introduce labor-saving techniques. It is also important to note the role of London. In 1800 the city had a population of almost 1 million, almost double the size of any other European city. The city's population growth provided a huge, growing, and secure market for food producers and manufacturers, and this effective demand was the basis for further investment in agricultural and industrial production.

The very early Industrial Revolution was centered in Manchester. One compelling reason is the importance of the culture of innovation: between 1600 and 1800, a protoindustrialization system of economic organization was established, there was a capacity for continuous innovation, and there was a large middle class of small capitalists able to employ a range of entrepreneurial talent, social networks, and local cultures that allowed a constant improvement in products and processes. The city's location was important: it was close to the coal that provided the raw power to animate the incessant spinning and weaving machines. But other places were also close to coalfields, so this single locational factor has limited explanatory power. Manchester became the first industrial city because it was the first truly innovative city.

REFERENCES

Allen, D. W. 2011. *The Institutional Revolution: Measurement and the Economic Emergence of the Modern World*. Chicago: University of Chicago Press.

Allen, R. C. 2009. *The British Industrial Revolution in Global Perspective*. Cambridge: Cambridge University Press.

Hall, P. G. 1998. *Cities in Civilization*. New York: Pantheon Books.

The Second Cycle

The second and third cycles were also experienced in other parts of the developed world, especially Germany and the United States (see Photos 9.3 and 9.4). New industrial regions were established in North America and Europe. The city of Paterson, New Jersey, was a center first of textile production and then the making of locomotive engines.

Pittsburgh, Pennsylvania, typifies the second and prefigures the third cycle. It was surrounded by bituminous coal; the state produced almost a quarter of the nation's coal, over 10 billion tons in 200 years. By the mid-nineteenth century, over 1,000 factories in Pittsburgh were using coal as their power source, including at least 65 glass factories. During the Civil War, the city became a center for armaments manufacturing. Scottish-born industrialist Andrew Carnegie

TABLE 9.3 The Four Kondratieff Cycles				
	TEXTILES	**IRON AND STEEL**	**MASS PRODUCTION**	**HIGH TECH**
	1787–1845	1846–1895	1896–1947	1948–2000 (?)
Key innovation	Power loom	Steel making	Electric light, automobile	Transistor, computer
Key industry	Cotton, iron	Steel	Cars, chemicals	Electronics
Industrial organization	Small factories	Large factories	Giant factories	Large and small factories
Labor	Machine minders	Craft labor	Deskilled	Segmented
Geography	Towns	Towns	Conurbation	New industrial regions

PHOTO 9.3 Paterson, New Jersey, was a site of textile production in the US industrial revolution.

PHOTO 9.4 Secondary sector: steel-making plant in Baltimore.

founded the first steelworks there in 1873, and by 1910 the city was producing a third of the nation's steel. Aluminum was first made in Pittsburgh in 1888. The city grew from a population of only around 46,000 with a metropolitan population of 100,000 to a city population in 1920 of almost 600,000 with a metro population of almost 1.5 million. Pittsburgh was the epicenter of US industrialization, and even today, major employers include Alcoa, U.S. Steel, and Wheeling–Pittsburgh Steel. As the economy matured, the city of Pittsburgh was only the central point in a wider suburbanization of people and jobs throughout a wider urban region. With the creation of large, integrated industrial establishments such as large-scale steel works, industry, jobs, and people were widely distributed throughout a forty-mile radius from downtown Pittsburgh. As firms increased in size, the plants grew larger. The result was a complex suburbanized metropolitan economy. The rapid and heavy industrialization also created unhealthy pollution.

The Third Cycle

The third Kondratieff cycle was based on automobile manufacturing, giant factories, and standardized deskilled manufacturing processes. The exemplar city was Detroit as it became home to car manufacturing and new forms of industrial production. At the beginning of the twentieth century, cars were luxury items, handcrafted and designed for the wealthy. Detroit was only one of many car-making cities. Even before producing cars, Detroit had been a manufacturing center; by 1900 it had almost 1,000 machine shops making ships, stoves, engines, and mining machinery. There was a pool of skilled labor and a network of local

financiers. Detroit was soon producing two-fifths of the nation's car output, concentrating on the cheaper end of the market (hence the moniker "Motor City," or "MoTown").

Henry Ford transformed the manufacture of cars. He was born in 1861 and grew up close to Detroit. In 1879 he went to work for the Michigan Car Company, which was building ten cars a day. In 1899 he formed the Detroit Automobile Company, but it was not a success. Ford was too much of a perfectionist and had yet to hone his market sensibilities. Fine-tune them he did, however, when he founded the Ford Motor Company in 1903 with twelve shareholders, all from Detroit. In 1908, the first Model T appeared on the market. It was a simple yet robust model that went through numerous design improvements to become one of the first mass-produced cars. Assembly lines and mass production had been developed over the previous century in a range of manufacturing industries, including bicycles and watches, as well as the Chicago meat-processing and packing industry. Ford did not invent mass production, but he refined and improved it. Car production became standardized, precise, and continuous. Mass production allowed reduction in the final price: the cost of a Model T in 1908 was $805, but by 1924 it was down to $290. With each price decrease, new markets were created, and so mass production not only met demand but also created new demand. Cars became less a luxury item and more a regular purchase, especially in the rural hinterland, where farmers used them for a variety of purposes. By 1920, every second car in the world was a Model T Ford. To reduce labor turnover, Ford also paid high wages for the time. A well-paid industrial workforce was essential to maintaining worker allegiance and limiting labor turnover. Unions soon formed and their collective bargaining power further increased wages, although there were periodic clashes between management and workers, especially during business downturns as management sought to reduce costs by cutting wages. Ford factories were instrumental in creating the highly paid blue-collar sector and unionized workforce of the new industrial city.

Detroit was the forerunner of a form of production named after Henry Ford. Fordist production is controlled

9.4 Inequalities: Disability in the Workforce

Hundreds of millions of people suffer from discrimination in the workplace. It is a persistent and multifaceted problem. Discrimination stifles opportunities, wastes the human talent needed for economic progress, and accentuates social tensions and inequalities. Discrimination in the workplace may occur based on religion, age, caste, HIV status, gender, nationality, and disabilities.

According to the International Labour Organization, people with physical or psychological disabilities make up an estimated 1 billion, or 15 percent, of the world's population. More than 80 percent are of working age. However, the right of people with disabilities to decent work is frequently denied. Compared to nondisabled persons, they experience higher rates of unemployment and economic inactivity. There are many obstacles that may prevent persons with disabilities from entering and advancing in the workforce. These include discriminatory laws and policies; inaccessible workspaces or offices; inaccessible transportation and information and communications technology; and stereotypes, negative attitudes, and mistaken assumptions about persons with disabilities, including their working capacity.

The exclusion of persons with disabilities from the labor market is a waste of potential, with an estimated loss of global gross domestic product of between 3 and 7 percent.

There is a growing body of evidence that inclusive policies not only enhance the reputation of businesses as corporate leaders but also boost profitability. In 2019, the Valuable 500 was launched. It is a global business collective of large companies committed to disability inclusion. Companies that have pledged a commitment to diversity include the BBC, Apple, Sony, McDonald's, and Google. In 2020, McDonald's in India launched a new packaging, called "EatQual," that makes eating its burgers easier for people with limited hand mobility. Yet there is much more to do to reach an inclusive workforce.

REFERENCES

International Labour Organization. 2015. *Disability Inclusion Strategy and Action Plan 2014–2017*. Geneva: International Labour Organization.

International Labour Organization. 2022. *EmployAbility: Tapping the Potential of Persons with Disabilities in Asia and the Pacific*. Geneva: International Labour Organization.

The Valuable 500. "Our Work" and "Companies," 2022. https://www.thevaluable500.com/.

by a small number of very large companies operating under oligopolistic conditions. Fordist production is mechanized and repetitive, with long production lines. Under Fordism there were outbreaks of labor unrest when the business cycle softened demand and management implemented layoffs and wage cuts. But by and large, a stable system of capital–labor relations was established as a relatively affluent working class arose. The high point of this wave lasted from 1945 to the mid-1970s.

Global Shift

There is a changing global geography to the third cycle of manufacturing mass production. Since the 1970s, there has been a **global shift** as manufacturing moved from the high-wage economies of the developed world to the lower-wage economies of the developing world. As manufacturing became more mechanized and routinized, employers shifted operations to areas of cheap, semiskilled, and unskilled labor. Transport improvements, such as container shipping and integrated transport connected at ports, made it cheaper and easier to ship goods. As transport costs plummeted, attention shifted to moving manufacturing plants to cheaper labor regions of the world. First Japan, then South Korea, Singapore, Hong Kong, and Taiwan, became centers for textile and shoe production. These newly industrializing countries were centers of rapid economic growth and were sometimes collectively known as **Asian Tigers**. The regional economic complexes then became the basis for a move up the **value-added chain** to more complex manufacturing, such as cars and electronics. As labor costs increased, low-value manufacturing, such as textile sand shoes, shifted to even cheaper areas such as Bangladesh, China, Indonesia, and Vietnam, while high-value manufacturing such as computer chips and electronics moved to Taiwan, Thailand, and, more recently, China.

As manufacturing becomes even more mechanized and transport improvements flattens the world, these industries move to even cheaper accessible areas. This global shift has had three major consequences. The first was the decline of manufacturing employment in Europe and North America. They experienced **deindustrialization** as manufacturing jobs were lost through a combination of efficiency improvements and global shift. For cities and regions based almost entirely on manufacturing, such as Detroit, the net effect was massive job loss and economic decline. Across the Global North, from Birmingham, Alabama, to Birmingham, UK, and from Newcastle, UK, to Newcastle, Australia, there was a

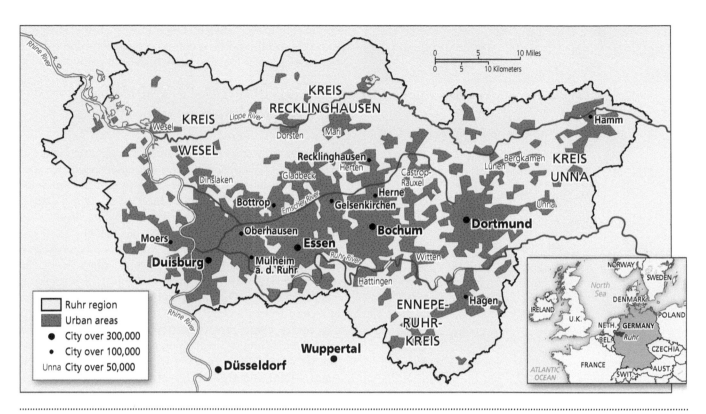

MAP 9.1 The Ruhr, one of the major industrial regions of Europe.

massive loss of manufacturing employment and a decline in industrial regions and cities. Map 9.1 shows the Ruhr in Germany, a major industrial region that experienced deindustrialization. There was also a decline in organized labor and an effective hollowing out of the blue-collar middle class.

The second was that the destination regions of this global shift now became places of sustained economic growth. Japan, then South Korea, and more recently China have witnessed rapid economic growth based on the increase in manufacturing.

Third, there was also the creation of new class and gender identities. In the Global North there was a reduction in male, blue-collar employment opportunities and a decline in organized labor. Meanwhile, in regions and countries that saw a manufacturing upswing because of global shift, a female working class was created, and economic growth enabled an enlarged and more prosperous middle class. Much of the labor in China's new manufacturing, for example, is young female workers. The global shift resulted in new forms of class and gender identities.

In summary, then, global shift meant, on the one hand, a decline of older industrial regions and cities and of the organized labor class in Europe and North America, and on the other hand, in the developing world, the creation of new industrial cities and regions, the emergence of a new female working class, and the great enlargement of the middle class.

The Fourth Cycle

The fourth Kondratieff cycle is associated with the development of high-tech information technology industries. The Kondratieff distinction between inventions and innovations is apparent in this fourth cycle. The inventions associated with the information technology sector occurred in a steady stream throughout the twentieth century, but became a wave of high-tech innovation that transformed the economy and society only by the very end of the twentieth century (Table 9.4). The information technology wave took more than a century to turn inventions into the wave of innovations that occurred from 1980 onward. By 1994, the World Wide Web had become a powerful information portal for business and personal computer users.

Each of the previous waves has an exemplar city or region: Manchester, Pittsburgh, and Detroit. For the fourth wave it is Silicon Valley, a high-tech cluster centered on Palo Alto, California (see Map 9.2), where personal computing was developed for mass markets. Nearby Stanford University played a major role as a source of smart computer scientists and as the force behind the establishment of Stanford Industrial Park, where two Stanford graduates established Hewlett Packard in 1953. Technological innovation was funded by a steady stream of government defense contracts and facilitated by corporate initiatives, such as Xerox's Palo Alto Research Center.

TABLE 9.4	Inventions in Computer Technology
APPROXIMATE DATE	**INVENTIONS**
1890	Punch cards and mechanical calculating machines
1939	Vacuum tubes and electromechanical calculating
1952	Transistors and electronics
1964	Mainframe computing
1971	Microprocessors
1981	Personal computers
1994	The Internet

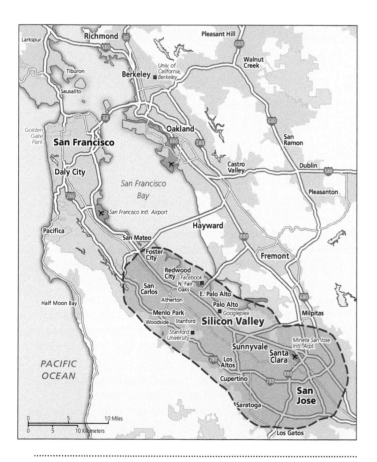

MAP 9.2 Silicon Valley.

Over time, the area became home to law firms and venture capital firms that facilitated and funded the startup and establishment of innovative companies. Major computing companies such as Adobe, Apple, Cisco, Google, Hewlett Packard, Intel, and Yahoo! are all headquartered in the region (Photos 9.5 and 9.6). Over time, the initial early advantages solidified into a major competitive edge for this region as a center for innovative high-tech computing.

9.4 Services

In mature economies there is a shift from manufacturing to services. In the United States, for example, service employment now accounts for almost 78 percent of the gross domestic product. Services are defined as selling, assistance, and expertise rather than making a tangible product. In some cases, the distinction is clear; when you buy a car, you are buying a good, whereas when you hire someone to clean your house, you are buying a service. At other times, the distinction is fuzzy; the classic example is a restaurant, where you buy food but also a service. The term "services" is best considered loose and hazy, with slippage at the edges. It also covers a wide range of activities, from healthcare and financial consultancy to computer information companies. The sector includes a range of wildly differing job experiences. At one end are Wall Street brokers working in high finance and international currency dealing, all making high wages and lucrative annual bonuses that fuel the local housing markets (Photo 9.7). At the other end are the nighttime contract cleaners of the offices that house these executives.

Evolution of the Service Sector

Part of the services sector broke away from the secondary sector. In the nineteenth and early twentieth centuries, the factory contained both the assembly line and the business service sector that administered the whole process of buying raw material, hiring workers, and selling finished products. These sectors have been hived off into separate divisions and companies. We can think of a simple model of contemporary multinational manufacturing business. At the base is the routine assembly plant that needs all those things noted in the standard location models of industry, such as cheap labor and low taxes. As economic globalization has created a flatter world, these plants can be located in a wide range of countries. Another level of the company that has separated out from the production plants is the research and development sector that tests new products. This sector needs to be close to pools of highly skilled labor, knowledge centers, and the amenity-rich locations that attract such workers. Then there is the company headquarters, which needs a metropolitan location to maximize face-to-face business contacts and be close to business services such as advertising, financing, legal services, and other important services and decision makers. Consider the case of Boeing, a company with almost 142,000 workers around the world, almost 60,000 of them in the state of Washington. In 2001, the company

PHOTO 9.5 Google headquarters.

PHOTO 9.6 Apple headquarters.

moved its headquarters from Seattle to Chicago and separated its corporate headquarters from its production base. According to the company, the move was made to secure ready access to global buyers and financial markets. It was also lubricated by subsidies of over $60 million from the state of Illinois and the city of Chicago, which was in competition with Denver and Dallas–Fort Worth.

Characteristics of a Service-Sector Economy

A service-dominated economy tends to have a number of characteristics. It produces a more polarized job market, with high-paying jobs with full benefits on the one hand

PHOTO 9.7 Tertiary sector: Wall Street is a center of financial services.

and minimum-wage employment with few benefits on the other. At the core of the new service economy are highly paid, knowledge-based professionals, analysts of data, trends, and ideas, such as business consultants and investment bankers. Paid generously, they have employment security, good working conditions, and generous benefits. They often work long hours with brutal deadlines, but they are firmly located in the middle to upper-middle class, and they can ensure their children's similar economic success by being able to buy good private education or make the right residential choices to ensure the best public education. At the very upper levels of this sector is a world of affluence. It is the world of the corporate jet and the million-dollar-plus stock options that provide a deep financial cushion against the vagaries of life. Outside this core group, there are two peripheries. A semiperiphery consists of full-time workers with less income, status, and prestige. Their jobs are less secure, and too often they are able to hang on to middle-class status only if more than one person in the household

9.5 Green Jobs

The International Labour Organization defines green jobs as decent jobs that directly contribute to environmental sustainability, either by producing environmental goods or by making more efficient use of natural resources. A green economy is seen as reconciling two competing agendas of environment and economy and offers a pragmatic middle ground. Green jobs reduce the consumption of energy and raw materials, limit greenhouse gas emissions, minimize waste and pollution, protect and restore ecosystems, and enable enterprises and communities to adapt to climate change. Some examples of some of the fastest growing green jobs include:

- clean car engineers,
- water quality technicians,
- urban growers for the sustainable food and restaurant industry,
- recyclers,
- green builders and green design professionals,
- biofuels jobs,
- ecotourism,
- socially responsible investing, and
- developing clean technology.

Many green jobs are in the environmental goods and services sector and directly benefit the environment or conserve natural resources. However, a green job does not necessarily have to be limited to jobs in and around the environment. A job that improves the environmental impact of production processes in a car manufacturing facility is such an example.

The green economy is likely to be a major source of job growth in the future of work. For example, transitioning away from fossil fuels to renewable energy could create around 24 million jobs, largely offsetting any job losses in the traditional energy sectors such as oil and coal extraction. In the United States, the clean energy sector is now creating new jobs twelve times faster than almost any other sector in the American economy. In Latin America and the Caribbean, the clean energy sector has the potential to generate 15 million net jobs in the region by 2030.

However, it is important to note that many of the jobs in energy are traditionally dominated by men; efforts will be needed to ensure these green energy jobs are available to women.

REFERENCES

International Labour Organization. 2018. *World Employment and Social Outlook 2018: Greening with Jobs.* https://www.ilo.org/global/research/global-reports/weso/greening-with-jobs/lang--en/index.htm.

International Labour Organization. 2019. *What Works: Promoting Pathways to Decent Work.* Geneva: International Labour Organization.

Stoknes, P. E. 2021. *Tomorrow's Economy: A Guide to Creating Healthy Green Growth.* Boston: MIT Press.

is working. Beyond this, a periphery comprises people on short-term or part-time contracts. They may work from home on their computer and relish the flexibility of such work or be picked up each day at the street corner by landscapers to manicure the lawns of the wealthy.

The shift from a manufacturing-based economy to a service-based economy also has profound effects on income distribution. A strong manufacturing base allows low-skilled workers to obtain relatively good wages. The service economy, in contrast, provides high-paying jobs for those with marketable skills, but more limited opportunities for low-skilled workers. For those lacking high educational attainment, the opportunities shrink to the low-wage service sector.

This shift also goes hand in hand with an increase in female employment. In the United States in 1950, the female participation rate in the formal economy was only 30 percent, but by 2022 it had increased to almost 55 percent. The workforce has been feminized in the shift to a more service-based economy.

As service employment increases, there is the danger of what is called **Baumol's disease**, after the economist William Baumol, who first noted that while innovation and efficiency occur rapidly in the manufacturing sector, leading to price reduction, they are less pronounced in the **service sector**. You can make a car quicker and more efficiently, but it still takes a string quartet the same time to play a Beethoven quartet. As the service sector comes to dominate an economy, it is subject to rising costs. While prices for goods decline as a result of manufacturing efficiencies, the costs of services keep rising in absolute and relative terms. In his book *The Cost Disease*, Baumol suggests that while we can work to reduce service costs, they are the price of a successful economy.

Quaternary Sector

The service sector has grown so much in size and sophistication that different subsections can be identified. Some argue for a distinct new sector, the **quaternary sector**, that involves use of knowledge and information. It includes communications, consultancy, and research and development. This sector is concerned with the increased and strategic use of information and knowledge.

Although an easy demarcation between the tertiary and quaternary sectors is not possible, we can make a broad distinction between the highly skilled workers in financial services and less-skilled service workers, such as a restaurant server. Although both may work equally hard, the levels of compensation and work condition are often very different.

Some even make a further distinction and define a **quinary sector** that is composed of key decision makers, the small group of people perched atop the information- and knowledge-based economy who make the executive decisions. Even in the quaternary sector there is world of difference in wages, working conditions, and social status

between someone at a desk in a bank inputting data into a spreadsheet and the bank's chief executive officer, who is part of the top 1 percent of the population that has done so much better than the vast majority.

The terms quaternary, quinary, and **advanced producer services** are attempts to designate a particularly specialized and often highly paid sector of the economy. As economies grow and become more sophisticated, shifting from primary through secondary to tertiary, there is a greater reliance on knowledge and information. It is not enough just to make things; it is also important to identify markets, assess consumer preferences, keep records of transactions, mobilize bank loans, influence government policy, and undertake advertising campaigns. The more an economy develops, the greater the reliance on the quaternary sector of knowledge- and information-based services.

Global Shift in Services

From the early 1970s, global shift affected mostly the manufacturing sectors. In more recent years, there has been a significant increase in the global shift in service employment. A new round of economic globalization, made possible by changes in technology, is sending a range of service employment overseas from the developed world to the developing world. Back offices in Bangalore, India, now process home loans for US mortgage companies, while many insurance claims made in the United States are routinely processed in offices situated in New Delhi. The economics are simple. Software designers in the United States cost $7,000 a month, while experienced designers in India cost only $1,000 a month. US companies now routinely outsource work previously done at home. In the 1970s and 1980s, engineers came to the United States and Europe; now the jobs come to them.

Selected cities in developing countries are developing as centers of service employment. In part, they develop as hubs of corporate national headquartering, which generate service employment. In China, for example, Beijing and Shanghai are the favored sites for headquarters of foreign companies. Corporations are attracted to Beijing because of the closer proximity to key decision makers, an important criterion for working effectively in a centrally planned economy. In Shanghai there has been a shift from manufacturing to service employment. In 1990, manufacturing accounted for 60 percent of the city's gross domestic product, while services only accounted for 31 percent. By 2022, the respective figures were 26 and 73 percent. While still an important manufacturing center, Shanghai has emerged as a predominantly service-based urban economy.

The Cultural-Creative Economy

An important and growing sector of the economy is the **cultural industry** sector, which includes the artistic industries of music, dance, theater, literature, the visual arts, crafts, and many newer forms of practice such as video art, performance art, and computer and multimedia art. There are also the "cultural-products sectors," which include high fashion, furniture, news media, jewelry, advertising, and architecture.

The work of creative artists enhances the design, production, and marketing of products and services in other sectors. Cultural industries concentrate in world cities like Los Angeles and Paris and other large cities, contributing immensely to their economies. The concept of "path dependence" can be used to analyze cultural production as well as technological innovation. Path-dependent theories claim that small historic events or locational advantages can affect macroeconomic consequences that privilege certain paths to development and limit others. This idea sheds light on the prestigious success that Paris enjoys in high fashion, New York in advertising, and Los Angeles in motion picture entertainment. Their leading roles in these industries have long enjoyed wide recognition, and their advantages over potential competitors relate not only to the quality of their products, but also to the "symbolic images"—such as authenticity and reputation—that those products carry.

Creative talent drives economic development. Richard Florida argues for a new social class, the creative class, which leads the new creative economy. The creative class divides into two groups of people. The first makes up the "super-creative core," a broad and diverse group including scientists, engineers, university professors, poets, novelists, artists, entertainers, actors, designers, and architects, as well as nonfiction writers, editors, cultural figures, think-tank researchers, analysts, and other opinion makers. The other group consists of "creative professionals" who work in a wide range of knowledge-intensive industries such as high-tech sectors, financial services, the legal and healthcare professions, and business management. The former group produces new forms or designs that are readily transferable and widely useful, while the latter group engages in creative problem-solving.

There is a vigorous debate about the rise of the so-called creative class, its contingency and flexibility, its scalability from a few well-known case studies, and its ideological underpinnings as a form of urban economic policy. A convincing argument could be made that the poor are the real creative class, because of their ingenuity and resourcefulness in the face of limited income and restricted employment opportunities.

9.5 Summary

A global trend is evident for each of the three economic sectors. There is an increasing intensification and commercialization of agriculture. Manufacturing is becoming more automated and also relocating toward a select group of developing economies in the Global South. In most high-income countries in the Global North, there is a decline of manufacturing employment and a rise in service employment. Around the world, the more dynamic economies are shifting from agriculture to manufacturing and from manufacturing to services.

Cited References

Baumol, W. 2012. *The Cost Disease: Why Computers Get Cheaper and Health Care Doesn't*. New Haven, CT: Yale University Press.

Carson, R. 1962. *Silent Spring*. Boston: Houghton Mifflin.

Cronon, W. 1991. *Nature's Metropolis: Chicago and the Great West*. New York: W.W. Norton.

Florida, R. 2014. "The Creative Class and Economic Development." *Economic Development Quarterly* 28:196–205.

Maertens, M., M. Zeller, and R. Briner. 2011. "Agriculture Land-Use in Forest Frontier Areas: Theory and Evidence from Indonesia." *Environmental Research Journal* 5:505–522.

Turner, S. 2012. "'Forever Hmong': Ethnic Minority Livelihoods and Agrarian Transition in Upland Northern Vietnam." *The Professional Geographer* 64:540–553.

Select Guide to Further Reading

Borras, J., and J. Franco. 2010. "Towards a Broader View of the Politics of Global Land Grab: Rethinking Land Issues, Reframing Resistance." Initiatives in Critical Agrarian Studies Working Paper 001. Transnational Institute and Land Deal Politics Initiative. http://www.tni.org/paper/towards-broader-view-politics-global-land-grabbing. Accessed January 19, 2012.

Byres, T. J. 2016. "In Pursuit of Capitalist Agrarian Transition." *Journal of Agrarian Change* 16:432–451.

Evenson, R. E., and D. Gollin. 2003. "Assessing the Impact of the Green Revolution, 1960–2000." *Science* 300:758–762.

Fiorino, D. 2018. *A Good Life on a Finite Earth: The Political Economy of Green Growth*. New York: Oxford University Press.

Hall, T. 2010. *Earth into Property: Colonization, Decolonization, and Capitalism*. Montreal: McGill–Queens University Press.

Hawkens, P., A. Lovins, and L. H. Lovins. 2010. *Natural Capitalism*. New York: Earthscan.

Horlings, L. G., and T. K. Marsden. 2011. "Towards the Real Green Revolution? Exploring the Conceptual Dimensions of a New Ecological Modernization of Agriculture That Could 'Feed the World.'" *Global Environmental Change* 21:441–452.

Hutton, T. 2014. *Cities and the Cultural Economy*. New York: Routledge.

Li, X. 2020. "Cultural Creative Economy and Urban Competitiveness: How One Matters to the Other." *Journal of Urban Affairs* 42:1164–1179.

Lyon, S., J. Aranda Bezaury, and T. Mutersbaugh. 2010. "Gender Equity in Fairtrade–Organic Coffee Producer Organizations: Cases from Mesoamerica." *Geoforum* 41 (1): 93–103.

Morgan, K., T. Marsden, and J. Murdoch. 2006. *Worlds of Food: Place, Power and Provenance in the Food Chain*. New York: Oxford University Press.

Pearce, F. 2012. *The Land Grabbers: The New Fight over Who Owns the Earth*. Boston: Beacon.

Pimentel, D. 2005. "Economic and Environmental Costs of the Application of Pesticides Primarily in the United States." *Environment, Development and Sustainability* 7:229–252.

Scholtz, C. S. 2006. *Negotiating Claims: The Emergence of Indigenous Land Claim Negotiation Policies in Australia, Canada, New Zealand, and the United States*. New York: Routledge.

Sen, D., and S. Majumder. 2011. "Fair Trade and Fair Trade Certification of Food and Agricultural Commodities: Promises, Pitfalls, and Possibilities." *Environment and Society: Advances in Research* 2:29–47.

Sinclair, R. 1967. "Von Thünen and Urban Sprawl." *Annals of the American Association of Geographers* 57:72–87.

Walker, R. T. 2021. "Geography, Von Thunen, and Tobler's First Law." *Geographical Review*. 112:4, 591–607. https://doi.org/10.1080/00167428.2021.1906670.

10

A Global Economy

In this chapter, we focus on the creation of a global economy. In many of the current discussions of globalization, one common assumption is that it is a recent phenomenon. In this reading, globalization goes hand in hand with the construction of the modern world. However, while its pace has increased, globalization is long established as a continuing process of the human occupancy of the globe. Our focus in this chapter is the growing economic interaction between peoples and societies spread around the globe. We will consider this interaction through three related topics: space-time convergence, waves of globalization, and the question, "How flat is the world?"

10.1 Space-Time Convergence

Space-time convergence is a collapse in the time it takes to cover distance. This convergence, both a cause and an effect of widening economic transactions, reduces the cost of transporting goods and people and pulls places closer to one another. A series of space-time convergences caused by new developments in transportation and communication effectively brought the world closer together. In terms of transportation, the very first was the invention of the wheel, which allowed more goods to be effectively transported. The wheel was invented around 5,000 years ago, appearing simultaneously at a number of sites in the Middle East and Europe. The earliest wheeled vehicles had to move over rough surfaces that made movement slow and difficult. The full flowering of the wheel's advantages had to await the creation of smoother transport surfaces.

A Widening World

For centuries, the easiest way to move people and goods long distance was across water. Natural wind could power sailing ships. Initially, most seaborne trade was limited by navigational knowledge, though the early Polynesians managed to populate islands across the vast Pacific using their intimate knowledge of winds, stars, and tides. The invention of the wet compass (in which a magnetic needle floated in liquid) in China over 2,000 years ago and the dry compass in Europe around 1280 enabled sailors to plot their direction more accurately, thus reducing the time and therefore the cost of long-distance sea travel. Equipped with a compass, European sailors could now venture out more confidently from the enclosed seas of the Baltic and the Mediterranean to travel across the wider oceans. This had geopolitical consequences. The Roman Empire, for example, was essentially restricted to the extended basin of the Mediterranean because it had limited ability to sail across the wider seas. Later traders, however, could use compasses to venture across the Atlantic and around the Horn of Africa. Maritime empires such as the Spanish, Portuguese, and British were established and developed because the compass made it easier to move across distant seas and expansive oceans.

Improvements in navigation made it quicker, safer, and hence cheaper to sail across the oceans. Sea travel also avoided the frictions of distance imposed by traveling through different societies, separate countries, and competing empires. The neutrality of the open sea aided long-distance trade and commerce. Because most long-distance trade for the past 500 years occurred by sea, economic activity is still concentrated in coastal regions and in (current and former) port cities. London and New York, for example, grew as major cities in part because their economies were invigorated by their port status. Amsterdam grew from a small, watery town to a major city as the Dutch commercial empire spread its tentacles across the globe. Port cities also took on a more cosmopolitan feel, filled as they were with exotic goods and foreign peoples. The big commercial ports connected with a world wider than their immediate hinterlands; they had an openness to the wider world. When Peter the Great wanted to modernize Russia, he created the new city of St. Petersburg in 1703, located on the Baltic and therefore more open to foreign influences and international trade. He moved the capital there, from the interior city of Moscow, to create a window on Europe.

The earliest cargo of long-distance maritime global trade was limited to luxury goods and prized items. When transport is costly, it is economically feasible to concentrate on only high-priced items. Early trade between Europe and East Asia, for example, focused on such expensive goods as spices, highly prized for their ability to season food in an era before refrigeration. In the seventeenth century, a pound of nutmeg purchased in the Spice Islands of Southeast Asia cost 60,000 times more when sold in a London market.

The spice islands of Banda and Run became vital resources fought over by Portugal, Holland, and England; they also become significant features of European maps of the region. Photo 10.1 is from a map of Southeast Asia in a 1636 Mercator–Hondius atlas. The Spice Islands are disproportionately large, a reflection of their economic importance at the time.

The Age of Steam

When the age of steam replaced the age of sail in maritime transport, there was further shrinking of space-time. Around 1890, steamships moved more tonnage than sailing ships. The increasing use of steam allowed bigger ships to carry more goods more quickly and more cheaply. In 1850, it took between eighteen and twenty-four days to sail from Liverpool to Gibraltar. A steamship could make the same journey in seven to nine days. Not only was steam quicker, but also it was more reliable, less dependent on the vagaries of the wind. Steam meant faster and more predictable transport times. Highly prized goods and commodities no longer dominated international trade as the declining transport cost per unit "smoothed out" the globe as a transportation surface for ever-cheaper goods and commodities.

Transport developments created new geographies that made the operation of **comparative advantage** more feasible. Comparative advantage refers to the ability of a region or country to produce goods at lower cost than other areas. It works best when the unit costs of transport are low. With a smoother transportation surface and cheaper transport costs, the mechanisms of global comparative advantage generate new economic geographies. It thus became more efficient in the late nineteenth century to grow wheat in

PHOTO 10.1 The Spice Islands.

the United States than in Britain because the cost of land in particular, and in some cases labor, was so much less. Steamships made it profitable for grain grown in the United States and other distant places, such as Australia, to be shipped to meet demand in Europe.

Canals

Before the railways, inland canals were an important transport artery for reducing overland transport costs. The longest canal in the world and one of the oldest is China's Grand Canal, also known as the Jing-Hang Canal. It stretches over 1,000 miles and connects the Yellow River in the north of the country with the Yangtze in the South. Work first began over 2,500 years ago, as it was extended and renovated during different dynasties. It continues to play an important role as a trade artery for the Chinese economy.

The nineteenth century saw the development of canals in Europe and North America. Take the case of the Erie Canal, begun in 1817 and completed in 1825: it cut a straight, watery line through the 363 miles between Buffalo and Albany and connected the Great Lakes with New York City (Map 10.1). The canal compressed space-time. Before the canal opened, it could take up to twenty days to travel between Buffalo and New York City. After the canal opened, it took less than four days. The cost of moving a ton of freight plummeted from $100 to $5. For 20 miles on either side of the canal, the equivalent of a day's wagon ride, the local economy was transformed. Along the path of the canal, villages turned into towns and towns into cities as local economies were now linked to larger national and global markets. Goods, people, and ideas flowed along the route of the canal.

But just as the canal was creating its greatest impact, a new form of transportation compressed space and time even more. The coming of the railways further reduced travel time from Buffalo to New York City. Goods moved along the canal at around fifty-five miles a day; the railway covered that distance in an hour. The railway could also enable travel year-round, whereas the canal froze over in wintry upstate New York. By 1850 more freight was being carried on rail than on the canal. The canal soon became obsolete, replaced as a major means of transport by the railroads, and

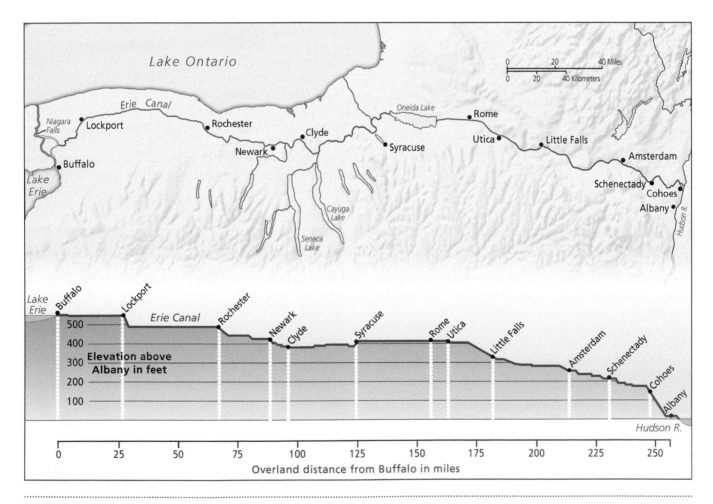

MAP 10.1 The Erie Canal.

towns along the canal soon withered. Some towns, such as Syracuse, which were located on both the canal and the railway line, grafted new railway-connected growth onto canal-based growth and thus continued to thrive.

Railways

Railways annihilated distance, brought cities closer together, and made interaction easier and cheaper. The geographer Donald Janelle plotted the travel times between Boston and New York. Figure 10.1 shows the declining travel time between the cities with the introduction of new transport modes and their subsequent improvement. Over time, Boston and New York came closer together, making the movement of people, goods, and ideas between them easier and quicker.

Space-time convergences can bend physical space. London's space-time map is shown in Map 10.2. Consider the case of the connection between Stranraer and London. Because of the lack of plane connections, the relatively poor railway connections, and the lack of motorways, it can take up to nine hours to travel between the two places. The result is that a space-time map warps the physical distance. London is closer in space-time terms to Paris than it is to Stranraer. The expansion of highways, airports, and transportation corridors, such as the London to Paris Channel Tunnel, has further reduced traveling times between these two cities (Photo 10.2).

Arnold Horner measured the convergence of rail times between cities in Europe over the period 1964 to 1999. More recently, David Banister has argued that the desire to move farther faster is ultimately unsustainable because of its energy costs. Rather than reducing travel times and increasing distance, he promotes alternative plans that reduce travel distances.

Continuing Convergence

Space-time compression continues apace. Developments in maritime trade include the increasing size of ships and the growing use of containers to pack the goods. The large box containers allow for easier loading on and off the ships; hence, the unit cost of transport declines as more goods are shipped more quickly. As the

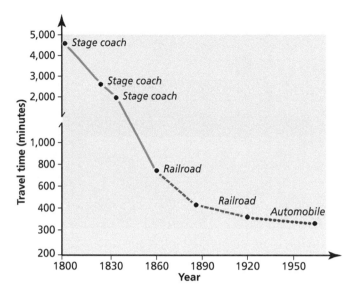

FIGURE 10.1 Space-time convergence between Boston and New York.

size of the ships increases, new types of port facilities are required. Unlike older, smaller ships, the newer container ships could not sail up narrow, shallow rivers. By the 1970s, many port cities worldwide could no longer handle the big container ships. Shipping therefore moved seaward to deepwater

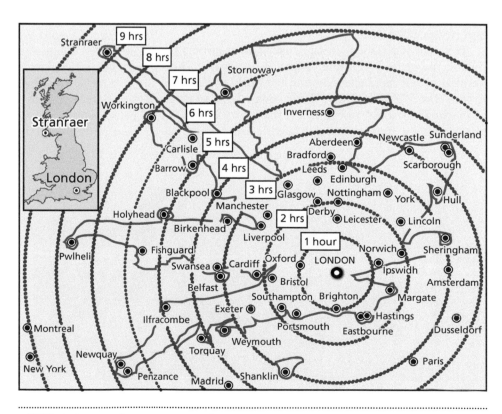

MAP 10.2 Space-time connections of London.

PHOTO 10.2 Traffic leaving London.

ports, and old upstream ports were abandoned. In places as varied as London's Docklands, Baltimore's Inner Harbor, and Sydney's Inner Harbor, these sites—large areas of vacant land at the heart of metropolitan regions—sometimes became the setting for new postindustrial residential and commercial activity. It is a phenomenon not restricted to the Global North. A similar process can be seen in cities such as Buenos Aires, Cape Town, Haifa, Hong Kong, Mumbai, and Shanghai. In this way, each space-time convergence destroys old geographies and creates new ones. Space-time convergence effectively shrinks and reshapes the globe as economic transactions are brought closer together. The world is flattened and distorted as it is shrunk.

In recent years, the increasing use of air transport has shrunk the globe even more (Photo 10.3). In 1873, Jules Verne depicted rapid travel around the world in the novel *Around the World in Eighty Days*; today, with good air connections, it could now take less than forty hours. Air transport is still more suitable for moving people and high-

value goods. Expensive perishable items, such as fresh flowers and fruits, are particularly suitable for transport by air. Flowers grown in Colombia are transported by air to flower markets in North America because the cost of the final item makes the relatively high-cost air freight affordable. There is a distinct seasonality to this trade, as fresh fruit grown in the Southern Hemisphere is shipped to northern markets in winter. Air freight allows a never-ending supply of seasonal produce. The growing trade in horticultural produce between Kenya and British supermarkets, for example, was possible because produce with a limited life span was shipped through Nairobi's airport. Although air freight allows quick travel along the path of global food chains, unstable fuel prices make it vulnerable to rapid price changes.

Information and Communications Technologies

As an economy matures, the transfer and exchange of information becomes as important as the movement of goods. The inventions of the telephone, radio, television, and later the Internet have allowed people all over the world to communicate instantly, shrinking space-time. Increasingly, peoples' daily activities happen in a virtual world where they can shop from their home or workplace or carry on business transactions from home. With increasingly integrated information and communications technology, ideas travel instantly and space-time has become more flexible. In turn, people can act on information more quickly. A commodities investor who receives a news alert about a hurricane in the Caribbean on her cell phone could sell stock shares on bananas within minutes.

And here we come to the importance of cost rather than time. Once the telephone was invented, messages could be transmitted very quickly. The most important variable became not space-time, but **time-cost compression**. In 1919, it cost almost $483 (in 2023 prices) to make a ten-minute call between New York and Los Angeles. One hundred years later, the cost is negligible (Figure 10.2). Calling long distance is no longer an expensive proposition. The same process occurs with Internet connections. As the real cost of communication falls, space-time-cost decreases. Each successive wave of space-time-cost convergence creates new markets and new economic geographies. Space-time convergences do not occur either randomly or regularly. They tend to be bunched together as distinct waves of globalization.

PHOTO 10.3 Baltimore/Washington International Airport. Space-time convergence: cheap air travel shrinks the globe.

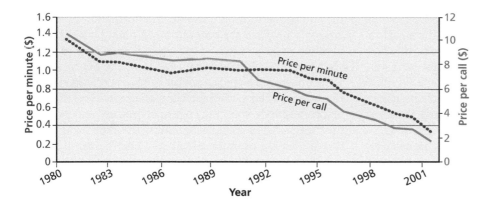

FIGURE 10.2 Decline in telephone costs.

10.2 Waves of Globalization

We can imagine the world as a series of places in a continual process of globalization and reglobalization. Just as waves sweeping onto a shore affect the topography of the beach, which in turn shapes the flow of subsequent waves, so the tides of globalization have impacted places, and the character of places has in turn affected the subsequent flows of global contact. We can identify at least three distinct waves of globalization.

The First Wave

The **first wave of globalization** followed the Columbian Encounter of 1492; it inaugurated a truly global economy. Before, the world was largely a system of regional empires and separate hemispheric ecologies. There was a substantial distinction between the Eastern and Western Hemispheres, what are often called the Old and New Worlds. After 1492, a global system took shape.

A major consequence of the encounter between the Old and New Worlds after 1492 was the demographic holocaust that we discussed in Chapter 5. The population estimates of the New World around 1490 range from 54 million to 112 million. By 1650, that population had fallen to just over 5 million. The causes included explicit forms of ethnic cleansing and the introduction of new diseases from the Old World against which the Indigenous people had no immunity.

The demographic holocaust was just one consequence of the Columbian Encounter. Others followed. The decline of the local populations led to the need for imported labor to dig in the mines and tend the estates. The slave trade shipped close to 20 million people from Africa to the New World. Then there was the biotic exchange. New World crops such as corn and potatoes enriched the diets of the whole world, leading to population increases in Europe.

Turkeys, tomatoes, sunflowers, and peanuts were transplanted from the Western to the Eastern Hemisphere, while horses, hens, cattle, pigs, sugar cane, and wheat moved in the opposite direction. Major environmental changes were wrought as a variety of plants and animals were dispersed around the world. When we pick tomatoes in Europe or ride horses in America, we are participating in a globalization process that began when Christopher Columbus landed in the New World. The Columbian Encounter bridged the hemispheric divide in a series of transactions and exchanges of people, plants, viruses, and animals that activated a global world.

The Second Wave

A **second wave of globalization** took place from approximately 1865 to 1970. From 1865 to 1914, economic globalization was facilitated by low tariffs, an international labor market, and relatively free capital mobility. Critics of the time called it "British internationalism." British overseas investments were twice those of France and five times those of the United States. It was an economic globalization centered in London. Other countries, such as France, Germany, and the United States, were also involved through their smaller trading empires. The global system was neither fair nor equitable. It was a system dominated by merchants, manufacturers, and investors in the rich countries seeking raw materials, markets, and investment opportunities.

This second wave of globalization was associated with major shifts in space-time convergences triggered by the appearance of the railways, the telegraph, and the steamship. It took less time to travel large distances. People, ideas, and messages could move more quickly. News of events was more quickly transmitted across the globe, in effect shrinking the globe to a more compressed space.

Railways had been introduced before 1865, but were limited to partial networks in only a few countries. After 1865, there is a widening and deepening of the railway network around the world. In 1870 there were only 125,000 miles of track in the world; by 1911 this had increased to 657,000 miles. Far away from the railroad tracks, life may have remained unchanged, but the whistle of the train announced that the world had sped up.

Ocean distances were reduced by the construction and enlargement of canals. The Suez Canal reduced the time it took to sail from western Europe to East Asia by linking the Mediterranean with the Red Sea. This **highway to India** was completed in 1869. The British invasion of Egypt in 1882 was in large part an attempt to control the canal, so vital for maintaining British links with their colony in India. But as ships increased in size and the region became susceptible to political instability, the Suez Canal became a less important transit site. On the other side of the world, the Panama Canal was cut through a narrow isthmus to reduce the distance between the Atlantic and Pacific. After unsuccessful attempts by the French, the United States built the Panama Canal, and it opened to traffic in 1914. To control the canal, the United States appropriated property along its route. The Canal Zone became a source of friction between the United States and an increasingly more independent-minded Panama. In 1999, the Canal Zone reverted to Panamanian control. As the container ships got bigger, more traffic was diverted away from the canal. The growing trade between the United States and East Asia, for example, was occurring through the ports on the West Coast of the United States. Almost a third of the world's container ships were too large to pass through the existing canal. Larger locks that can handle bigger ships opened to traffic in 2016 (see Photo 10.4). The older locks could only cope with ships each carrying 4,800 containers. These new locks allow the passage of much larger ships, carrying 12,000 containers with 200,000 tons of freight.

The period from 1865 to 1914 marked a widening and deepening of economic integration around the world. London was the pivot of the international trading system,

the sun of the trading universe. The formation of a world economy involved the economic and often political incorporation of larger areas of the world. Large parts of Africa and Asia, and even faraway islands in the South Pacific, were annexed to a global economic order. Overseas territories provided cheap raw materials, secure markets for manufactured goods, sources of national prestige, and pawns in the great game of global geopolitics. The 1865–1914 period was also an **age of imperialism**, as a neo-mercantilist doctrine emerged that argued for protection of home industries and the possession of overseas colonies both to ensure cheap raw materials and secure markets and to deny access to competitors.

This second wave of globalization has a major disjuncture, the modern "Thirty Years War" from 1914 to 1945, marked by two world wars and a global economic depression. It was a time of dislocation and economic disintegration, but the experience of world wars led many countries to seek the basis for a renewed integration. The charter of the United Nations was first drawn up in 1945 to provide a forum for international dialogue. The Council of Europe in 1949 was followed by the European Coal and Steel Community in 1952, the European Economic Community in 1958, and then the European Community in 1967. From the ashes of the war there arose a conscious attempt to create a unified Europe. In the new postwar world, rules and organizations were established to stop the world economy from slipping into the downward spiral caused by the national economic protectionism that occurred during the Great Depression. Delegates from forty-four countries met in 1944 in Bretton Woods, New Hampshire, to establish a new economic order that stressed free trade. They created the World Bank and the International Monetary Fund (IMF) to integrate national economies into a global system. The United Nations, IMF, and World Bank were systems of international regulation that provided the backbone of a globalized world, a world now centered more on the United States. While the UN is headquartered in New York, the IMF and the World Bank are only a few blocks from the White House in Washington, DC.

The Third Wave

A **third wave of globalization** emerged after 1970. It is associated with an even more marked space-time-cost compression, more global production chains, and a deregulated global financial system, creating a "flatter" world with easier and hence quicker worldwide economic links. The icon of the flat world

PHOTO 10.4 Panama Canal: the new locks that opened in 2016.

10.1 The Flow of Remittances

Remittances are the funds that temporary and permanent migrants send back to their home countries. As both international migration and capital mobility have increased, so has the level of remittances. As it becomes easier, and much cheaper, to send relatively small amounts of money back home, modest remittances contribute to the global flow of money.

By 2021, close to $590 billion in remittances was sent from people in one country to another. The biggest sender was people living in the United States, with almost $48 billion. The next largest was Saudi Arabia, with $26 billion. The largest receivers include India ($87 billion) and China ($53 billion). However vast these sums are, they constitute less than 5 percent of each country's gross domestic product (GDP). Table 10.1 lists a sample of the countries where remittances constitute a relatively large proportion of the GDP. What are some of the main characteristics of this subset of countries?

In the case of El Salvador, for example, emigration increased substantially from the late 1970s as bloody civil war wracked the country. Millions fled, and many of them sent money back to their families, villages, and neighborhoods. In 1978, traditional agricultural exports were responsible for most of the hard currency in the country; remittances accounted for only 8 percent. By the first decade of the twenty-first century, remittances made up 70 percent and traditional exports only 12 percent. In many small counties, as well as regions and villages in large countries, remittances are a vital source of much-needed cash and capital, fueling local housing construction and paying for education and the provision of social services.

In some cases, remittances are an integral part of international economic relations. Almost 10 million Mexicans, 15 percent of the entire Mexican-born workforce, are now working in the United States. Many of these migrants send significant parts of their earnings back in the form of remittances. By 2020, the flow of total remittances was close to $53 billion. The flow varies over time. Economic downturns reduce immigration and hence the flow of remittances. The longer migrants stay in a country, or if they bring in their own families or start new ones, the more the flow tends to

TABLE 10.1 Inflows of Remittances, 2010 and 2020		
COUNTRY	**REMITTANCES AS PERCENTAGE SHARE OF GROSS DOMESTIC PRODUCT**	
	2010	**2020**
Bangladesh	11.8	6.6
Bosnia	12.7	9.3
El Salvador	15.7	24.0
Guyana	17.3	8.0
Haiti	15.4	21.4
Honduras	19.3	23.6
Jamaica	13.8	22.2
Jordan	15.6	8.9
Lesotho	24.8	25.1
Moldova	23.1	15.8
Nepal	22.9	24.1
Tajikistan	35.1	26.7
Tonga	27.7	39.0

Source: Based on World Bank and IMF data.

drop off. In countries such as Saudi Arabia, where economic migrants are not encouraged to stay despite being vital to the economic functioning of the country, the level of remittances remains steadily high.

Remittances sent back home are used to pay for education. There is also the flow of what we may term reverse remittances, as some of the more affluent families send money to their children studying abroad, especially for higher education.

is the **shipping container** (Photo 10.5). First developed in 1956, the container permits the easy and cheap global transportation of trailer-sized loads from ships to trains and trucks without unloading the containers' contents. It is not the shipping container that created economic globalization, but globalization that created the container. The container both flattened the world and emerged from this flattening. A series of decisions between 1968 and 1970 by

the International Organization for Standardization created global standards of container size, markings, and corner fittings. The result was a truly global form of cheap transportation. Today, close to 20 million containers make over 200 million journeys a year.

The steep fall in transport costs has made everyday things from around the world readily available. It has flattened the world as low transport costs enable relatively

PHOTO 10.5 Container ship in Da Nang, Vietnam.

cheap products to be made in one country and then sold in another halfway around the world. Blueberries grown in Chile, T-shirts manufactured in Vietnam, plastic toys made in China, and white wines bottled in New Zealand can all be purchased in the same store in Syracuse, New York. The global trade in goods is an important part of our everyday lives. And global trade now has the power to influence, affect, and even destroy local and national economies. Distance and transport costs are no longer the barrier they once were. A flatter world is a smaller world.

The Flow of Capital

There are distinct phases of global capital mobility. First, there was a period of marked upswing from 1880 to 1914, when there were few restrictions on the movement of capital. Much of the capital flowed from London markets to developing economies around the world. The First World War ended this period. A brief resurgence from 1925 to 1930 was followed by the Depression and the collapse of international capital markets. Second, a period of stability emerged after the Second World War. Capital flows were relatively small and regulated. The system ended in 1971 and ushered in the current period of increased capital mobility. Financial deregulation promoted global capital flows. The amount of direct foreign investment has increased in most countries. Four qualifying points need to be made:

1. The level of foreign direct investment is highly variable at the aggregate and national levels. Capital outflows constituted over 5 percent of world GDP in 2007, but this figure rapidly fell in the wake of the 2008 collapse. By 2021, it was still only around 1 percent.

2. The bulk of foreign investment occurs between already-rich countries. It is a form of diversified investment rather than development investment. That more investment goes to capital-rich rather than capital-poor countries is termed the Lucas paradox, after the economist who authored a famous paper, Robert Lucas. The disparity from classic theory, which says that capital should flow to places that need it, is the result of institutional structures, heightened sense of risk, and asymmetric information. Many developing economies are considered too much of a risk. The paradox is that the places that need the most foreign investment may be getting less of it.

3. Be careful what you wish for. Increased capital mobility makes it easy to attract investment but comes with the danger that it may lead to rapid and destabilizing capital flows. A smooth surface for capital flows means that a highly mobile capital can undermine national economic planning. Consider the case of Argentina, where the government racked up deficits to reduce unemployment. Investor confidence was shaken, and capital outflows increased. As the currency depreciated, even more capital flowed outward, creating capital shortages and the potential for runs on the bank. The government introduced capital controls and forced companies to repatriate foreign profits.

4. Even when capital flows into developing economies, it comes at a price. Consider the case of an economy with an important commodity that attracts a lot of capital from overseas. Carmen Reinhart and Vincent Reinhart looked at capital bonanzas in 181 countries. They found that bonanzas have become more frequent because of fewer restrictions on capital mobility and are associated in medium- and lower-income countries with a higher incidence of banking, currency, and inflation crises. As they note, "A bonanza is not to be confused with a blessing."

10.3 Global Shift

Global shift is used to describe the movement of industrial employment from the industrial regions and cities of the Global North to those of the Global South. Since the 1970s there has been a redistribution of manufacturing employment from western Europe and North America to Asia, Latin America, and other parts of the world, involving the deindustrialization of the advanced economies and a rapid

10.2 When Debt Becomes Unsustainable

Governments borrow money all the time. They do so to mobilize financing for growth and development. When governments borrow, this is known as public debt or public-sector debt (in contrast to private debt, which is taken out by individuals, for example, in home mortgages).

Many countries in the Global South have borrowed money to build the expansive infrastructure needed to industrialize and modernize. Borrowing money gave governments extra resources to build roads and water and sanitation systems and to invest in health systems, education, and other infrastructure. Debt is not necessarily bad, if it is short term and a country can pay back its loans.

However, many lower-income countries are struggling with debt distress. A country experiences debt distress when debt servicing becomes so large that a country uses a high proportion of its foreign exchange earnings to service the debt and still searches for more loans to enable it to meet urgent and pressing domestic obligations. Some countries attempt to "restructure" or modify their debt arrangements. This can involve getting lenders to agree to reduce the interest rates on loans or extend the dates when payments are to be made. Countries may also be forced to implement austerity measures—which consist of cuts to public-sector spending such as health, education, social services, and other public goods. Austerity measures could also reduce public-sector salaries, impose new labor laws, and increase taxes. Unfortunately, in many countries in debt distress, the impact of austerity measures has meant declining incomes, increased poverty and malnutrition, and increased rates of mental depression.

Today, over half of the world's seventy-six lower-income countries struggle with high levels of debt. Thirty-eight countries in sub-Saharan Africa are in debt distress; so, too, are Afghanistan, Bangladesh, Nepal, Haiti, Honduras, Cambodia, and Myanmar. Countries that experience debt distress have debt that is unsustainable, threatening to undermine economic progress and increase inequalities.

REFERENCES

United Nations. 2020. *World Economic Situation and Prospects 2020*. New York: United Nations. https://www.un.org/development/desa/dpad/wp-content/uploads/sites/45/WESP2020_FullReport.pdf.

United Nations Inter-agency Task Force on Financing for Development. 2019. *Financing for Sustainable Development Report 2019*. New York: United Nations. https://desapublications.un.org/file/538/download.

industrialization of selected cities in a small group of developing countries, including China, Singapore, Taiwan, and South Korea (Photos 10.6 and 10.7).

Smooth global transport networks reduce the cost of transporting goods around the world, and the development of routine manufacturing processes allows manufacturing to become established in areas of unskilled labor. Economic globalization allows companies to search for cheap labor outside their national borders. Economic globalization allows corporations to relocate to minimize wage costs and effectively deterritorialize. The old adage that what is good for General Motors is good for America no longer applies, because General Motors now makes cars in China and Europe. Although it is listed as a US company, Nike's interests do not necessarily parallel US interests. What is good for Nike shareholders is good for Nike shareholders; it may not be good for US workers or US consumers, however.

Global shift is moving up the manufacturing chain. In the early 1990s, China's economic growth was dominated by traditional labor-intensive manufacturing sectors such as textiles, clothing, and footwear. More recently, however, growth has been more noticeable in the capital-intensive, high-tech sectors such as machinery and electronics.

As manufacturing jobs are routinized and deskilled, they become capable of being offshored. Production is being pried away from pools of skilled and higher-paid labor in traditional manufacturing regions. This shift has two consequences. The first is the rise of a new working class in the newly created industrial regions. A female working class emerges as young women fill many of the new jobs in the maquiladoras along the US–Mexico border

PHOTO 10.6 Abandoned factory in Syracuse, New York.

or in the industrial cities of coastal China. The second is the decline in the economic and political strength of the organized working class of the old industrial regions of Europe and North America.

The global shift in manufacturing employment was most pronounced from the 1970s to the 2000s. Since 2010 there has been a further redistribution as some manufacturing jobs relocated back to the United States. From 2010 to 2012, 525,000 manufacturing jobs were added in the United States, and 50,000 of those arose from overseas firms moving operations there. One reason is the narrowing wage gap between China, Thailand, and Vietnam and the United States as wages increase in those countries and decline in the United States. The gap in wage cost between China and the United States, for example, shrank from $20 per hour in 2000 to an estimated $7 per hour in 2015. Rising productivity and increasing innovation are making manufacturing in the United States a more attractive proposition. We may be at the beginning of another global reshifting, especially for the advanced manufacturing sector. The more unskilled routine manufacturing will continue to shift to areas with lower labor costs.

Big-box retailers reinforce the world's flattening by forcing the global shift in manufacturing. Traditionally, manufacturers made things and then sold them to retailers, who sold them to customers. Retailers were simply in the middle. With the rise of the big-box retailer and a more flexible production system, with many small companies producing small runs of goods relatively quickly, power shifted from the large number of small manufacturers to the small number of larger retailers. The very large retailers have such enormous power that they can force producers to compete with each other to meet strict cost limits. Manufacturers are willing to do this because access to a big retailer ensures a large customer base. The lower price per item is offset by the higher turnover of items.

Transnational Corporations

A major change in the global economy has been the increasing power, size, and reach of **transnational corporations (TNCs)**. Today, TNCs control as much as two-thirds of the world trade in goods and services. Some TNCs have sales or revenues that exceed the GDP of countries. For example, Exxon has been valued at $63 billion; its assets are comparable in economic size to the economies of Chile and Pakistan. TNCs are major drivers of globalization and economic growth. It is critical to understand their role in the global economy.

There are four types of TNCs:

1. multinational, with a decentralized organizational structure that operates as a series of semi-independent entities;

2. international, controlled from one central headquarters with overseas operations as appendages;

3. global, in which a centralized hub implements parent-company strategies to reach a global market; and

4. an integrated network of complex, globally integrated operations.

We can get some idea of the range by considering three TNCs: Nike, Starbucks, and Toyota. Nike is a sportswear and equipment company that exhibits characteristics of the integrated network. By 2022, Nike had an annual revenue of $44.5 billion. Nike does not make any goods directly. There are no Nike factories. Instead, the company buys goods from factory owners and then sells them to consumers. In Nike's early years, these factories were in Japan; by the early 1980s, as labor costs increased in Japan, most Nike shoes were made in South Korea, and the city of Pusan became the capital of Asian shoe manufacturing. Since then, Nike has further reduced its costs by having shoes made in China, Indonesia, and Vietnam, where labor costs per worker were only $100 a month. Indonesia is now one of the largest suppliers of Nike shoes: seventeen factories employ 90,000 workers, producing around 7 million pairs of shoes. In southern China, the center of shoe manufacturing is the city of Guangzhou. Just outside the city, one shoe factory, used by Nike, makes 35,000 shoes a day.

Toyota is an automotive company headquartered in Japan but with factories around the world and a global workforce of over 366,000. It strongly exhibits characteristics of the international and integrated network. Toyota emerged in the 1930s and produced its first car in 1935. It made cars in Japan but wanted to sell them abroad. The problem was that many countries, especially those with their own auto industries, put up tariff walls to protect domestic producers. To scale these walls, Toyota established manufacturing plants in selected countries. It became a multinational company headquartered in Japan but with manufacturing plants around the world. Unlike Nike, it retained direct control of the factories producing its products. Over the years they established a reputation for reliable, well-made cars that ranged from budget to luxury. A successful global brand was established, and in 2008 it became the world's largest automaker and the fifth largest company in the world, with significant market shares in North America, Asia, and Africa and smaller shares in Europe. It makes around 10.5 million vehicles a year, with manufacturing and assembly plants in twenty-six different countries. Toyota has fourteen major plants in the United States. Although a global company, Toyota is still anchored in Japan. The company is headquartered there, and the very senior management tends to be Japanese. As late as 2021, almost 9.4 million vehicles were produced in Japan. The explicit desire to become the world's number one automaker was ambitious. This drive to global dominance is, according to many analysts, the root of its recall problems in 2009–2010, when over 8.3 million vehicles were recalled for either accelerator or braking problems or both. The rapid increase in production perhaps outstripped quality control and the ability of engineers and managers to carefully scrutinize new model developments for design problems and associated safety issues. The problems were widely reported—a global company also gets global publicity when things go wrong.

The first Starbucks opened in 1971 with a small storefront cafe in Seattle. Its globalization is relatively recent. The first store outside the United States opened in Tokyo in 1996. By 2020, Starbucks had 32,660 stores in fifty-two countries and a global workforce of over 130,000. Starbucks cafes have remarkable similarity around the world. There are few nods to local particularity or national differences. In some cases, the generic nature creates backlash. There was early difficulty breaking into European markets; Italy, a country with millions of dedicated coffee drinkers, only opened its first Starbucks in 2018. The COVID-19 pandemic forced many stores to close and changed consumer preferences to drive-through and pickup. Company revenues have declined from a peak of $26.5 billion in 2019, and in 2021, 400 stores were closed.

10.3 TNCs, Climate Change, and Carbon Neutrality

As part of efforts to switch to renewable energy, an emerging trend among countries, cities, and businesses is pledging to become carbon neutral. The concept "carbon neutral" or having a net-zero carbon footprint refers to achieving net-zero carbon emissions by offsetting any carbon released with an equivalent amount of sequestered or offset carbon.

There are three main ways to achieve carbon neutrality. The first is reducing or limiting energy usage and emissions from transportation (by walking, using bicycles or public transport, avoiding flying, using low-energy vehicles), as well as from buildings, equipment, animals, and processes.

The second way is to obtain electricity and other energy from a renewable energy source. This can be done either by generating energy directly from the renewable source (e.g., installing solar panels on the roof) or by selecting an approved green energy provider and using low-carbon alternative fuels such as sustainable biofuels.

The third way involves buying carbon offsets. Examples of sequestering carbon include planting trees, funding carbon projects that would prevent future greenhouse gas emissions, and buying carbon credits. An example of a carbon credit could be investing in a sustainable forestry project in the Pacific Northwest or a biogas project from farms in the region. Carbon offsets enable organizations to reduce their environmental impact by supporting projects that reduce, absorb, or prevent carbon and other emissions from entering the atmosphere. A carbon offset is created when one ton of greenhouse gas is captured, avoided, or destroyed to compensate for an equivalent emission made.

A growing number of businesses have pledged to reach carbon neutrality by the year 2050 or sooner. Being carbon neutral is increasingly seen as good corporate social responsibility. Dell, Google, and PepsiCo are Fortune 500 corporations that have pledged carbon-neutral initiatives. In 2006, Google began installing thousands of solar panels on its Bay Area corporate campus, enough to provide approximately 30 percent of the energy needs for that campus. The company has pursued strategies to achieve carbon neutrality, such as investing in wind farms in several locations, including a wind farm in Iowa and two wind power plants in South Dakota. By 2017, Google announced that it reached 100 percent renewable energy. In September 2019, Google's chief executive announced plans for a $2 billion wind and solar investment, the biggest renewable energy deal in corporate history, making Google the largest corporate buyer of renewable power.

Businesses, especially more powerful TNCs, that invest in renewable energy can save money, show commitment to sustainability, and leverage their capital and brand recognition to further investment and interest in renewable energy. It may be TNCs, rather than governments, that lead the way with regard to climate change.

REFERENCES

Funk, M. 2014. *Windfall: The Booming Business of Global Warming*. New York: Penguin Press.

Reisser, W., and C. Reisser. 2019. *Energy Resources: From Science to Society*. New York: Oxford University Press.

What Starbucks did was less to meet a demand than to create a demand. Starbucks embodied a global coffee culture, sometimes building on existing cultural traditions and in other places creating them anew. Before the pandemic, Starbucks was a third place between home and workplace, where socializing and telecommuting occurred as much as coffee drinking—a meeting place as much as a selling place. While in some places it signified US cultural dominance, in others it was perceived as a symbol of modernity. Whether its physical stores can flourish in a postpandemic world is an intriguing question as it transitions away from a distinct place and more toward a point of service.

Today, TNCs are among the world's biggest economic institutions and are able to exert influence over policies and economies in many countries. TNCs have driven the creation of complex production chains and supply chains.

Global Production Chains

Production chains now stretch and snake their way across the globe looking for competitive advantage. The process is driven by competition. At one end, consumers seek greater value, and retailers, competing with one another to provide the best value, force producers into cheaper labor areas and more efficient production zones. At the other end of the chain, producers need to keep finding ways to reduce costs, improve techniques, and hire cheaper labor. As distance shrinks, relocation to ever-cheaper labor areas continues to flatten the world.

The geography of manufacturing is complex, with global supply chains linking iron ore fields in Australia to steel plants in China to fabrication sites throughout Asia, Europe, and North America. Factories are now linked into a dense circuitry of connections as products designed in one country are assembled in another, with parts drawn from suppliers in many other countries.

Textiles are a major sector. Textile manufacturing is a common gateway industry for newly industrializing countries, because the entry costs are low. Textile manufacturing can be established relatively cheaply; it is easier and cheaper to provide sewing machines than a car-making plant, and the skill requirement for workers is relatively modest compared to more complex manufacturing systems. For many nations, textile manufacturing is an easier gateway from a primary-based to a manufacturing economy because it requires relatively smaller amounts of capital, lower skills, and less technological sophistication than advanced manufacturing.

As an example of one production chain, consider the manufacturing of clothes for retailers in the United States and Europe. One clothing retailer, Hennes & Mauritz (H&M), a Swedish company with stores in Europe

10.4 Inequalities in Supply Chains

Chances are your chocolate bar was made possible by child labor. In Western Africa, cocoa is a commodity crop grown primarily for export; cocoa is the Ivory Coast's primary export and makes up about half of the country's agricultural exports in volume. As the chocolate industry has grown over the years, so has the demand for cheap cocoa.

Most cocoa farmers earn less than $1 per day, an income below the extreme poverty line. As a result, they often resort to the use of child labor to keep their prices competitive. Approximately 2.1 million children in the Ivory Coast and Ghana work on cocoa farms. This form of child labor endangers the health and safety of the child.

The children of western Africa are surrounded by poverty, and many begin working at a young age to help support their families. Some children end up on the cocoa farms because they need work and traffickers tell them that the job pays well. Other children are sold to traffickers or farm owners by their own relatives, who are unaware of the dangerous work environment and the lack of any provisions for an education. Often, traffickers abduct the young children from small villages in neighboring African countries, such as Burkina Faso and Mali. In one village in Burkina Faso, almost every mother in the village has had a child trafficked onto cocoa farms. Once they have been taken to the cocoa farms, the children may not see their families for years.

Most consumers are not aware that labor is embedded in many supply chains including chocolate, coffee, cell phones, clothing, and cotton. Consumers can respond with their wallets: when choosing between brands, find out what you can about their supply chains and support companies that practice transparency.

REFERENCES

Food Is Power. n.d. "Child Labor and Slavery in the Chocolate Industry." https://foodispower.org/human-labor-slavery/slavery-chocolate/.

International Labour Organization. 2017. *Global Estimates of Child Labour: Results and Trends 2012–2016*. Geneva: International Labour Organization. https://www.ilo.org/wcmsp5/groups/public/@dgreports/@dcomm/documents/publication/wcms_575499.pdf.

and North America, keeps its prices low by contracting in low-wage areas of the world, such as Bangladesh, where 4 million people work in 5,000 textile factories, most of them women paid around $38 a month to work. H&M contracts with 900 factories worldwide to produce a constantly changing design portfolio. The company has been successful in keeping inventory low; the **just-in-time production** system ensures that goods are made to meet demand. Stores often receive daily supplies. High turnover means that profits can be made through selling many items rather than one; hence, the price of individual items can be reduced, which in turn aids turnover. Workers, however, toil for low wages in often poor working conditions. Between 2005 and 2013, more than 1,800 workers died in Bangladesh in factory fires and building collapses in the textile sector. One building collapse in 2013 killed 1,127 workers.

While supply chains are global, there are also regional manufacturing clusters such as the garment factories in Dhaka. The small town of Poole in Dorset, England, is home to the two biggest makers of tiny electric motors that rotate on bearings made of gas molecules. The motors are used to make printed circuit boards. Warsaw, Indiana, is the center of medical implant makers. While global supply chains highlight the wide spread of a globalized economy, these clusters reveal the importance of spatial agglomeration for the regional economy.

On the Trail of the Flip-Flop

We can follow one global chain by looking at the simple flip-flop (Map 10.3). Caroline Knowles followed the flip-flop trail. It begins in Kuwait, where oil is pumped from underground and onto a supertanker that then makes its way to a South Korean petrochemical factory, where it is turned into plastic. The plastic is then shipped to a factory in China, where it is transformed into flip-flops. Hundreds of thousands of flip-flops are then shipped to Djibouti and then taken to markets in Addis Ababa, Ethiopia, a major market for China's flip-flops. Even when they are abandoned in a giant landfill site, "scratchers," people eking out a living by recycling the rubbish, salvage them along with other materials. To complete the chain, one of the scratchers is saving the money gained by selling the recycled materials to buy an airline ticket so that she can travel to Dubai and maybe get a job as a housekeeper.

Measuring Risk

Doing business in a global economy can be risky. TNCs tend to avoid countries that are politically unstable. Countries are now ranked on their vulnerability to crisis and collapse. The Fund for Peace's Fragile States Index, the Center for Systemic Peace's State Fragility Index, the

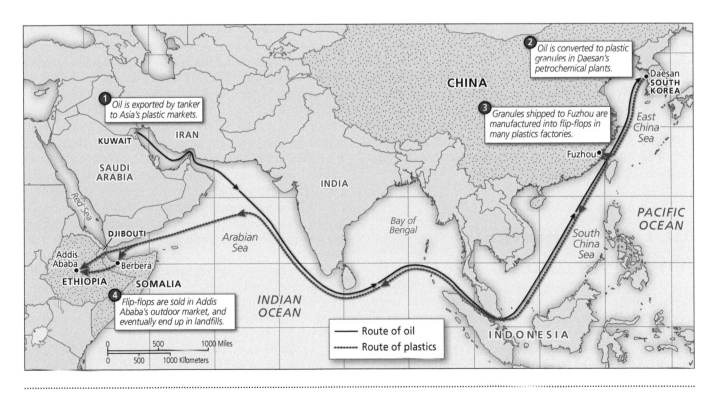

MAP 10.3 The flip-flop trail.

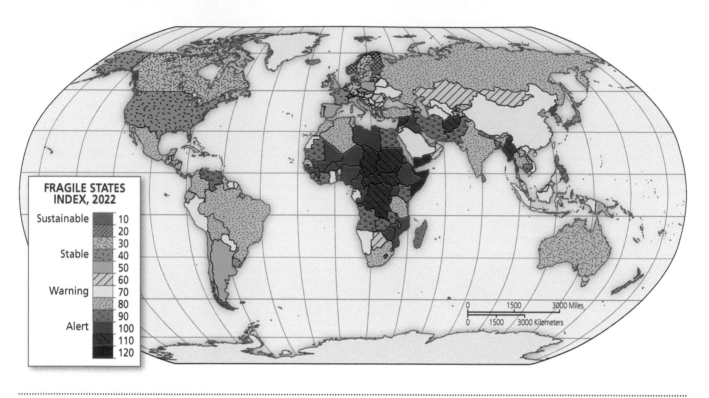

FRAGILE STATES INDEX, 2022

Sustainable — 10, 20
Stable — 30, 40, 50
Warning — 60, 70, 80
Alert — 90, 100, 110, 120

0   1500   3000 Miles
0   1500   3000 Kilometers

MAP 10.4 The Fragile States Index.

Economist Intelligence Unit's Political Instability Index, and the PRS Group's International Country Risk Guide all measure risk and state fragility. The latter, for example, publishes monthly rankings for 140 countries measured along twelve factors, including socioeconomic conditions, internal conflict, corruption, and ethnic tensions. There is also the Political Instability Task Force's global model, which focuses on political structures and relationships via four independent variables: regime type, infant mortality, conflict-ridden neighborhoods, and state-led discrimination. Map 10.4 shows the map of politically fragile states produced by the Fund for Peace.

10.4 International Nongovernment Agencies

Certain **nongovernment agencies** are important actors in the global economy. Three supranational agencies, in particular, have great influence: the IMF, the World Bank, and the World Trade Organization (WTO). The IMF was initially set up to enhance world trade by providing monetary and currency stability. By 2022, the IMF had 190 members. Each member country contributes a certain sum of money as a credit deposit. Collectively, these deposits provide a pool of money from which members may draw. The IMF acts like an international credit union for states. A country's deposits determine how much it can draw on, known as "special drawing rights," and its voting power. Rich countries dominate the IMF, because they contribute most of the funds and set the agenda through their voting power. The IMF wields its power through its **surveillance** system and the strings it attaches to lending. Surveillance (the term used by the IMF) involves monitoring a member's economic policies and evaluating their economy. In one example, South Korea asked for IMF assistance on November 21, 1997, in the wake of depleted foreign reserves and rising short-term loans. The IMF agreed to lend $56 billion and, in exchange, asked for an economic adjustment package that depressed economic growth, caused unemployment, cut government spending, led to a decline in living standards, and granted foreign financial institutions more access. The IMF imposed market principles and freer international trade on Korean society; its argument was that South Korea's problems were caused by a lack of transparency in financial dealings, a bloated public sector, and a job market that ensured jobs rather than productivity.

The World Bank was created in 1944. Its initial name, the International Bank for Reconstruction and Development, describes its role (Photo 10.8). It was a fund established to aid the reconstruction of Europe, and its first loan was $250 million to a war-damaged France. Once Europe got back on its feet, the bank widened its remit. The first

PHOTO 10.8 Headquarters of the World Bank, Washington, DC.

loan to a developing country was $13.5 million given to Chile in 1948 for a hydroelectric scheme.

By tradition, and a continuing example of the domination of the Global North, the IMF is always headed by a European and the World Bank by someone from the United States. The World Bank now has almost universal membership. Like the IMF, it was and still is dominated by rich countries in the Global North. In response to criticism, both internal and external, from both the developing and the developed world, from both private investors angry at delays and administrative snafus and aid workers upset at corruption and inefficiency, the World Bank shifted its emphasis in 1989 to eradicating world poverty with more carefully assessed and monitored development schemes.

The WTO is the latest version of what was originally called the General Agreement on Tariffs and Trade, established in 1948 as a forum for stimulating world trade by reducing customs duties and lowering trade barriers. It was a forum for getting rid of protectionism. The first round of multilateral discussions, held between 1948 and 1967, led to some tariff reductions. The long, drawn-out nature of the discussions led many commentators to suggest that the acronym stood for "General Agreement to Talk and Talk." The most decisive round was the Uruguay round, 1986 to 1994, which established new rules for trade in services and intellectual property, new forms of dispute settlement, and trade policy reviews, as well as the creation of the WTO in 1995. The WTO now has 164 members that account for 97 percent of the world's trade. China became a full member in 2000. Membership in the WTO is vital to countries seeking access to global markets. But this access comes at a price. For the stronger economies, free trade can mean access to new markets, especially for banking and cultural economies, but it can also mean the decline of traditional industries. For the weaker economies, membership signifies access to the large consumer markets of the rich world, but can mean an inability to control the fate of traditionally protected industries. Membership in WTO implies a willingness to shape national economic policy around the principles of neoliberalism and free global trade.

10.5 The Promise and Reality of Neoliberalism

Neoliberalism, an economic ideology that promotes deregulation, minimal or small government, low taxation, and free trade, has circulated widely around the world. It has a number of strands, a long history, and now pervasive adoption by governments and economic elites all around the world. Classical economists of the eighteenth century, such as Adam Smith, believed in the invisible hand of the market to make things work. More recently, it is part of a broader shift in economic ideologies away from the New Deal Keynesianism that dominated from the 1930s to the late 1970s. Neoliberal theorists included Friedrich Hayek (1899–1992), who, in his 1944 book *The Road to Serfdom*, argued against centralized government planning. Milton Friedman (1912–2006), a Nobel Prize–winning economist, also lauded the virtues of free markets and small, limited governments. His book *Free to Choose*, published in 1980, linked individual freedoms with functioning markets and restrained governments.

In this restated neoliberalism, unregulated markets were the solution and governments were the problem.

The neoliberal agenda promotes the privatization of public goods and services, reduced taxes (especially for the wealthy), reduced governmental commitment to the welfare of its ordinary citizens, and an emphasis on enhancing corporate profitability and improving business competitiveness. This agenda's core proposition is that deregulated markets will increase economic growth and raise living standards. Neoliberalism as a political theory has the benefits of being simple, easily understood, and even easier to articulate. As a political practice, however, its greatest difficulty lies in implementation when it comes up against the brute realities of mature markets dominated by powerful sectional interests. These interests promote free trade when it meets their needs but not when it undercuts their power and market share. The tenets of neoliberalism are breached regularly in the face of political realities. Elaborate subsidies and tariff barriers, for example, maintain agricultural producers in the developed world. While efficient Japanese carmakers want free trade, inefficient Japanese rice farmers do not. The US financial services sector wants to penetrate foreign markets, yet the US agricultural lobby funds the erection of import-blocking tariffs.

The neoliberal agenda is promoted by global systems of governance. Both the IMF and the WTO, for example, promote neoliberal economic policies. Around its central ideological core, neoliberalism has subsidiary ideas about accountability, choice, competition, incentives, and performance. It is also a political process that recasts citizens as consumers, reimagines states less as providers of public services and more as promoters of private growth, and shifts governance issues from citizen entitlements to consumer choices. In full-blown neoliberalism, markets trump the state, capital wins out over organized labor, consumers replace citizens, and market choice replaces citizen rights. The private market and the accompanying individual consumerism are enthroned as the means and measure of success.

Yet there are sites of resistance and narratives of dissent. There has been the development of a global discourse of human rights, environmental protection, and shared projects of poverty reduction. An alternative global vision to neoliberalism is emerging.

10.6 Globalization and Its Discontents

A Flat World?

The journalist Thomas Friedman wrote an influential book with the provocative title *The World Is Flat*. He argued that economic growth and development in key countries such as China, India, and Brazil were evidence that the competitive global playing field between high-income countries and emerging markets was leveling and that the large differences created by uneven development were declining. He was optimistic that free trade and global markets would eventually raise all living standards.

But is the world really flat? Is the smoothing of the differences enough to justify using the term **flat world**? Two counterarguments can be made. The first is the simple fact that only some of the world is flat, and by "flat" we mean a smoothing of the tyranny of economic distance and a lessening of the friction of national economic differences. Production chains snake their way only into selected countries. Nike athletic shoes are made from components in some countries, including China, Indonesia, and Vietnam, but not in Myanmar, Pakistan, Somalia, or Zimbabwe or most other developing and poor countries. Starbucks has few stores in sub-Saharan Africa. There are limits to economic globalization. The cheapness of labor is not the only issue for a mobile capital. Reliable infrastructure, cheap transport links, the rule of law, and a stable political system are all prerequisites for long-term global shift, and much of the developing world lacks one or many of these elements. The world is not completely flat, but parts of it are, where distance is almost frictionless and economic transactions are easily connected to other parts of the world. In other parts, however, distance is overcome only at great transactional costs. We live in a misshapen world, parts of which are flat, others curved, and yet others pockmarked by disconnected ravines and inaccessible peaks.

The world's topography is more varied than flat. At the international level, there are still many countries only loosely connected to global flows of investment or global shifts in manufacturing. And even within the more connected, flatter countries, there is often a marked demarcation between the more inaccessible rural areas and the metropolitan regions where most of the global economic activity is concentrated. Many of the rural areas of the developing world are far from markets and lack the necessary infrastructure to ensure easy and cheap transportation, the hallmark of a flat world. The flat areas, where much of the world's global economic activities take place, are like landing strips in the middle of rugged terrain, often constituting small parts of larger countries and, in total, only a tiny part of the Earth's economic surface.

The second point is that economic globalization is about creating hierarchies as well as flat surfaces; it generates spatial difference as well as homogeneity. Economic globalization is not just the product of a consumerist culture and a competitive capitalism continually seeking to squeeze costs. Since the 1970s, there has also been a globalization of finance. Capital now flows in larger and more viscous currents across the world. In this case, the flattening of the world is also tied to the networking of the world.

One particularly dynamic sector of services is the knowledge-based industries, so-called advanced producer services such as advertising, banking, and financial services for the larger corporations. Together these sectors constitute the dynamic edge of mature capitalist economy. Since 1980, in the developed world, a region's success rests less on manufacturing employment and more on the extent to which it can generate, retain, and attract knowledge-based employment. This sector is heavily concentrated in the global cities. Flat surfaces are ideal to move goods cheaply and quickly, but to do business you need nodes of concentrated activity. A global system of cities, connected by flows of capital, ideas, information, and skilled personnel, "landscapes" the world into concentrated nodes of heightened connectivity and accumulated control. We will discuss this topic more in a later chapter.

In summary, while there has been a flattening of the world, a flat world is not yet a reality.

The differential flattening that does occur has redistributional consequences. Globalization has different effects on capital and labor and on different sectors of labor. While capital is freer to move across the surface of the globe, labor is much more restricted. The low cost of international transport and the growing ease of international trade, crucial requirements of economic globalization, have allowed capital to be more easily disassociated from national interests and local community concerns. Capital is free to roam the space of the world in search of lower wages and higher returns, while labor is more immobile. Retailers can move their production contracts to another factory in another country, but organized labor is more fixed in place. The result is an uneven bargaining arrangement. Globalization has liberated capital but largely restricted workers to national markets.

There are different levels in the flows of globalization that lead to different geographies of globalization. At the top level of the flat world, capital flows with fewer and fewer restrictions. Here are the global rich and the highly skilled, especially in the field of global business. At the bottom are the unskilled and those without capital, whose movements are much more circumscribed; they live in a more rugged topography.

Long ago, distance severely restricted economic interaction. The cost and difficulty of movement meant that most contact was between peoples and areas close together. The world that people inhabited was shaped more by the local than the global. The geographic-historical long view would perceive a world gradually coming together, not as a continuous process but in a series of fits and starts with some areas now pulled together, now pulled apart. World historical geography is in large part the story of global space-time convergence and successive waves of globalization.

The Backlash

Globalization is under attack. The electoral victory of politicians promoting an economic nationalism, the Brexit vote that led to the United Kingdom's exit from the European

PHOTO 10.9 Antigovernment, anticapitalist, anti-IMF protest in Buenos Aires.

Community, and the rise of an aggressive nationalism in mainland Europe and around the world are all part of a backlash to globalization (Photo 10.9). Why the discontent? A deeper examination of global integration may shed some light.

The roots of today's global economic order were established just as World War II was coming to end. In 1944, delegates from the Allied countries met in Bretton Woods, New Hampshire, to establish a new system around open markets and free trade.

New institutions such as the IMF, the World Bank, and a precursor to the WTO were established to tie national economies into an international system. There was a belief that greater global integration was more conducive to peace and prosperity than economic nationalism.

While capital could now survey the world to ensure the best returns, labor was fixed in place. This meant there was a profound change in the relative bargaining power between the two—away from organized labor and toward a footloose capital. When a company such as General Motors moved a factory from Michigan to Mexico, it made economic sense for the corporation and its shareholders, but it did not help workers in the United States. Freeing up trade restrictions also led to a global shift in manufacturing. The industrial base shifted from the high-wage areas of North America and western Europe to the cheaper-wage areas of East Asia: first Japan, then South Korea, and more recently China and Vietnam.

As a result, there was a global redistribution of wealth. In the West, as factories shuttered, mechanized, or moved overseas, the wages of blue-collar workers declined in real terms. Meanwhile, in China prosperity grew, with the country's official poverty rate falling from 84 percent in 1981 to under 1 percent in 2022.

Political and economic elites in the West argued that free trade, global markets, and production chains that snaked across national borders would eventually raise all living standards. But as no alternative vision was offered, a chasm grew between these elites and the mass of blue-collar workers. The backlash against economic globalization is most marked in countries such as the United States where economic dislocation unfolds with weak safety nets and limited government investment in job retraining or continuing and lifetime education.

The flattening of the world allowed for a more diverse ensemble of cultural forms in cuisine, movies, values, and lifestyles. Cosmopolitanism was embraced by many of the elites but feared by others. In Europe, the foreign other became an object of fear and sometime resentment. But evidence of this backlash to cultural globalization also exists around the world. The ruling BJP Party in India, for example, combines Hindu fundamentalism with political nationalism. There is a rise of religious fundamentalism around the world in religions as varied as Buddhism, Christianity, Hinduism, Islam, and Judaism. Old-time religion, it seems, has become a refuge from the ache of modernity. Religious fundamentalism held out the promise of eternal verities in the rapidly changing world of cultural globalization.

There is also a rising nationalism, as native purity is cast as a contrast to the profane foreign. In Europe, from Bulgaria to Poland and the United Kingdom, new nationalisms have a distinct xenophobia. Politicians such as Marine Le Pen in France recall an idealized past as a cure for the cultural chaos of modernity. Politicians can often gain political traction by describing national cultural traditions as under attack from the outside.

Globalization has now become the catchword to encompass the rapid and often disquieting and disruptive social and economic change of the past twenty-five years. Globalization has provided enormous economic benefits but also has generated economic inequalities, political uncertainties, and cultural anxiety.

Cited References

Banister, D. 2011. "The Trilogy of Distance, Speed and Time." *Journal of Transport Geography* 19:950–959.

Friedman, T. 2006. *The World Is Flat: A Brief History of the Twenty-First Century.* New York: Farrar, Straus and Giroux.

Horner, A. 2000. "Changing Rail Travel Times and Time-Space Adjustments in Europe." *Geography* 85:55–68.

Janelle, D. 1968. "Central Place Development in a Space-Time Framework." *Professional Geographer* 20:5–10.

Knowles, C. 2014. *Flip-Flop: A Journey through Globalization.* London: Pluto.

Lucas, R. 1990. "Why Doesn't Capital Flow from Rich to Poor Countries?" *American Economic Review* 80:92–96.

Reinhart, C. M., and V. Reinhart. 2008. "Capital Flow Bonanzas: An Encompassing View of the Past and Present." National Bureau of Economic Research Working Paper 14321. http://www.nber.org/papers/w14321.pdf.

Select Guide to Further Reading

Baldwin, R. 2018. "If This Is Globalization 4.0, What Were the Other Three?" *World Economic Forum*, December, 22 1–4. https://www.weforum.org/agenda/2018/12/if-this-is-globalization-4-0-what-were-the-other-three/.

Crosby, A. A. 1972. *The Columbian Exchange: Biological and Cultural Consequences of 1492.* Westport, CT: Greenwood.

Dicken, P. 2015. *Global Shift: Mapping the Changing Contours of the World Economy.* 7th ed. New York: Guilford.

Dumenil, G., and D. Levy. 2011. *The Crisis of Neoliberalism.* Cambridge, MA: Harvard University Press.

Engel, C., and J. Wang. 2011. "International Trade in Durable Goods: Understanding Volatility, Cyclicality, and Elasticities." *Journal of International Economics* 83:37–52. http://dx.doi.org/10.1016/j.jinteco.2010.08.007.

Friedman, M., and R. Friedman. 1980. *Free to Choose.* Orlando, FL: Harcourt.

Goldstone, J. A., R. H. Bates, D. L. Epstein, T. R. Gurr, M. B. Lustik, M. G. Marshall, J. Ulfelder, and M. Woodward. 2010. "A Global Model for Forecasting Political Instability." *American Journal of Political Science* 54:190–208.

Harvey, D. 2007. *A Brief History of Neoliberalism.* New York: Oxford University Press.

International Labour Organization. 2017. *Global Estimates of Child Labour.* Geneva: International Labour Organization. https://www.ilo.org/wcmsp5/groups/public/@dgreports/@dcomm/documents/publication/wcms_575499.pdf.

Jackson, T. 2017. *Prosperity without Growth: Foundations for the Economy of Tomorrow.* 2nd ed. New York: Routledge.

Khanna, P. 2016. *Connectography: Mapping the Future of Global Civilizations.* New York: Random House.

Knowles, R. D. 2006. "Transport Shaping Space: Differential Collapse in Time-Space." *Journal of Transport Geography* 14:407–425.

Lewis, D., N. Kanji, and N. S. Themudo. 2020. *Non-governmental Organizations and Development*. London: Routledge.

Mann, C. C. 2006. *1491: New Revelations of the Americas before Columbus*. New York: Knopf.

Mann, C. C. 2011. *1493: Uncovering the New World Columbus Created*. New York: Knopf.

Marsh, P. 2012. *The New Industrial Revolution: Consumers, Globalization and the End of Mass Production*. New Haven, CT: Yale University Press.

Sheriff, C. 1997. *The Artificial River: The Erie Canal and the Paradox of Progress*. New York: Hill & Wang.

Short, J. R. 2001. *Global Dimensions: Space, Place and the Contemporary World*. London: Reaktion Books.

Short, J. R. 2004. *Global Metropolitan: Globalizing Cities in a Capitalist World*. London: Routledge.

Sobel, D. 1995. *Longitude: The True Story of a Lone Genius Who Solved the Greatest Scientific Problem of His Time*. New York: Walker.

Steger, M., and R. K. Roy. 2010. *Neoliberalism: A Very Short Introduction*. New York: Oxford University Press.

Tang, X. 2020. *Global Space and the Nationalist Discourse of Modernity*. Stanford, CA: Stanford University Press.

The Cultural Organization of Space

There is an intimate relationship between space, place, and **culture**. Space is given meaning and place is rendered unique through cultural formations. Culture is expressed in space and embodied in place. Part 4 considers the cultural organization of space through an examination of population, religion, and culture. Chapter 11 explores the geography of religion, Chapter 12 examines the geography of language, and Chapter 13 considers the geography of culture.

OUTLINE

11

The Geography of Religion

Religion is an important element of culture and an important subject for cultural geography. Many people define themselves primarily in terms of religious affiliation. Religious beliefs order society. In this chapter we will look at a number of elements in the geography of religion and how religion and religious beliefs structure the cultural organization of space.

11.1 The Geography of the Major Religions

Three broad types of religion can be identified: the **animist religions** of hunting-gathering societies, the distinctive religions of Asia, and the religions of **monotheism** that emerged from the Middle East. We will look at each in turn but with a gentle reminder that there are similarities and connections as well as differences and ruptures between the three categories.

The Animist Tradition

The cosmologies of hunting-gathering societies are rooted in an animistic conception of the world. The essence of the animistic cosmologies of hunting-gathering societies is that the Earth is alive and nurturing. Nature is a spiritual resource as well as an economic entity, part of the fabric of connections between all things. Space is sacred, and territory is the fundamental source of meaning and identity.

In Cherokee tradition the word for Earth, *elohi*, also means history, culture, and religion. All living things and all places are bound in a web of material and spiritual connections. Collective and individual identity is bound to specific places, plants, and animals. There is a kinship between animals and humans, a bond between the hunted and the hunter, planter and planted. A deep and profound religiosity imbues the economic relations of everyday life. Ritual and performance in dance and song, as well as material culture such as shelter, clothing, and decoration, link people, places, and living things in a tight nexus of material and spiritual connections.

Consider the case of the Indigenous peoples of the central Australian desert. Before the coming of the whites in the late nineteenth century, the total Arrernte precontact population is estimated to have been between 8,000 and 10,000. Food was scarce. The landscape was a complex system of land titles based on lineage and family ties. Responsibility for "managing" specific sites often lay in the hands of senior elders, both men and women (Photo 11.1). The complex, interlocking titles were less monopoly controls, as we understand private property under a capitalist system, and more relational and totemic, allowing individuals and groups a broad range of claims, responsibilities, and bargaining options that enabled long-term occupancy of a harsh environment with irregular water and variable food sources. The

cosmology bound people to the land in intricate webs of meaning that also sustained long-term economic usage.

The cosmologies of these early societies did not prevent major ecosystem transformation and species collapse. Calvin Martin looked at the effect of early contact with Europeans on the Micmac tribe in the Gaspé Peninsula of Canada. Prior to the arrival of Europeans, the beaver was hunted in a sustainable way; there were millions of beaver, and their hunting was bound by rules, regulations, and religious invocations that severely limited the hunting of "relatives in fur masks." With the coming of the French and the Dutch, a vigorous trade was promoted. Europeans wanted furs from the Micmac and other tribal groups, for which they traded knives, axes, guns, and alcohol. The increasing competition for furs, the decline of traditional constraints caused by the deadly epidemics of the Columbian Encounter, and the waning influence of the traditional shaman led to a wholesale slaughter of beaver by Micmac hunters, who hunted no longer for immediate needs but instead without restraint. It became an unrestrained slaughter as the beaver was turned from a relative in a fur mask to a commodity.

As hunting-gathering societies were overwhelmed by agriculturalists and later urbanization and industrialization, the world lost much of the rich variety of the animist traditions. Map 11.1 show the main religions across the world and reveals how the traditional animists are now restricted to the more peripheral and marginal lands of desert, tundra, and rainforest. It is a fast-disappearing world.

ANCIENT ANIMISMS AND MODERN ENVIRONMENTALISM

In a rapidly modernizing world, the original animist traditions are not unchanging but transformed and reinterpreted. Sam Gill, for example, examined the cosmologies of Native American animist traditions. He found little original evidence of a Mother Earth figure. The myth appears in nineteenth-century Anglo-American interpretations of Native American religions. It was blended into and drew on Christian beliefs in the Virgin Mary and understandings of Old World Neolithic cultures. However, it then became part of Native American writings as Native Americans used the myth to express their attachment to the land in a way that was understandable for the dominant Anglo-American audience. A product of Anglo-American cosmologies was incorporated into Native American beliefs and then projected back to influence Anglo-American perspectives.

Animist traditions, real and invented, are now used to inform the contemporary **environmental movement**. Here is the oft-cited 1854 speech of Chief Seattle of the Suquamish in response to the US government's offer to buy their land:

But how can you buy or sell the sky? The land? The idea is strange to us. If we do not own the freshness

PHOTO 11.1 Sacred site in Alice Springs, Australia.

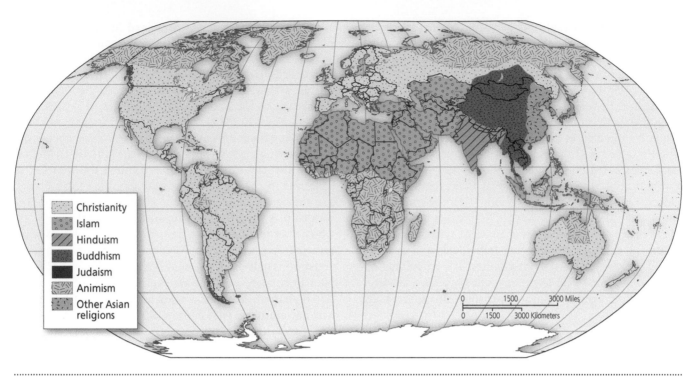

MAP 11.1 Main world religions.

of the air and the sparkle of the water, how can you buy them?

Every part of the earth is sacred to my people. Every shining pine needle, every sandy shore, every mist in the dark woods, every meadow, every humming insect. All are holy in the memory and experience of my people.

We know the sap which courses through the trees as we know the blood that courses through our veins. We are part of the earth and it is part of us. The perfumed flowers are our sisters. The bear, the deer, the great eagle, these are our brothers. The rocky crests, the dew in the meadow, the body heat of the pony, and man all belong to the same family.

It is a moving, eloquent speech. But it is an invention; we have no idea what the chief actually said. The "speech" was rewritten at numerous times by different people with different agendas—some religious, some ecological, and some a combination of the two. It was written into the script of a film made by the Southern Baptist Convention in 1972. The speech has entered the **religio-ecological** mainstream almost as a sacred text in its own right. It makes three important points that connect to present-day ideas about sustainability:

The environment is a spiritual resource more than an economic entity.

There is web of connection between all things.

Occupancy of the Earth is a religious bond with responsibilities and obligations.

The speech is less a lament for the present and more a hope for the future. It also implies that humans have a responsibility to be stewards of the natural world, a key idea that we now hear echoed in the demands for action on climate change or to protect biodiversity. The ecology of the animists is now an important element in the contemporary religio-ecological movement and acts as a counterpoint to the "unsustainable" discourse that believes resources exist only to be used by humans.

The Religions of Asia

The **Indic region** of northern India, centered on the Indus valley, is the hearth source for Buddhism, Hinduism, Jainism, and Sikhism (Photos 11.2 and 11.3). The two largest are Hinduism and Buddhism. Hinduism spread across the Indian subcontinent into much of Southeast Asia. Later, Buddhism spread along the trade routes of the Silk Road, finding expression and acceptance in China, Korea, and Japan and replacing Hinduism in much of Southeast Asia.

HINDUISM

Hinduism has more than a billion adherents and is the dominant religion in India, though not the only one. It is also found in Sri Lanka and Nepal and is celebrated around

11.1 Alternative Visions of Place

The Huichol Indians live in a mountain area in western Mexico. Their total numbers are around 25,000 to 50,000. Although deeply impacted by the Columbian Encounter, they still retain many of their traditional belief systems. Huichol is still spoken, and there are two local governments: one answers to the local people, while another represents the Mexican state.

The Huichol interact with the modern world. Their crafts cater to a tourist market. But they also practice a form of animism that imbues the area with primary spiritual significance. The landscape is a sacred geography whose protection is vital to the integrity of their worldview. The local mountains are not inert geological features but sacred portals to an understanding of the infinite, sites of cosmic significance. One mountain, Cerro del Quemado, known to the Huichol as Wirikuta, is considered the birthplace of the Sun.

Two mining companies, Canada's First Majestic Silver and the Mexican Real Bonanza, have a very different conception of the area. They see it is as a place to extract silver. In 2009 they purchased mineral rights from the Mexican government and will most likely use cyanide in open pit mining.

This is a form of struggle witnessed throughout the world. On the one hand is a small group of people with indigenous beliefs and uncertain land rights; on the other are powerful international corporations looking to extract commodities. The Huichol have sought to resist the mining proposals, while the mining companies have sworn to protect the sacred sites, maintain international environmental standards, and employ local people. Many of the non-Huichol local people support the mining proposal for the jobs it will generate. Behind the conflict are very different and competing ideas of the environment, as a commodity to be exploited and as a sacred space to be revered. As one Huichol said of the mountain,

This is the center of the world. It is our church. The mining company has promised that the mountain will look the same. But it won't be the same if you take away its soul.

REFERENCES

Booth, W. 2012. "Where Cosmic and Commercial Collide." *The Washington Post*, February 14, 2012, A1 and A11. Quote is from page A11.

Liffman, P. M. 2010. *Huichol Territory and the Mexican Nation: Indigenous Ritual, Land Conflict and Sovereignty Claims.* Tucson: University of Arizona Press.

---

the world because the Indian diaspora took their religion in their migrations to the Middle East, Australia, western Europe, and North America.

Hinduism is one of world's oldest religions. It is a belief system that stretches back to the early civilizations of the Indus valley over 6,000 years ago. It diffused south across India into Southeast Asia and the islands of what is now Indonesia. It is a monotheistic religion, because it believes there is one life force, but it is also polytheistic with numerous deities. There are three male gods, Brahma, Vishnu, and Shiva, and a female deity, Devi, the mother goddess. There are a host of other gods such as Ganesh, a man with the head of an elephant who is a god of strength; Lakshmi, the goddess of wealth; and Hanuman, the monkey king, who is a god of cunning. Hinduism also has a strong belief in reincarnation.

Over thousands of years of development, Hinduism has combined different strands of belief, from the earliest folk traditions of the worship of local deities to the very recent associations with Indian Hindu nationalism. There is no one foundational text, in contrast to the Bible for Christians or the Koran for Muslims. Sacred texts include the Vedas, a large number of texts written around 3,500 and 5,000 years ago; the 700-verse *Bhagavad Gita*; and many texts of the Upanishads, written between 1000 BCE and 1500 CE. There is no single source of religious authority.

Hindus believe in an ultimate reality, Brahman. Numerous gods link this ultimate reality with the populace. The most popular include the four-faced Brahma, the creator god; the blue-skinned Vishnu, the preserver; and Shiva, the destroyer or transformer, with an eye in the center of his forehead.

Hinduism is not an unchanging set of beliefs; it has been touched and transformed by historical changes. The increasing urbanization of India from 500 to 200 BCE, for example, challenged previous rituals. A Hindu modernism emerged in the nineteenth century that focused on social issues, while the contact with the British Raj impacted a wider understanding of Hinduism.

As one of the world's oldest religions, Hinduism provides the best case study of the malleability of religious beliefs and practices. Religions have a tight core of beliefs but these are changed and transformed over time and in contact with new historical realities, changing social forces, and other beliefs and doctrines. Religions do not remain pure and untouched but rather are marked by their place in time and their time in place.

Hinduism is an ancient, complex religion that defies easy definition. It has changed and transformed over the millennia. However, there are a number of key beliefs. The core belief involves leading a righteous life, realizing wealth, fulfilling

PHOTO 11.2 Jain Temple in Mumbai, India.

significance and resources devoted to marriage celebrations in Hindu culture. The final stage is supposed to be marked by a detachment from material life to focus on spiritual matters.

Hindu temples are places of festivals, ritual, and community celebrations (Photo 11.4). The highest point in the temple represents Mount Meru, the sacred five-peaked mountain considered the center of the spiritual universe. In the temples of south India in particular, the outside walls are richly decorated. These temples become distinctive features of the cultural landscape, a visual reminder of the role of faith in that community. Rituals are also practiced at home and include worship at family shrines, yoga, meditation, and the chanting of mantras. There are also festivals celebrated throughout the year. Many of them are local and regional, while a few, such as the annual festival of lights celebration known as Diwali, are found throughout the Hindu culture region.

There are a number of pilgrimage circuits that link holy cities and sites. One of the holiest is held every three years and the location rotates around the four cities of Allahabad, Haridwar, Nashik, and Ujjain. The river

PHOTO 11.3 Buddhist Temple in Myanmar.

desires, and attaining liberation. The classic stages of life are student, householder, retired, and renunciation. The socially important household stage, when people take up their role in society, starts with a marriage that explains, in part, the

PHOTO 11.4 Hindu Temple in Kuala Lumpur, Malaysia.

11.2 Pollution, Climate Change, and the Ganges River

Known as *Ganga Ma* (Mother Ganges), the Ganges River is revered as a goddess whose purity cleanses the sins of the faithful and helps the dead on their path toward heaven. Hindus believe that if the ashes of their dead are deposited in the river, they will be ensured a smooth transition to the next life or be freed from the cycle of death and rebirth. It is said that a single drop of Ganges water can cleanse a lifetime of sins. In cities along the Ganges, daily dips are an important ritual among the faithful.

Despite its spiritual importance, the physical purity of the river has deteriorated dramatically. The Ganges River begins as pristine, clear waters in the Himalayas. But pollution, untreated sewage, and use by hundreds of millions of people transform the river into a toxic sludge by the time it reaches the Bay of Bengal, some 1,500 miles later. The Ganges winds through twenty-nine cities with populations over 100,000.

Today, half a billion people live in the basin of the Ganges and more than one hundred cities dump their raw sewage directly into the river. Coliform bacteria measurements have consistently been at levels too high to be safe for agricultural use, let alone for drinking and bathing. Industrial pollution and nutrients from agricultural lands also add to the river's pollution load. The UN estimates that as much as 80 percent of sewage discharged into two major tributaries that feed the river is still untreated. As much as 1.5 billion gallons (6 billion liters) of untreated sewage is dumped into the Ganges *each day*.

Climate change may amplify existing water problems in the Ganges River. In terms of water quality, droughts may reduce river water flow, increasing the concentration of pollution in certain areas, such as the heavily industrialized Kanpur region. Increased water pollution may increase waterborne illness, a significant issue, because in India, 80 percent of health problems are caused by waterborne diseases.

In terms of water quantity, climate change may alter the river's flow. Because the Ganges depends on monsoon rains and ice melt from the Himalayas for its flow, the river will be impacted by increased temperatures and shifting rainfall patterns. Ice melt may come earlier, or all at once, causing major flooding. In the last twenty years, the monsoon season has arrived earlier, lasted longer, and caused record rainfall and flooding. In 2019, flooding displaced almost 1.2 million people from their homes. Monsoon floods in May 2022 triggered numerous flash floods and landslides that displaced over 9 million people in India and Bangladesh. It is likely that flooding from the monsoons is only going to worsen in the future.

REFERENCES

Chandrashekhar, V. 2019. "As the Monsoon and Climate Shift, India Faces Worsening Floods." *Yale Environment 360*. https://e360.yale.edu/features/as-the-monsoon-and-climate-shift-india-faces-worsening-floods.

Hamner, S., D. Pyke, M. Walker, G. Pandy, R. K. Mishra, V. B. Mishra, C. Porter, and T. Ford. 2013. "Sewage Pollution of the River Ganga: An Ongoing Case Study in Varanasi, India." *River Systems* 20:157–167. https://doi.org/10.1127/1868-5749/2013/0058.

Ganges is worshipped as the goddess Ganga, and devout Hindus believe that bathing in the river washes away sins and helps liberate them for the endless cycle of life and death. It is a problematic place for pilgrimage because it is one of the most polluted rivers in the world, a symbol of how sacred space is transgressed by the material world.

Partly in response to Western colonial presence, the loose collection of beliefs that constitute Hinduism became more codified. More recently, it has become the basis of identity and nationalist politics in India.

A significant social aspect of Hinduism, which was reinforced under the British, is the **caste system**. These are social categories that determine not only position in the social hierarchy but also appropriate occupation, behaviors, and social customs such as eligible marriage partners. There are many different castes, such as the high-caste Brahmins, whose traditional occupations include priests and teachers. The Kshatriya formed the political and military elite. The Vaisya were involved in agriculture but now are landowners, merchants, and bankers. The Sudras are the "worker bees" of the system, involved in many occupations such as barbers, gardeners, and potters. At the very bottom of the hierarchy are the Dalits, the untouchables, who number around 200 million people in India. They were restricted to the impure occupations such as leather workers, sweepers, manual laborers, and sewage workers. They were considered unclean and often physically segregated. Across India many Hindus became Muslims to avoid the debilitating and limiting caste designation. The conversion was almost universal in the rural parts of east Bengal, what is now Bangladesh.

The caste system is more complex than the simple categories outlined here. There are many more subcategories and further divisions of subcategories. There is no simple mapping of caste onto wealth. There are many poor Brahmins and rich Vaisya and affluent Dalits. The caste system is strongest at the village level. In Madhya Pradesh, for example, Dalits are prevented from entering most temples and are often segregated at the edge of the village. As people move to the cities, the entrenched caste system that operates strongly at the village level is breaking down—not quite to the easier mobility of modern secular societies, but still toward a greater fluidity. However, only 5 percent of marriages in India cross caste boundaries. Since 1950 the Indian government has actively worked to improve the employment opportunities and living conditions of the Dalits. They are now found in the upper levels of the government and government-controlled enterprises. They still face serious discrimination in public life, with high dropout rates from schools and more limited access to public services. Violence against Dalits continues, though it is on the decline. Urbanization and policies of affirmative action have eroded much, but not all, of the bias in the caste system. There is also variation across India. The caste system was more challenged in Kerala, for example, part of the more radical political tradition in a state that regularly elects Communist state governments.

BUDDHISM

Buddhism emerged as a critical offshoot of Hinduism, and like Hinduism, it also developed in northern India, around the sixth century. It spread across East and Southeast Asia. The transformation of Hinduism into Buddhism is embodied in the life of Buddha, who was born a prince in northern India around 563 BCE. Spared the vicissitudes of life by his wealthy family, he instead chose poverty, ill health, old age, and eventual death. He distanced himself from the caste system of Hinduism and taught a middle way between the asceticism of a retreat from society and the indulgence of society driven by desire. Buddha preached the pursuit of enlightenment.

A fundamental belief of Buddhism is that desire in any form, whether for wealth, fame, or sensory pleasure, causes misery. Bliss is achieved through not having desire. Retreat from the material world through permanent or temporary periods as a monk or nun is a common practice. Buddhism slowly lost influence in India but gained adherents in East and Southeast Asia. There are two main forms: Mahayana Buddhism, found in China, Japan, and Korea, and Theravada Buddhism, which ties more closely to the Buddha's teaching of the search for nirvana. Theravada Buddhism is particularly strong in Sri Lanka, Myanmar, Thailand, and Laos, and in these countries most young men try to spend a portion of their time as monks learning to meditate while begging for food.

In parts of Southeast Asia, Buddhism was influenced by British colonialism. After the British gained control of Burma in 1824, Christian missionaries began a conversion program. Buddhist monks, fearful that their religion would crumble, encouraged people to study Buddhist texts and to meditate, a practice formerly limited to monks. Buddhism was reshaped and meditation became a standard practice in response to the threat from Christian proselytization. Most people in the West now think of Buddhist meditation as a centuries-old tradition rather than a product of nineteenth-century resistance to British colonialism and its contact with modernism.

This provides yet another example of the constantly changing nature of religious practices and beliefs and their evolution as they come into contact and sometimes compete with other religious faiths.

OTHER ASIAN RELIGIONS

Hinduism and Buddhism are the two main religions of South and East Asia, but there are others. Sikhism combines Hindu mysticism and Muslim monotheism. It is indigenous to this region and was founded in the fifteenth century in the Punjab. By 1800, a vast Sikh empire stretched over Pakistan and India, with a capital in Lahore. The entry of the British undermined this empire, and by 1900 the Sikh heartland had become the Punjab region of British India. In the early twenty-first century there are around 25 million Sikhs, with the vast majority in the Indian state of Punjab; smaller numbers in Haryana, India; and a diaspora in Asia, North America, and Europe.

Jainism is an ancient religion with around 5 million followers. Its main teachings include nonviolence, a detachment from the world, and a commitment to honoring other forms of life. Jains are vegetarian to avoid inflicting unnecessary suffering on animals.

Shinto is the largest religion in Japan, intermittently followed by around three-quarters of the total population. It is considered Japan's traditional religion and is constituted in thousands of public shrines and temples devoted to the worship of numerous gods (Photo 11.5). Shinto is the religion of

PHOTO 11.5 Shinto Temple in Japan.

Japan's traditional agricultural society, and many of the rituals were originally festivals to placate the gods to ensure a good harvest and to thank them for the bounty.

Shintoism provides a good example of an important trait of all religions: a hard core of true believers and regular adherents and an often much larger periphery of occasional practitioners. There are, for example, Catholics who go to Mass and confession every week and many others who may go to religious services only during Easter and Christmas, to bury someone, or to attend a wedding. They are all Catholics, but some are in the **core of believers**, while the others are in the periphery. In Japan, only a very small minority of Shinto practitioners are in the core. Most use Shinto temples just for weddings and occasionally visit its shrines to make an offering for a successful business venture or social project; it is not an integral part of everyday life as, say, Buddhism is in Thailand or Myanmar. There is a wide continuum to religious beliefs, from the ardent to the casual and from the committed to the irregular.

Confucianism is as much a philosophy as a religion. It developed from the teachings of Confucius (551–479 BCE). It stresses family loyalty, social harmony, and respect for hierarchy. It cultivates virtue, education, and social ethics with an emphasis on tradition and stability. It is a major influence not so much as a religion, but as an organizing social worldview in China, Singapore, Taiwan, Korea, Japan, and Vietnam. Its influence is profound. The recent spectacular economic growth of East Asian countries since the 1970s is in no small part a result of the Confucian legacy of good government and social harmony. All the countries mentioned have experienced spectacular rates of growth, starting with Japan and more recently in China and Vietnam. That is one reason why the success of East Asia has been difficult to replicate in other parts of the world that do not share the Confucian legacy. The social order of Confucianism does play an important part in their economic success, based as it is on hard work, family sacrifice, and a belief and trust in the role of government. When the Communists took control of China in 1949, they initially discouraged Confucianism, but now his teachings are widely taught and admired. As part of their projection of soft power, the Chinese government now supports Confucius institutes in countries around the world.

Taoism, also sometimes known as Daoism, is a traditional religion that originated in China. The main goal is that people should become one with the rhythms of the world. The three main beliefs are inaction, simplicity, and living in harmony with nature. There is no god figure as in many other religions. A central tenet is the need to live with the shifting forces that shape our cosmos. A key text is the *Tao Te Ching* written over 2,400 years ago. Today, there are around 12 million self-identified Taoists in China, although a much larger number, almost 173 million, claim they pursue Taoist practices. Taoist temples can be in many cities around the world, especially where there is or was a significant Chinese population, such as Hanoi, Kuala Lumpur, and San Francisco.

While these religions developed in Asia, they have spread around the world. There are Buddhists in America, Hindus in Britain, and Sikhs in Canada. Along the way, the rituals and practices have subtly transformed. The practice of yoga, long an element of Hinduism, has now been severed from its religious roots in the West to become a physical exercise. So when you see people carrying yoga mats in the cities of the West, they are less likely to be religious devotees and more likely to be people coming and going from physical exercise class.

The People of the Book

The **Semitic region** of the Middle East is the hearth area of three of the worlds' significant religions: in order of historical appearance, Judaism, Christianity, and Islam. They are the religions of the Book, where religious beliefs are encoded in written texts. Their texts often contain the same myths and stories. Some stories are sacred to all three **Abrahamic religions**: Adam and Eve, Abraham, Cain and Abel, Jonah and the whale, Moses, and Job. The stories helped to define each religion as both special in God's eyes and supporting community identity in the face of its other two rivals. Robert Gregg has looked in detail at some of the shared stories and their varied tellings. In the Jewish telling, Jonah was a cautionary tale of a complex figure: an angry prophet who did not repent, and later a rehabilitated agent of God. For the Christians, he was a rescued prophet who prefigured Jesus and forecasted Christ's victory over death. In the Muslim telling, he was an angry man and then a penitent one who excelled in glorifying God.

Let us briefly consider the geography of these three religions.

JUDAISM

Judaism emerges from the early city-states of Mesopotamia. Abraham and his family clan left the city of Ur and set out for Canaan around 3,000 to 2,500 years ago. Urban sophisticates, they left civilization to establish settlements in what was considered wilderness. Abraham eventually came to believe in a monotheistic religion, the foundation of Judaism. The Judaic theology has roots deep in Mesopotamian beliefs and writings. *The Epic of Gilgamesh* is echoed in the story of Genesis, including the idea of a great flood. The Jewish universe was one ruled by an all-powerful God, accessible by prayer and incantation and present in appearance and action. Abraham's faith was tested in the searing experience of his willingness to sacrifice his son. The Jews, who saw themselves as the chosen people of God, were continually tested in their faith. Famine stalked the land and they were taken into slavery in Egypt, to build the great cities and temples of the empire on the Nile. Led by their prophet Moses, they managed to escape, as described in the story of Exodus, and displaced the Canaanites in a brutal campaign described in the Bible as a form of ethnic cleansing. The Jewish kingdom soon become rich because it was strategically located on the routes between major trading regions. Hebrew emerged as the dominant language, and a vast temple complex was constructed in Jerusalem.

And then a series of disasters tested the people's faith in the love of God. The country came under the rule of nearby empires, first the Assyrians and then the Babylonians. The Jews resisted full incorporation into the Babylonian Empire; when they revolted, the rebellion was crushed and, in 586 BCE, the Temple in Jerusalem was destroyed. Many Jews were taken into exile until freed by the Persian king in 538 BCE. The Jews returned and rebuilt the temple. Its rededication, centuries later, is the basis of the Hanukkah celebration. The Romans destroyed the temple complex in 70 CE, and the Jewish state disappeared for close to 2,000 years, until 1948.

The Old Testament Bible, then, is a historical geography of the Middle East, albeit embroidered and tailored for specific purposes. Four distinct sources make up the Old Testament. The J source is the earliest and draws on Mesopotamian and Canaanite sources, the E source was written two centuries later, the D source is a book of laws for more centralized worship at a time when the priests were gaining power, and the P source was written at the same time as the Babylonian exile. Ecclesiastes, written around 400–200 BCE, draws on ancient wisdom literature in which a character offers important life lessons, but also expresses the skepticism and muted joy and mournfulness that perhaps reflects the experience of exile. In keeping with the idea of the Middle East as a cauldron of shared and different cultural traditions, Ecclesiastes also contains Persian words and Greek themes.

After the destruction of the temple in 70 CE, Jews dispersed throughout the Middle East and later Europe and the New World. Centuries later, from around 1940, Jews from across much of eastern and central Europe either left or were forced to leave as Nazi Germany conquered Europe or were killed in the horror of the Holocaust. Jews in the Middle East largely escaped the Holocaust that blighted European Jewry. But their position changed dramatically after the creation of Israel in 1948. Wars between Israel and the surrounding Arab countries made life difficult for Jews of these countries because they were now associated with **Zionism** and Israel. Jews left them for Israel or the West, especially the United States. Outside Israel, the region lost a Jewish presence that had survived for thousands of years.

Judaism is not a universalizing religion like Christianity. It does not seek out converts or promote expansion into other faith communities. Judaism is the religion that connects Jews with their God; it is not promoted as a universal religion open to everyone. It is the religion of a people, not a religion for all the people.

Jews and significant Jewish communities are found in many countries of the world, including the United States, where 36 percent of the world Jewry core population lives, often playing significant roles in local economies and political life. The "Israel lobby," for example, generally taken to mean a commitment to Israel, is an important element in US presidential politics and plays a key role in crucial electoral states such as Florida. However, you can see from Map 11.1 that Judaism is the dominant religion only in Israel, where just over 70 percent of the population of Israel is Jewish. Israel is now the recognized homeland for Jews. In 1897, only 50,000 Jews lived in Palestine and they accounted for less than 1 percent of world Jewry. By 1950 it was 11.6 percent, and now, with 6.5 million Jews in Israel, it is 41 percent. Between 1989 and 1991, 1 million migrants came to Israel from Russia. Since Israeli Jews are younger and have higher birth rates than non-Israeli Jews, by 2050 the majority of all Jews in the world will live in Israel.

CHRISTIANITY

Jesus was a Jew claimed by Christians as the son of God. He grew up and died in the Middle East, a region then situated on the edge of the Roman Empire. His short life and resurrection became the basis for the New Testament. This minor Judaic sect soon spread along the trade routes of the Roman Empire, first between seaports along the Mediterranean and then later inland. Armenia became the first Christian country, around 303 CE. The Romans initially persecuted the Christians and suppressed their religion, but after 313 it was officially recognized, and in 380 it became the official religion of the empire. The Catholic Church embodied the Roman Empire; it is still headquartered in Rome, and its faith and influence spread across the territories of the former Roman Empire. Christians were persecuted because their religious faith undermined civic responsibilities. Yet when the Roman emperor Constantine converted to Christianity in 312 CE, it had the backing of political power and became the dominant religion throughout Europe. It built on the older, more place-based religions, often using their sacred sites as locations for places of Christian worship. Many important cathedrals in Europe were sited on pre-Christian sacred sites.

After the fall of Rome and the passing of the medieval period into the Renaissance, Christianity was exported around the world as part of European colonialism and imperialism. The Spanish took their Catholicism with them when they invaded the New World, thus transforming the cultural landscape of the Americas (see Photo 11.6). The British settler societies of North America and Australia brought their Protestant religion with them, self-consciously believing in its superiority over local religious beliefs and practices. Christianity was one of the more expansionist religions, associated as it was with the powerful colonial forces of the day. In some countries, it was resisted. British control in India and the Middle East did little to change

PHOTO 11.6 St. Joseph's Cathedral in Antigua, Guatemala, first built in 1594.

the religious beliefs of the majority of the colonized people. In others, it was incorporated into complex blends of the old and the new. In parts of South America, the old gods still appear, suitably modified in Catholic worship; in Haiti, Catholic rituals are part of ceremonies that also draw on older African practices brought there by enslaved people. In still others, it was enthusiastically adopted. In late-nineteenth- and early-twentieth-century Korea, Christianity gave meaning and purpose to the lives of ordinary people denied both in a very hierarchical and feudal Confucian system.

ISLAM

Islam emerged from desert oases in an Arabia sandwiched between the competing powers of the Byzantine and Persian Empires. Around 610, a merchant, Muhammad ibn Abdullah, sleeping in a cave near Mount Hira, received a revelation that was written down in what became the Quran, the words spoken to him by God. Muslims accept the Quran as the literal word of God; that is why devout Muslims recite the Quran in Arabic and are particularly sensitive to criticisms of it. In 622, Muhammad, with his followers, left Mecca for Medina; this date is taken as the start of the Muslim calendar, much like the birth of Christ is the starting point of the Christian calendar. A persecuted cult quickly became a religious and military force, first by taking control of Mecca in 630 and then by extending its reach in one of the fastest imperial expansions in the premodern world. Like Christianity, Islam was a universalist doctrine, in which faith, not ethnicity, was important. And like early Christianity it combined secular and religious

power. Whereas in the Christian world there was a steady separation, especially in modernity, between the religious and political spheres—rarely a neat or clean break, and in some cases still a site of struggle and contestation—this distinction did not materialize in Islam, where the religious and the political were conceived as one and the same.

Soon after the prophet's death in 632, an Islamic empire had taken control over former Persian and Byzantine territory; a century later, by 732, Islamic power had swept through the Middle East, across North Africa, and up through Spain and into France. Like all rapid and successful imperial expansions, it was fueled by an economic incentive to gain monopoly control over taxation, goods, and people, as well as by a religious fervor to spread the word of Islam to the entire world. The designation of a region known as the Middle East and North Africa reflects the geography of Muslim expansion from the heartland in Arabia.

As Islam spread from its core region, it was influenced by local traditions. Thus, Islam in Southeast Asia lost some of its desert austerity as it was grafted onto local traditions of animism, Hinduism, and Buddhism. To this day, Islam in Indonesia is more socially moderate than the fundamentalist Islam of Saudi Arabia.

Early on, Muslim places of worship were forbidden to show human figures, which led to artistic displays of Koranic writing and dazzling geometric forms as decoration (Photo 11.7).

11.2 The Geographies of Religious Belief

Map 11.1 is a useful guide to the global distribution of religion. However, it is a broad-brush approach that assumes religious homogeneity subsumed under very wide categories.

Religions change, adapt, and modify as they diffuse from their original hearth area. Religion is spread in a variety of ways. Hierarchical diffusions can involve the conversion of a king, emperor, or major leader who then influences others to convert. Contagious diffusion is often the result of direct proselytizing, while relocation diffusion occurs

PHOTO 11.7 The interior of the Blue Mosque, Istanbul.

through the migration of traders, missionaries, and settlers. All three of these types of diffusion can occur as a religion diffuses, and all three produce complex spatial patterns.

From the sixth century BCE, Buddhism spread along the Silk Routes to Central Asia and China and along the trade routes to southern and Southeast Asia. The geographic spread led to the evolution of different forms as the religion interacted with differing local traditions.

Islam spread throughout Southeast Asia and along the east coast of Africa by Arab traders. In 1135, for example, the Langkasuka Kingdom, situated along the important trade routes of the Malacca Strait, converted to Islam and became the Kedah Sultanate. Eventually, most of the kingdoms and empires along this maritime trade route converted to Islam, and in the early twenty-first century, Islam remains the dominant religion in Malaysia, Indonesia, and the southern Philippines.

There is also the migration of believers. The colonial settlers of North America and Australia, for example, brought their religions with them. Christian churches were built in all the towns and cities. Later, more recent migrants brought their religious beliefs so that many large cities in Australia, Canada, and the United States now contain mosques and temples as well as churches.

The diffusion of religion can take the form of preachers and missionaries spreading the word. An early convert to Buddhism, King Asoka (ca. 268–232 BCE), sent out monks to surrounding areas to spread the message of the Buddha. Sufi priests are reckoned to be a possible source of the diffusion of Islam in what is now Malaysia. Catholic priests brought Roman Catholicism to the New World, British Protestant missionaries sought to convert the people under the rule of the British Empire and young missionaries for the Church of Latter-day Saints still seek to spread their faith across the world.

There is also the more informal diffusion as religious ideas spread and diffuse through travel, social media, and the Internet. These ideas range from the more formal religions to smaller cults and sects. We should be careful with words, however, because many people who see their beliefs in a positive way describe them as a religion, while seeing others as a sect or cult, with all the negative connotations that these words often imply.

Religions transform and change as they spread, adapting to and incorporating local existing belief systems. We can gain some idea of this by looking at the sites of religious worship. Islam spread across Eurasia and Africa from its heart in the Middle East. As one of the later monotheistic religions, it encountered a rich mosaic of existing beliefs and practices, some of which were incorporated. Mosques, for example, were often constructed using existing traditions. Photos 11.8 and 11.9 show the Great Mosque of Xi'an in China and the adobe Djenne Mosque in Mali. Both are Muslim mosques but they employ regional designs and use local building materials. The Great Mosque of Xi'an, built in 742 CE and constructed in its present form in 1384 CE, employed traditional Chinese architectural forms. The Djenne Mosque, built of mud in the thirteenth century, is designed in the regional Sudano-Sahelian style. These two mosques provide vivid witness to how religions adapt and adopt as they diffuse spatially.

The spread of religion is intimately associated with the distribution of power. The Arab empire that emerged after the death of Muhammad was able to use its power to spread Islam. Colonial empires, such as the Spanish and British, promoted specific religious beliefs while marginalizing others. Animist religions, for example, were long regarded by imperial powers and settler societies as idolatrous and were inhibited or banned. Specific religious beliefs are sometimes adopted by the state as the official state religion, while others are prevented from flourishing. The United States is one of the few countries where there is no official state religion. Elsewhere, religion is often intimately connected to state power, and the state enforces and embodies its rule through regulating religious practices.

PHOTO 11.9 Djenne Mosque, Mali.

PHOTO 11.8 Minaret in Great Mosque of Xi'an, China.

We can see something of the complexity of religious beliefs in one example, the case of the island of Bali in Indonesia. Indonesia is one of the largest Muslim countries in the world, but the dominant religion on Bali is Hinduism. This ancient religion reached the islands of Indonesia over 2,000 years ago when Hindu and later Buddhist kingdoms were established across the archipelago. Later, Islamic sultanates expanded from their coastal enclaves to gain control. Hindus left Java and other nearby islands to escape Muslim rule, and some moved to Bali. Hinduism was grafted onto animist religions and local belief traditions to become the dominant religion for centuries. When Indonesia gained independence from the Dutch in 1947, the Islamic-based government made it illegal to practice a religion that was not monotheistic. To meet the new constitutional requirement, Bali Hinduism developed into a complex amalgam of ancient animism, core Hinduism beliefs, and adaption to the required monotheistic requirement. Bali Hindi is also deeply associated with the wonderfully expressive culture of Bali, rich in song and dance. Across the island, Hindu shrines dot the landscape (Photo 11.10).

PHOTO 11.10 Hindu shrine in Bali.

The Sunni–Shia Split in the Middle East

If we change the scale from the global to the world regional and more national level, then a different picture emerges and we can see the many varieties of Christianity, Buddhism, and Islam. Sharpening our focus from the global to world regional allows us to see marked differences even within the same religion. Take the case of Islam in the Middle East. It is divided into two main sects, Sunni and Shia. Both share the basic tenets of Islamic teaching: a belief in Muhammad as the messenger of God, daily prayers, pilgrimage to Mecca, alms, and fasting during Ramadan. Their argument lies in different interpretations of the succession after the death of the Prophet Muhammad. The Shia believe that the succession should be hereditary and follow family bloodlines. The first caliphs after Muhammad were Sunni choices from outside his immediate family. The fourth was Ali, Muhammad's son in law, and thus more acceptable to Shiites. When he was killed at a battle in 661, a rift developed between the two branches and was further widened when Ali's son, Hussein, was killed at the battle of Karbala in 680 CE, or 61 in the Islamic counting. It is now a Shiite sacred site.

Sunni Islam is the dominant form, with more than 1.3 billion followers. Shiites are a minority, with less than 250 million, concentrated in Iran, southern Iraq, southern Lebanon, and eastern Saudi Arabia and neighboring Bahrain (Map 11.2). The split has grown in recent years, embodied in two major and competing powers in the region, Saudi Arabia and Iran. Saudi Arabia is one of the richest Sunni Arab countries, where fundamentalist teachings see Shiites as non-Muslims; and Iran has been the only officially Shiite state since the Islamic Revolution in 1979. There is conflict between the two countries, and their proxies in Iraq, Syria, Lebanon, and Yemen, for geopolitical dominance, especially in weak, dysfunctional, or collapsing states such as Yemen and Syria that provide space to project national interests. The rise of a more fundamentalist Sunni Islam across the Middle East has endangered minority Shia communities in countries such as Pakistan and Afghanistan. The Middle East remains a contested space between the two branches of Islam, their primary states, their proxies, and their terrorist offshoots.

The Case of the United States

Religious beliefs and practices are also expressed spatially at a national level. Map 11.3 plots the majority church affiliation in every county in the United States. Distinct

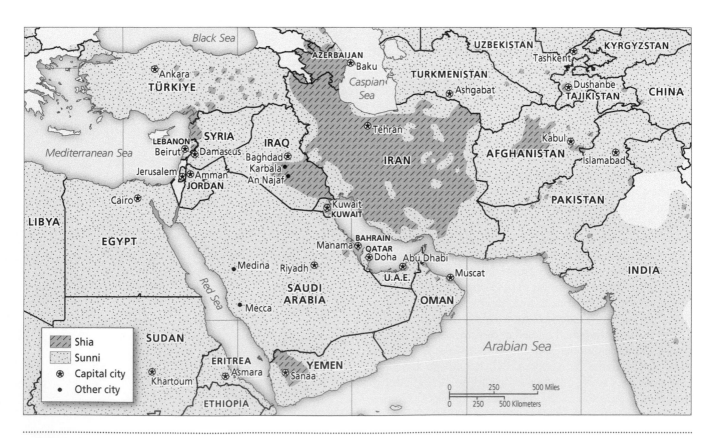

MAP 11.2 The Sunni–Shia divide in the Middle East.

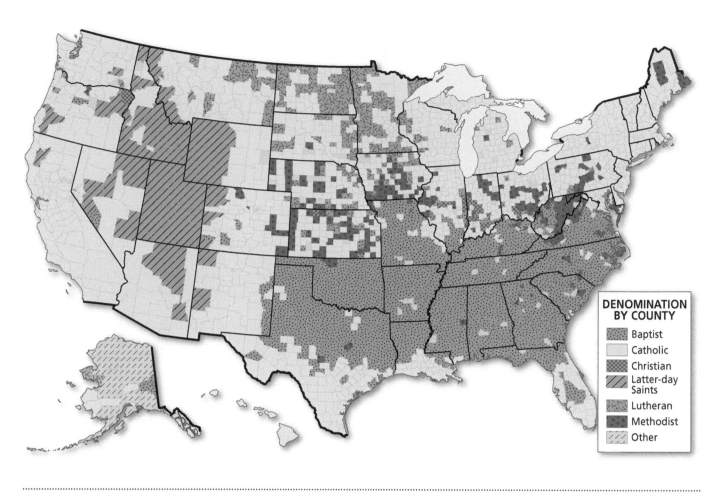

MAP 11.3 Majority religious affiliation in every county in the United States.

DENOMINATION BY COUNTY

- Baptist
- Catholic
- Christian
- Latter-day Saints
- Lutheran
- Methodist
- Other

regional groupings are evident: Mormons in the mountain West, grouped around Salt Lake City; a Baptist heartland region south of the **Mason–Dixon Line**, the so-called **Bible Belt**; and a concentration of Lutherans in the upper Midwest. In part, these regional differences reflect the settlement patterns, as Mormons trekked west to establish their way of life far from mainstream America, the Scots-Irish poured over the Appalachians in the eighteenth century, and Germans and Scandinavians later settled in the upper Midwest. There are also small outliers in select rural counties such as the Mennonites. Different groups coming at different times brought their religions with them. These religious affiliations have had wider impacts on the tone and practices of local communities in these regions. Using a slightly different religious classification, geographer John Bauer looked at regional changes in religious affiliation from 1980 to 2000. While there was great stability, there were also some marked changes as the strength of the Catholic region deepened and extended in the West, a trend associated with the increasing size of the Hispanic population.

The Place of Religion

Traditional maps that plot the national distribution of the major religions are now inadequate to convey the complexity of religious devotion. When there are mosques in Detroit and ashrams in Memphis, the spatial geography of religion is more complex than can be summarized in simple maps that shade in entire areas with a single category of religious belief.

At the regional and local levels there are profound differences in religious affiliation and the degree of churchgoing. In the United States we can make a distinction between the Bible Belt, where fewer than 10 percent consider themselves irreligious, and New England and the Pacific Northwest, with 30 percent. The Bible Belt is not just an area of Protestant churches but also an area of an explicit religiosity compared to the rest of the country, often combining social and political views in a social conservatism that impacts political affiliations. The context of place is important in influencing religious practices. In more religious communities such as the Bible Belt, social pressures may make it more difficult not to go to church, whereas in the more irreligious Pacific Northwest, going to church, since it is not a widely practiced ritual, may be less frequent even for people of faith.

At the very local level, religion often reflects and shapes local communities. Local churches, for example, are important community resources, providing sites of worship and witness, public spaces for activism, and places of devotion and celebration. In contemporary American Evangelicalism, the suburban megachurches also offer a space with meaning amid the dispersed character of suburban sprawl.

And it is often at the very local level, where different religious groups share the same public space, that both ecumenical cooperation and religious conflict can occur. In recent years Muslims in Europe and in some parts of the United States have felt threatened by a rise in anti-Muslim rhetoric. Controversies over building mosques and debates about the proper place for religious observances, such as dress codes, are all highlighted in local places shared by different religious practices.

Religion and Urban Public Space

One persistent theme is that religion declines with modernization. As societies become modern, more connected to the rationalities of science and technology, magic loses its hold, spirituality is replaced by materiality, and religion loses its appeal. Matthew Arnold's 1867 poem *Dover Beach* suggested that the retreat of religion and faith left us abandoned on

> *A darkling plain*
> *Swept with confused alarms of struggle and flight*
> *Where ignorant armies clash by night*

The city as the embodiment of the modern and the industrial and mechanical is often depicted as the locus for the breakdown of religion and faith. In fact, this trajectory is rarely so simple. We can consider three examples where faith within cities remains strong.

Tokyo is a very modern city, having been built and rebuilt twice in the twentieth century after earthquake damage and war bombing. Most Japanese report that though they think religious belief is important, their formal religious commitment is slight. Yet in this secular, modern city there are thousands of religious sites, from major temples to tiny shrines. The more popular ones are sites of festivals, holidays, and ritual occasions. Premodern religious sites are integrated into the urban landscape; Buddhist sites are associated with funereal rites, and Shinto shrines, drawn as they are from premodern harvest festivals, affirm life in weddings and other celebrations. The sacred spaces embody the tension between the continuing strength of ritual practice and the weakness of commitment to religion.

In Bangkok, Thailand, the many markets are filled with sacred sites, including shrines and spiritual markers (Photo 11.11). These religious places continue to exist, not despite the growing modernization and commercialization of society but because of it. The sacred space of the vernacular is transformed into more contemporary prosperity religions that align traditional Buddhism to the pursuit of individual material gain. Just as the cityscape is transformed by modernization, so the city's sacred geography is reimagined. Bangkok is the rule rather than the exception, as the religious and the secular coexist in modernizing cities around the world.

Orthodox Christianity was the state religion of Ethiopia until 1975, when a religious plurality was introduced. Now in the main city of Addis Ababa, urban public space is a contested site for interreligious rivalry and competition between Orthodox Christians, Protestants, and Muslims. Many taxicabs display overtly religious messages. Religious festivals provide the opportunity for **faith communities** to display their strength. During Epiphany, the Orthodox Christians display banners of green, yellow, and red, the colors of the Ethiopian national flag, across the city. Posters of saints adorn the streets and public

PHOTO 11.11 Shrine in downtown Bangkok.

squares, and young men wear T-shirts with the slogan "Ethiopia, A Christian Land." The Muslim faithful take to the streets in large numbers during their two annual holidays of Eid al-Adha and Eid al-Fitr to show their strength and commitment. And Protestants mount loudspeakers on vans and patrol the streets. Urban public space in the city is not a secular place, but the arena for displays of religious beliefs.

11.3 The Religious Organization of Space

Religion and the Environment

The very first religions were animist, firmly based in place, with sacredness located in specific sites. Some of these belief systems grew into more universal religions that effectively lifted the religious gaze above particular peoples and places to an all-seeing, all-knowing, single God on high. The transition was a gradual grafting rather than a sharp break. Most of the major cathedrals in Europe were constructed on pre-Christian sacred sites, and the key Christian festivals of Christmas and Easter are built on the observances of pre-Christian sun worshippers, who celebrated the winter solstice, and the festival rituals of pre-Christian agriculturalists who celebrated the start of the growing season. The cult of the Virgin Mary developed from pagan beliefs in the Earth Mother.

Many of the myths and stories of the animist and sacred books refer to environmental changes. The recurrences of flood myths are less fantasies than references to the melting of the ice sheets and subsequent flooding at the end of the Pleistocene. The flooding of the Black Sea around 7,500 years ago may be the basis for the story of Noah. Many of the stories of early religions are imaginative and elaborate retellings of actual environmental change and collapse.

The monotheistic religions of Judaism, Christianity, and Islam depict a sky-centered, universal God looking down over the entire world. The brutal defeat of the Canaanites by the Israelites, mentioned at various points in the Bible, is one of the first descriptions we have of the destruction of an Earth-based **polytheism** by a monotheistic religion. The Israelites' basis for this holy war was the fundamentalist religious doctrine that people should worship only one God. The Canaanites were depicted as corrupt and evil, and their fertility-based polytheism was replaced by the stringent demands of adherence to a single universal God. The war between the Canaanites and the Israelites played out again and again across time and over space as monotheistic religions displaced the more local, place-based belief systems,

whether it was in the destruction of local religious traditions in South America by the Catholic Church or the marginalization of Indigenous Native American beliefs by Christian missionaries in North America. Across the world, it is a similar tale of sky-based monotheistic religions displacing more Earth-based polytheistic beliefs. Barbara Ehrenreich makes a convincing case that many religions began more as festive events, ceremonies filled with music and dance, what she terms **ecstatic worship**. As the religions became connected with establishment power, the authorities sought to control and reduce these forms of ecstatic worship. Christian worship, for instance, was made more formal over the medieval period. When Christian missionaries arrived in the wake of imperialism and colonization, they similarly devalued the ecstatic forms of worship that they encountered.

Organized religions, with their gods and priests and rituals to make sense of the mysteries of life, also grew out of the need to justify and legitimize the social hierarchy in the early city-states. When these empires expanded and came into contact with one another, creative and destructive cultural collision ensued. Karen Armstrong updates the idea of a pivotal **axial age** that lasted from 900 to 200 BCE, during which the great religious traditions of the world came into being across the globe—Confucianism and Daoism in China, Hinduism and Buddhism in India, monotheism in Israel, and philosophical rationalism in Greece. Their universal messages of empathy and compassion grew out of the collective experience of living in cities as well as their awareness of people in other cities and regions.

In recent years there has been a renewal of interest in the idea of sacred space as part of modern environmentalism. Such environmentalism draws on the notion of sacred space to promote environmental protection and a sacred duty to protect and save the environmental bounty that the Creator provided. There is now a significant faith-based approach to environmental issues.

Sacred Sites

In place-based animist beliefs, the landscape was filled with spiritual significance. In the more universal religions, there is a more marked categorization of sacred and profane space. Most of the world was profane, a word whose etymological meaning is "in front of (i.e., outside) the temple." A small number of places were considered sacred sites: birthplaces of prophets, scenes of revelations, centers of religious authority and worship—places where the veil between the material and spiritual world was only lightly drawn. Jerusalem, for example, is a sacred site for the three religions of the Book: Christianity, Judaism, and Islam. The sharing of this sacred space by different religions is part of the political tension in the city (see Box 11.3). Sacred sites can be places of retreat as well as sites of engagement.

11.3 Jerusalem

Jerusalem has a long history. With a freshwater spring making it an oasis in a dry land and Mount Moriah forming a high point on the landscape, it became a sacred site for early peoples and religions. To this day it remains at the very heart of three of the world's great religions (see Map 11.4).

Jerusalem figured greatly in early Jewish tradition when Abraham took his son to Mount Moriah to sacrifice him. Under King David it became the political and religious center of a Jewish state. He chose the site because of its sacred appeal, but also for geopolitical reasons: it was centrally located among the twelve tribes of Israel but not directly linked to any one of them. King Solomon built a great temple around 1000 BCE. It was destroyed by the Babylonian Empire in 587/586 BCE, whose forces not only torched the city and temple but also carried off its wealthy citizens and many of its skilled craftsmen. Most of the Jewish elite was transported to Babylon.

The Jews had more luck with the next emerging empire, the Persians, whose great king Cyrus transported them back to Jerusalem and even gave them funds to rebuild the temple around 520 BCE (something to remember as the Israel–Iran conflict in the present era deepens). Jewish religious writings after this event are imbued with some of the beliefs of the Persian state religion of Zoroastrianism and its cosmology of a world divided into forces of good and evil.

The temple was refurbished and expanded, and Jerusalem grew, drawing pilgrims from all over Judea. The city came under the rule of the rising empires of the region, the Greeks under Alexander and then the Romans. When the Jews revolted against Roman rule, the city was demolished in 70 CE, the second temple was destroyed, and the survivors were exiled or sold into slavery. The Roman emperor Hadrian rebuilt the city but banned Jews.

Jerusalem also played a huge role in the Christian tradition as the place where Jesus was tried, crucified, and resurrected. When it came under Byzantine control around 313 CE, the Christian element was emphasized; the Church of the Holy Sepulchre was constructed, and the city became a place of Christian pilgrimage. In the medieval maps of Europe Jerusalem is often depicted as the center of the world.

The city plays an important role in Muslim theology, too. In 638 CE it came under Muslim control, and in 691 Abd al-Malik built the Dome of the Rock, one of the earliest mosques, on Mount Moriah. Jerusalem is considered the third most holy site after Mecca and Medina, and for a while devout Muslims prayed in the direction of Jerusalem before the orientation was shifted to Mecca.

The city went through a cycle of destruction and rebuilding as different Muslim dynasties changed their attitude to churches and synagogues. Apart from a brief period of Crusader rule, the city came under centuries of Muslim rule and was then part of the Ottoman Empire from 1517 to 1917. In 1538, Suleiman the Magnificent rebuilt the city walls, still extant, that define the Old City, divided traditionally into four quarters (Latin Christian, Armenian Christian, Jewish, and Muslim) and containing the Temple Mount, also known as the Noble Sanctuary.

In 1947, the UN voted to create a Jewish and Palestinian state in Israel. The State of Israel was declared in 1948, but

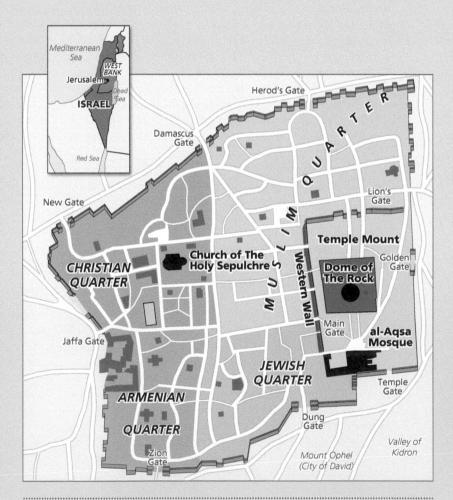

MAP 11.4 Sacred sites in Jerusalem.

Jerusalem sits on the border between Israel and Jordan. West Jerusalem was the capital of Israel, while the east was part of Jordan. During the Six-Day War of 1967, Israel captured the Old City and east Jerusalem came under Israeli control. The city has become more distinctly Jewish as settlers moved into east Jerusalem and there is a marked segregation in housing and use of public services, with separate Arab and Jewish bus routes.

Jerusalem is not just a place of streets and buildings; it is also a place of imaginings, longings, hopes, and fears. For Jews it is the place where Heaven and Earth meet, a site of pilgrimage, the center of the universe. For the Christians the city is also a place of pilgrimage because it is intimately associated with the life, death, and resurrection of Jesus Christ—a city made holy. The city plays a central role in the sacred geography of Muslim belief because it ranks only just below Mecca and Medina as a place of pilgrimage and just after Mecca in religious significance.

These sacred geographies superimpose and interconnect in the tight quarters of the Old City, a designated World Heritage Site. On the thirty-six-acre site a platform, known as the Temple Mount or Noble Sanctuary, is the site of the Second Jewish Temple; the Muslim Dome of the Rock, modeled in part after the Christian Church of the Holy Sepulchre to commemorate Muhammad's ascension to Heaven; and the Al-Aqsa Mosque, constructed in 705 on part of the old Jewish temple. At the edge of the platform is the Western Wall, also known as the Wailing Wall, all that remains of the second temple and one of the holiest sites for observant Jews.

Jerusalem is a holy place for three religions that have enjoyed periods of tolerance as well as interaction marked by conflict, strife, and outright hostility. The city remains at the center of conflict between Palestinians and the Israeli state.

Different sects of even the same religion have their own sacred sites. Catholics visit the shrine at Lourdes in search of cures for illness, but Protestants do not. The profane treatment of sacred sites can become a source of friction. In 1984, Indian troops attacked militants hiding out in the Golden Temple in Amritsar, one of the holiest places for Sikhs. The damage incensed the Sikh community and led to sectarian violence and the assassination of the prime minister who ordered the attack. Sectarian rivalry is expressed most vividly in the destruction of these special sites. In 2004, when terrorists exploded a bomb near the golden-domed mosque of Karbala in Iraq, they desecrated one of the most revered sites for Shia Muslims. The explosion triggered a bitter round of sectarian violence in Iraq between Shia and Sunni.

Pilgrimages

Some sacred sites become places of pilgrimage. Catholics in medieval Europe made pilgrimages to the tomb of St. James in Santiago de Compostela in Spain. Pilgrims continue to follow these trails across the Pyrenees and through northern Spain.

Part of the appeal of the pilgrimage is the search for spiritual revelation through physical endeavor. Pilgrimage involves a journey, a move into greater spiritual awareness through physical exertion. In many societies we have retained the sense of the efficacy of physical exertions in the gym and health club, but disconnected it from the spiritual dimension.

The world's largest pilgrimage is the hajj. One of the principal duties of every devout Muslim is to make a pilgrimage to Mecca at least once in their life. Millions converge on the city each year. They circle the Kaaba, a giant rock, seven times counterclockwise, walk four miles to Mina, and then a further eight miles to Mount Arafat, where they pray and then collect pebbles. The next day they walk back to Mina, where they throw their pebbles at the pillars that represent the devil, then return to circle the Kaaba seven more times. Because of increased accessibility and cheaper air travel, many more Muslims can now visit the city and make the pilgrimage. Crowds are huge, numbering close to 2 million at any one time and presenting major problems resulting from the large number of people in a restricted area over just a six-day period. During the pilgrimage of 2015, more than 760 people were killed and almost 1,000 were injured when crowd-control measures failed to stop a stampede after crowds coming from different directions collided.

Religious and Secular Space

There is a spatial variation in the importance of religion in people's lives. When people across the world were asked in 2016 what was the most important criterion of self-identity, 43 percent of those interviewed in Pakistan said religion. This contrasted with only 4 percent in the United Kingdom; 5 percent in France, Germany, and Spain; and 6 percent in Canada. Among the richest countries of the world, the United States had the highest percentage, 15 percent. And these differences have impacts. An explicitly Christian politician has a hard time being elected prime minister in the United Kingdom, while a committed atheist would have a similarly difficult time in the United States. But notice that only 15 percent of the people interviewed in the United States cited religion as the main defining criterion. So even in a nation noted for its religiosity, the large majority of people use other criteria such as national citizenship, being a world citizen, local community, race, or culture to self-identify.

Societies and local places vary in their degree of religious tolerance, from fundamentalist societies that allow only one form of devotion to be openly displayed to more multifaith societies that show more tolerance of religious diversity. But even in more tolerant societies a tension exists between freedom to practice in private and limitations in the public sphere. All societies regulate religious practices in the public sphere, some more than others. In a number of European countries, restrictions are now being placed on the building of mosques and the public displays of Muslim religious practices. In Egypt, Muslim men can marry Christian women but Christian men cannot marry Muslim women unless they convert to Islam.

In societies dominated by one religion, there is a conjoining of public and private space, a centrality of religious space in social and spatial organization. In medieval Europe, the tallest and most central buildings were churches and cathedrals. They towered over the surrounding city as a constant symbol of both religious expression and the central importance of religiosity in organizing human affairs. Myanmar remains a predominantly Buddhist society, and in its capital city of Yangon the most conspicuous building, set high on a hill and easily visible from all parts of the inner city, is the Shwedagon Pagoda, the most sacred site in the country. It was constructed over 1,500 years ago and rebuilt numerous times after earthquakes. It is the defining landmark of the city and the center of Buddhist sacred geography in the entire country. Its centrality also attracts political demonstrations. In 2007, Buddhist monks critical of the junta gathered in the temple as a symbol of resistance to military rule.

The rise of a secular public space is most evident in the founding of the United States, where the constitutional separation of church and state in effect means an explicit secularization of public space. In America and elsewhere, this division is still an arena for debate and conflict. Some fundamentalists call for the religious control of public spaces, such as Orthodox Jews in Israel asking for separate buses for men and women in public transport. Others, including avowed secularists, demand a more complete erasure of religion from public space, such as calls to ban religious dress in French public schools. The precise separation and demarcation of public and private space and how it maps religion and religious observance can be divisive issues in multicultural, multifaith societies.

Spatial Organization of Religion

There are also the territorial administrative units of religion. The Catholic Church, with its center in Rome, is divided into archdioceses, dioceses, and parishes. In more multicultural societies, the spatial organizing role of religion may be more opaque. An *eruv* is a Jewish ritual enclosure that marks out a space in which Jews observing the Sabbath can carry things and move between public and private space without transgressing Orthodox beliefs. In the Upper West Side of Manhattan, an area of 200 city blocks was separated using existing walls and fences and specially made cord and wire. Map 11.5 shows the map of another eruv in New Rochelle, New York, centered around the Congregation Anshe Sholom. An eruv is a silent presence, a marker for only some members of the community.

Religious Conflict

Religious conflicts occur for many reasons, but space and place are often important sites and reasons for religious conflicts.

Conflict can occur on the borders of faith communities. Map 11.6 shows the gradual nature of the shift from Muslim to non-Muslim from north to central Africa. The liminal division is most obvious in nations that straddle this line, such as the Central African Republic and Nigeria. The Central African Republic sits at the edge of the division with Chad and Sudan to the north and Congo to the south. Almost 80 percent of its 4.7 million people are Christian and the remainder are Muslim. Christians are mainly sedentary farmers inhabiting the south, while the more nomadic Muslims live in the north of the country. A civil war in 2013 was fought along this religious division, overlaid with struggles over land and political power. Muslims in the south of the country moved to its north or overseas to escape violence. The Muslim population of Bangui, the republic's capital, fell from 138,000 to 900. More than 1 million people were displaced by the civil unrest and violence.

There is also conflict when different religious groups share the same national space. Even in predominantly Islamic Pakistan there are Christians who officially make up 1.5 percent of the country's 180 million. Other estimates suggest closer to 5 percent. As the country has become more rigidly Islamic, the position of Christians has become more problematic. In 2013, more than 100 were killed by a suicide bomb attack in a Christian church. Christians are at the bottom of the socioeconomic hierarchy, living in slums and forced to work in menial occupations. One million live in Karachi, Pakistan's largest city. One of the largest Christian crosses in the world, 140 feet tall, was erected outside the entrance to the Christian cemetery in the city, where gravestones are often vandalized. The cross not only signifies belief but also is a symbol of hope for a beleaguered minority.

Conflict is heightened when religious differences are compounded by language and ethnic differences, such as in the conflict in Sri Lanka between Hindu Tamils and Buddhist Sinhalese. The Tamils live in southeast India and northeast Sri Lanka. After Sri Lanka gained independence in 1948, the Tamils faced discrimination, and in 1983 a militant Tamil group, the Tamil Tigers, attacked a military base.

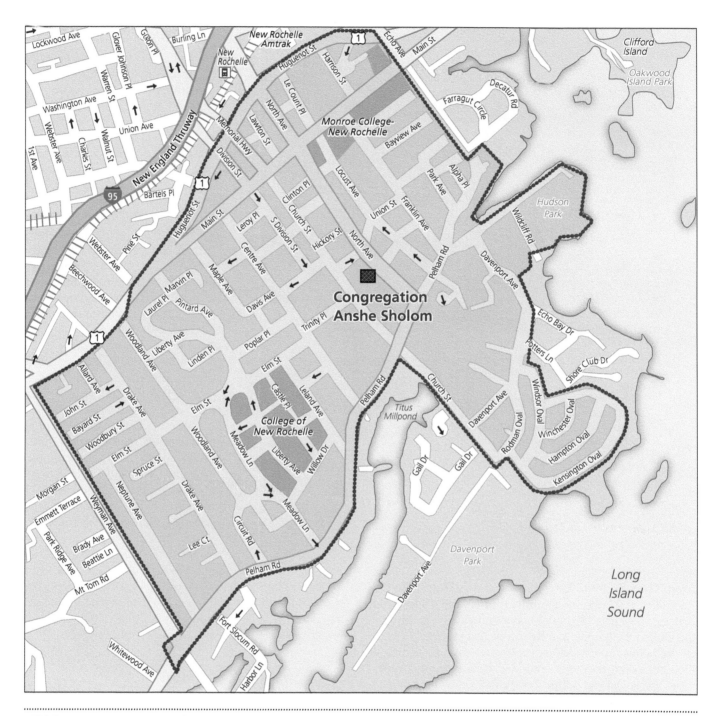

MAP 11.5 An eruv in New Rochelle, New York.

The resultant civil war was a brutal affair with war crimes on both sides, 100,000 casualties, and hundreds of thousands displaced as the war zone moved across the country. Worst hit was the Tamil area in the north. The civil war officially ended only in 2009 with the final defeat of the Tamil Tigers by the Sri Lankan military. In the very last stages of the war, 40,000 civilians died and 300,000 more were displaced and moved into camps.

Conflict can also occur between different sects of the same religion. In Pakistan, terrorists from the country's majority Sunni Muslim sect (80–90 percent of the population) have waged a campaign against the Shia minority (10–20 percent). Between 1987 and 2007, as many as 4,000 people were killed as Shia were targeted by Sunni terrorist groups.

In Northern Ireland, a long-running conflict between Protestants and Catholics known as the Troubles lasted

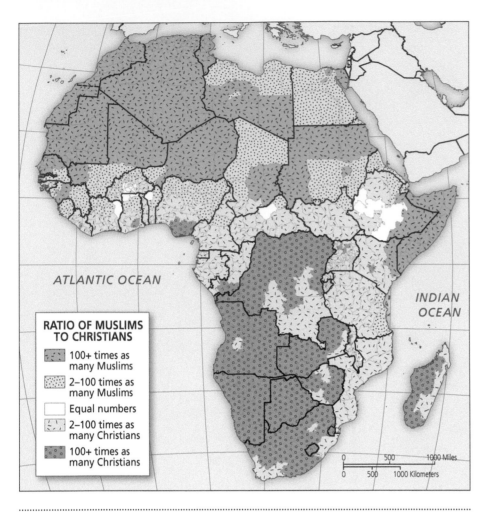

MAP 11.6 Muslims and Christians in Africa.

RATIO OF MUSLIMS TO CHRISTIANS

- 100+ times as many Muslims
- 2–100 times as many Muslims
- Equal numbers
- 2–100 times as many Christians
- 100+ times as many Christians

from 1960 to 1996. Over 3,500 people were killed by Republicans (Catholics committed to breaking away from Britain), Loyalists (Protestants committed to remaining in the United Kingdom), and British security forces.

Religious differences are rarely the only source of conflict, but when they are combined with economic differences and social and political marginalization, then religion can become the source of identity and mobilization. Religious differences are not monolithic, unchanging categories. Conflicts are not fixed in time or space, but emerge, rise, and fall in connection with internal and external forces.

11.4 Religion and Society

Religion is as much a social phenomenon as a spiritual one. Its practices and beliefs connect with wider social issues. Here we will consider just three: identity, globalization, and gender.

Identity

Religious belief is not a singular form of identity, but is also folded up with other forms of identity that have a spatial component. In Northern Ireland, for example, to be Catholic is often associated with those who support a connection with Eire rather than the United Kingdom, while to be Protestant is bound up with a commitment of being part of the United Kingdom.

Sometimes religion becomes part of national identity. Although Poland had a significant Jewish population before the Holocaust, it was primarily a Catholic country when the Communists controlled it from 1945 to 1989. Part of Polish resistance to the dominance of the USSR was in and through the practice of Catholicism. Being Catholic was part of Polish nationalism and an act of resistance to Soviet authority.

In some cases, national identity is a contested site between competing religions. India, for example, has one of the world's largest Muslim populations: officially 177 million, a figure exceeded only by Indonesia (204 million) and Pakistan (178 million). Yet this Muslim minority is threatened by the rise of a Hindu nationalism that tends to imagine India as a Hindu society. Conflicts between the two communities continue to fracture the society at national and local levels, especially because India's main adversary, Pakistan, proclaims itself as an Islamic republic.

Globalization and Religion

Religious beliefs are part of cultural identity—for some people, the principal source of their identity. So, when they feel their beliefs are under attack or threat, from secularism or other religions, the political backlash can be profound. We are in an interesting time with powerful competing forces. On the one hand, we are in the middle of a major wave of cultural globalization and the expression of religious diversity in shared national spaces; on the other hand, we are also witnessing a rise of religious fundamentalism, whether in an Islam that promotes sharia law based on the Koran, a Hinduism strongly associated with Indian nationalism, strict Judaism that sees occupation of the West Bank as part of God's covenant, or Christians who

11.4 Sustainability and Historic Preservation

Since the 1960s, many countries have established historic preservation laws that authorize funds for preservation activities, protecting landmarks and a variety of historically and architecturally significant buildings, sites, structures, districts, and objects.

Historic preservation is an important part of sustainability. Historic structures and buildings contribute to creating a community's sense of place and make an area special and unique. They are a legacy from the past that offer a source of inspiration. Religious buildings such as temples, mosques, shrines, and churches are especially valued.

Historic preservation is at the core of the United Nations' World Heritage Program. World Heritage Sites designated by the United Nations Educational, Scientific, and Cultural Organization (UNESCO) are considered to be of outstanding value to humanity. Sites may be designated natural (for example, Yellowstone Park) or cultural (Stonehenge in the United Kingdom) or a combination of both. Examples of cultural World Heritage Sites include ancient ruins, buildings, or historical structures.

By 2023, there were 1,157 sites, 900 of them cultural sites, 218 natural sites and 39 mixed sites spread across 167 countries.

Many World Heritage Sites are religious buildings, such as churches, temples, shrines, and other sacred spaces. These include Buddhist sites such as Angkor Wat in Cambodia, Islam's Blue Mosque in Istanbul (Photo 11.7), and the Cathedral of Notre-Dame in France. The old historic area of Antigua in Guatemala is a designated World Heritage Site composed of several buildings, including St. Joseph's Cathedral (Photo 11.6), which is often mentioned as an excellent example of Spanish colonial architecture.

A country that is home to a World Heritage Site is charged with preserving and protecting it for all time. However, uncontrolled urban development, unsustainable tourism, pollution and climate change, armed conflict, and a lack of political interest in investing in conservation and protection measures are increasing threats to many World Heritage Sites.

A full list of all UNESCO sites is available at https://whc.unesco.org/en/list/.

---

argue for the Ten Commandments as the guiding light of civil law. Geographer Roger Stump provides a geographical perspective on religious fundamentalisms with reference to the control of sacred space—the attempt to conjoin public and private space with religious structures and the association between territory and identity.

Religious belief provides many with a simple message in a complex world, an articulation of fixed primary identity when cultural change is everywhere, a platform and vehicle for political mobilization, and an anchor of meaning and relevance in a society of ever-accelerating transformation. Yet these permutations are also important generators of religious fundamentalism. At a more global scale, we are now in an era of both religious fundamentalism and religious diversity. The recent wave of globalization has spread and diffused different religions around the world.

Globalization also involves the movement of people and ideas, including religious beliefs and practices, around the world. Religions are diffused across the globe but become reterritorialized in specific sites. Diasporas are often constructed and maintained in and through church gatherings and religious festivals. For example, Filipino domestic workers in Italian cities are largely invisible in the public arena, but each year in the city of Padua, as well as in Berlin, Germany, and New Orleans, Louisiana, they celebrate their own festival in the streets. Filipino identity is performed by the annual religious celebration known as Santacruzan. By appropriating public space through a religious ritual, Filipino immigrants make themselves visible to reconfirm their

identity and renew their community. Likewise, immigrants from the Democratic Republic of Congo connect in their own Pentecostal and Catholic churches in Atlanta and London. The religious landscapes of both cities, altered by Congolese immigrants and their religious practices, are shaped by local circumstances. David Garbin describes the spatial appropriation and re-enchantment of the urban landscape as the Congolese often situate their churches at the margins, in former industrial areas and warehouse districts.

Diasporas are also recast as religious identities if the religious practices of immigrants become their defining feature, either in their own eyes or in those of the host society. When Turks living in Germany practice Islam, they are seen—and start to see themselves—primary as Muslims rather than as Turks or Turkish Germans. They become less Turkish and more Muslim. Distinctly "Muslim" spaces are created by immigrants from many different counties. In British cities, immigrants from a variety of Muslim nations such as Algeria, Afghanistan, Bosnia, Tunisia, and Iraq find collective religious homes and identity in the mosque and the madrasa, their shared religious identity now trumping their different national origins.

Social Differences: Gender

Religion embodies, defines, and reflects social differences. Religion can play an enormous role in differentiating men and women, because the social differences are often codified in religious beliefs and practices. The marginalization of women in religion has faced little scrutiny.

Many religions promote segregation between men and women and institutionalize gender differences in leadership roles, ordination, spatial behaviors, dress, and comportment (Photo 11.12). The Catholic Church, for example, does not allow women to be priests. In effect, men run the church and women play a secondary role. Men fill all the positions of real power, from pope, cardinal, and archbishop to bishop and priest. This is not limited to Catholicism. Around the world, most religious positions of authority are held predominantly by men. Yet the Pew Research Center found that worldwide, 97 million more women than men claim a religious affiliation, making their absence at the highest levels of decision-making in religious communities even more stark and inequitable.

In traditional Islam, women's lives are shaped by religious strictures on dress and public behavior, codified in states such as Saudi Arabia where women must remain veiled in public and are not allowed to drive.

PHOTO 11.12 Women in Istanbul, Turkey.

Many feminists decry formal religions as patriarchal structures that can lead to the oppression of women. The monotheistic religions do have a patriarchal character, often assigning women restricted social and religious roles, defining their functions for a private rather than public space.

Gender is perhaps the most obvious example of how religious beliefs and assigned spatial behaviors reinforce gendered social differences. However, a growing number of feminist faith activists are leading the way in re-examining religious teachings and laws and advocating for women's rightful participation in the interpretation of religious doctrine. If we are working against gender discrimination in every other arena, why do we accept it in religion?

Witness to Power

As social institutions, religions embody and reinforce, yet sometimes overturn and question, social hierarchies, gender differences, and socioeconomic power relations.

The antislavery movement in the United Kingdom and the United States was in large part driven by Christians appalled at slavery and its exploitation of fellow humans. Abolitionists felt compelled by their religious beliefs to work to end the slave trade. The more recent civil rights movement in the United States was in large measure advanced by religious groups, especially by African American religious institutions and leaders who felt that racial discrimination was evil. Religion has reinforced existing power arrangements but has also been a vehicle for undermining, contesting, questioning, and sharply criticizing authority.

From Buddhist monks in Myanmar protesting against the military junta to Catholic priests protecting the poor in the slums of South America and Muslim feminists in Iran fighting for women's rights, religious believers are confronting inequalities and using their faith as a guide and justification for profound social criticism.

Cited References

Armstrong, K. 2006. *The Great Transformation: The Beginning of Our Religious Traditions.* New York: Knopf.

Bauer, J. T. 2012. "U.S. Religious Regions Revisited." *The Professional Geographer* 64:521–539.

Ehrenreich, B. 2006. *Dancing in the Streets.* New York: Metropolitan.

Garbin, D. 2014. "Regrounding the Sacred: Transnational Religion, Place Making and the Politics of Diaspora among the Congolese in London and Atlanta." *Global Networks* 14: 363–383.

Gill, S. 1987. *Mother Earth: An American Story.* London: University of Chicago Press.

Gregg, R. C. 2015. *Shared Stories, Rival Tellings*. Oxford: Oxford University Press.

Martin, C. 1978. *Keepers of the Game: Indian–Animal Relations and the Fur Trade*. Berkeley: University of California Press.

Pew Research Center. 2016. "The Gender Gap in Religion around the World." http://www.pewforum.org/2016/03/22/the-gender-gap-in-religion-around-the-world/.

Stump, R. W. 2000. *Boundaries of Faith: Geographical Perspectives on Religious Fundamentalism*. Lanham, MD: Rowman & Littlefield.

UN Women. n.d. "Religion and Gender Equality." https://www.partner-religion-development.org/fileadmin/Dateien/Resources/Knowledge_Center/Religion_and_Gender_Equality_UNWOMEN.pdf.

Select Guide to Further Reading

Becci, I., M. Burchardt, and J. Casanova, eds. 2013. *Topographies of Faith: Religion in Urban Space*. Leiden: Brill.

Bergmann, S., ed. 2014. *Religions, Space and the Environment*. London: Transaction Publishers.

Collins-Kreiner, N. 2010. "The Geography of Pilgrimage and Tourism: Transformations and Implications for Applied Geography." *Applied Geography* 30:153–164.

Dwyer, C. 2016. "Why Does Religion Matter for Cultural Geographers?" *Social & Cultural Geography* 17:758–762.

Gottlieb, R. S., ed. 1996. *This Sacred Earth: Religion, Nature, Environment*. New York: Routledge.

Heine, S. 2012. *Sacred High City, Sacred Low City: A Tale of Religious Sites in Two Tokyo Neighborhoods*. Oxford: Oxford University Press.

Hopkins, P., L. Kong, and E. Olson, eds. 2013. *Religion and Place: Landscape, Politics and Piety*. Dordrecht, The Netherlands: Springer.

Ivakiv, A. 2000. "Toward a Geography of 'Religion': Mapping the Distribution of an Unstable Signifier." *Annals of the Association of American Geographers* 96:169–175.

Krech, S., ed. 1981. *Indians, Animals, and the Fur Trade: A Critique of Keepers of the Game*. Athens, GA: University of Georgia Press.

McMahan, D. 2008. *The Making of Buddhist Modernism*. Oxford: Oxford University Press.

Morin, K. M., and J. K. Guelke, eds. 2007. *Women, Religion and Space*. Syracuse, NY: Syracuse University Press.

Olsen, D. H., and D. J. Timothy, eds. 2021. *The Routledge Handbook of Religious and Spiritual Tourism*. London: Routledge.

Orsi, R. A., ed. 1999. *Gods of the City: Religion and the American Urban Landscape*. Bloomington: Indiana University Press.

Scott, J. S., and P. Simpson-Housley. 2021. *Mapping the Sacred: Religion, Geography and Postcolonial Literatures*. Leiden: Brill.

Shelton, T., M. Zook, and M. Graham. 2012. "The Technology of Religion: Mapping Religious Cyberscapes." *The Professional Geographer* 64:602–617.

Sopher, D. E. 1967. *Geography of Religions*. Englewood Cliffs, NJ: Prentice Hall.

Stump, R. W. 2008. *The Geography of Religion: Faith, Place and Space*. Lanham, MD: Rowman & Littlefield.

Sutherland, C. 2017. "Theography: Subject, Theology, and Praxis in Geographies of Religion." *Progress in Human Geography* 41:321–337.

Watson, S. 2005. "Symbolic Spaces of Difference: Contesting the Eruv in Barnet, London and Tenafly, New Jersey." *Environment and Planning D: Society and Space* 23:597–613.

Wellman, J. K., Jr., and K. E. Corcoran. 2013. "Religion and Regional Culture: Embedding Religious Commitment within Place." *Sociology of Religion* 74:496–520.

Wilford, J. 2012. *Sacred Subdivisions: The Postsuburban Transformation of American Evangelicalism*. New York: New York University Press.

Wilson, A. 2008. "The Sacred Geography of Bangkok's Markets." *International Journal of Urban and Regional Research* 32:631–642.

The Geography of Language

Language is an important part of culture that is intimately connected to place and space. We make sense of the world by naming places and spaces. Language is rooted in place but also emerges from spatial diffusion and contact. Language can be a hard medium in which to lock up, in sounds and words, our understanding of the world, but it is also fluid enough to change and reflect changing circumstances. Language embodies the fixity as well as the flow of cultural geographies.

12.1 The Distribution of Languages

Language of the Local

Language is connected to specific places and local geographies. Local places are the context and setting for a multitude of languages. In the premodern world, there were many more languages spread widely across the surface of the habitable globe. Map 12.1 depicts the complex linguistic map of Indigenous languages of the Pacific Northwest. Notice the mosaic of many language families across this single region.

We can focus on one particular group, known to themselves and others as the Nooksack. They are part of a larger group of people known as Salish, part of the Salishan language group shown in Map 12.1. The Nooksack were a **speech community** comprising a group of villages located around the Nooksack River. The population was estimated at around 1,125 in 1800. In 1855, they agreed to give up the title of their land in return for fishing and hunting rights and a guarantee of government services. Government attempts to keep them in reservations were fiercely resisted. Some of the Nooksack eventually gained formal title to their traditional

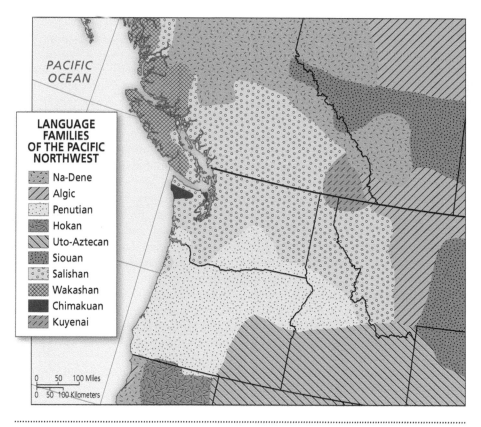

LANGUAGE FAMILIES OF THE PACIFIC NORTHWEST

- Na-Dene
- Algic
- Penutian
- Hokan
- Uto-Aztecan
- Siouan
- Salishan
- Wakashan
- Chimakuan
- Kuyenai

PACIFIC OCEAN

0   50   100 Miles
0   50   100 Kilometers

MAP 12.1 Indigenous languages of the Pacific Northwest.

connection to the land through patient documentation and discussion with Nooksack elders about place names.

There are profound differences between the names given by English settlers/speakers and the original Nooksack. Many of the English place names refer to people. Thus, Johnson Creek was originally Koy-yoti, meaning "former" or "in the past." Many of the English names suggested positive feelings to encourage settlement, such as Happy Valley and Eden Valley. The original Nooksack names, in contrast, focused on environmental factors and the flora and fauna used for food. Thus, the Nooksack names for what were called Ryder Lake and Ryder Creek in English referred to "many dead trees." Fairhaven Creek was originally Xwsisel7echem (yes, that's the name and not a typo) and it means, literally, "place of always finding salvage in the back" (as of finding a drifted canoe). To

land. In recent years, as part of their cultural survival, the tribe created an inventory of cultural sites. Allan Richardson and Brent Galloway uncovered the rich recover the original Nooksack names is to reveal a complex and sophisticated way of understanding and describing the environment.

12.1 Environment and Language

Indigenous peoples of the northern latitudes have a rich variety of words for snow. Igor Krupnik at the Smithsonian studied the vocabulary of ten Inuit and Yupik dialects. They found that Inuit speakers in Canada's Nunavik region have 53 words for different types of snow, including a word for wet snow that can be used to ice sled runners. Speakers of a local dialect around Wales in Alaska have 70 different terms for sea ice. When life is firmly based on local conditions, language must become a very fine filter of environmental understanding. The Sami people of Finland and surrounding countries have 180 words related to snow and ice and 1,000 separate words to describe reindeer. Indigenous languages are rooted in place and highly sensitive to the nuances of the local environment. They are a counterpoint to how the dominant languages refer to the natural world, and thus to identity.

Western society often views nature as separate from people, and this is reflected in the language we use—for example, referring to trees or water as "resources." For the Indigenous people of the Syilx Okanagan Nation, the Nsyilxcən language refers to the land and our bodies with the same root syllable.

REFERENCES

Armstrong, J. 2006. "Sharing One Skin." *Cultural Survival.* https://www.culturalsurvival.org/publications/cultural-survival-quarterly/sharing-one-skin.

Jerome, E. 2021. "How Language Affects Our Relationship with Nature/National Environmental Treasure," https://www.oursafetynet.org/2021/03/26/how-language-affects-our-relationship-with-nature/.

Roughly 6,000 to 7,000 languages are spoken around the world. Their distribution is uneven. The highest level of linguistic diversity is found in the equatorial region of the world, such as Papua New Guinea and southern Nigeria, which, respectively, have 820 and 516 distinct languages. In the tropical areas, climate allows year-round agriculture. Since there is less need for trade to secure food supplies, it is easier for groups to remain isolated and maintain their languages. Papua New Guinea is particularly diverse in languages not only because it is tropical, but also because the rugged terrain and steep mountains make movement difficult, allowing very small and localized speech communities to remain intact.

Where places are less self-sufficient and/or engaged in more extensive trade, languages of dominant groups are diffused across a wider area. The Hausa language, for example, spread across the savannah areas of West and West Central Africa. Today, more than 39 million people speak the language.

Trade links spread particular languages and enabled the emergence of multilingualism as people needed to communicate with those of other language groups. Even in the more isolated villages of Papua New Guinea, other languages are spoken and understood. In one village everyone over the age of ten speaks two languages and can understand at least one more. In other words, multilingualism was probably a very common feature of the premodern world.

Table 12.1 indicates the main language groups. Half of the world's population speaks only ten languages. Table 12.1 also shows the second language usage for these

dominant languages. Notice how some languages have little traction outside their core **cultural region**. The number of people who speak Japanese as a second language is less than 1 percent of those who use it as a first language. Similarly, while 207 million people speak Bengali as their first language, only 4 million more speak it as a second language. Compare this with English, which is more widespread outside the core area of England. We can make a distinction between dominant languages with large numbers of speakers and **global languages** that reach outside their primary language area. While both English and Spanish have similar numbers of primary speakers, making them a global language in terms of size, more people speak English as a foreign language than Spanish, making English an extensive global language. We will consider the practice of global English later in this chapter.

Map 12.2 shows the current distribution of dominant languages in the world. Notice how certain languages are highly localized in particular countries: Chinese in modern-day China and Hindi in India. Others are more diffuse, such as Arabic, spoken throughout the Middle East and across North Africa.

12.2 Language and Power

Language is a good marker and a permanent legacy of the spread of empires, which diffused language use around the world. The Roman Empire created the first pan-European language of Latin, the language of the dominant elites, which became the accepted linguistic current of the expanding empire. As the empire declined, flow and spread were then replaced by place and stasis. Local variations became more important, and we can make a broad distinction between the Romance languages at the center of the empire and the languages not descended from Latin, such as Old Norse, German, and English, that emerged at the periphery. We retain this linguistic legacy in the names of the days of the week. Days in the Romance languages have names that derive from the name of the Roman god Jupiter: *jeudi* (French), *jueves* (Spanish), and *giovedì* (Italian). In contrast, English still draws on the Old Norse name of Thor to designate the fifth day of the week as Thursday. The day is similarly named *torsdag* in Danish, Norwegian, and Swedish. The closer to the heart of the old Roman Empire, the stronger the linguistic legacy of Latin.

Empires spread language. The language areas shown in Map 12.2 are in large part maps of imperial expansion. Arabic is spoken across the Middle East and North Africa in a direct mapping of the spread of the Muslim empire out from its core region in Saudi Arabia. French, English, Spanish, and Portuguese are European languages that diffused to appropriated regions around the world. Spanish

TABLE 12.1 Ten Most Spoken Languages in the World		
	NUMBER (MILLIONS)	
	AS FIRST LANGUAGE	**AS FIRST AND SECOND LANGUAGE**
Chinese	1,146	1,343
Hindi	366	487
Spanish	358	417
English	341	573
Arabic	280	n/a
Bengali	207	211
Portuguese	176	191
Russian	167	277
Japanese	125	126
German	121	128

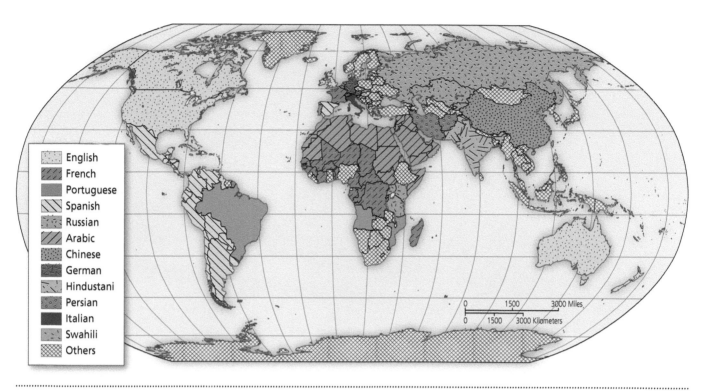

MAP 12.2 Dominant world languages.

spread into South America and Portuguese into South America and Africa.

Language and language use reflects colonial expansion and **imperial incorporation**. We can imagine a three-level linguistic model of colonization of increasing linguistic impact. First, in small trade colonies such as the Dutch trading port in Japan, there was limited linguistic contact outside the enclave. Second, in a situation of colonial exploitation such as mining and planation economies (e.g., the French in Indochina or the British in the Caribbean), segregated elites spoke the language of the imperial center. However, as locals learned the language of the powerful, the dominant language percolated down through the local society. Third, in settler colonies, immigrants from the home country brought their own languages, which became dominant and replaced a range of local ones. This is what happened with English in Australia, Canada, and the United States.

When the language diffused out from its home base, it was adopted but also altered. Many settler societies spoke English, but subtle differences evolved in English usage between Australia, England, and the United States. Similarly, as the French expanded their commercial and political empire and gained colonies, local elites soon learned to speak the language of the dominant. But the French of the center was used alongside the local languages, and the result was the emergence of various forms of spoken French (see Map 12.3).

Naming the World

The naming of things is how we humanize the world. Naming gives shape and meaning to our world; it turns space into place. Naming is never innocent of politics.

The naming of the Earth's surface is shaped by three basic toponymic eras, indigenous, colonial, and postcolonial. In the first, names were given by Indigenous people. Consider the case of Australia, where there is still legacy of Aboriginal names. Cities such as Wollongong remind us of an Aboriginal past. Then there are the colonial names. In Australia, the principal cities of Sydney, Melbourne, Brisbane, Adelaide, and Perth all refer to names from English aristocracy. We live in a postcolonial period in which we are aware of the Indigenous legacy as well as the colonial rewritings. In the central Australian city of Alice Springs, the long-silenced Aboriginal name of Mparntwe is now used.

In a neighboring country to Australia there is a vigorous debate about the national name. Many want the name New Zealand, derived from a cartographer working for the Dutch East India company, to be changed to Aotearoa, which, roughly translated from Māori, means "land of the long white cloud." This name is currently used on the nation's passports and currency and is employed in media broadcasting. It is part of a wider linguistic push in the country to have European colonial names changed to their original Māori names. While some resist the new name as

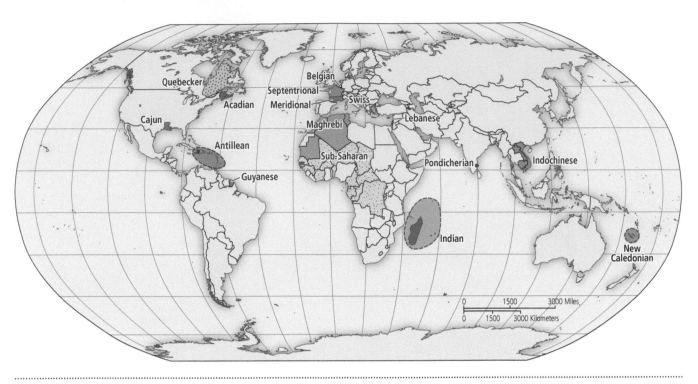

MAP 12.3 Dialects of the French language in the world.

12.2 Language Use and Climate Change

Since climate change first emerged on the public agenda in the mid- to late 1980s, public communication of climate change has been problematic. Part of the challenge is to communicate something that has invisible causes, with distant impacts, over a long time frame. Added to this are complexity and uncertainty, which can result in inadequate signals that indicate the need for change and allow people to disbelieve the science.

Initially, climate communication involved attempts to overcome uncertainty—whether climate changes was happening or not. Currently, communication efforts have changed from persuading people that climate change is happening to persuading people to act.

There has been a lot of research on how to use language (metaphors, words, strategies, and narratives) around climate change. For example, some research indicates using "alarming" language may have the opposite effect to what was intended; instead, the language of popular culture may be more effective than the discourses of politics or science. Words matter: when scientists use the word "enhance" to mean increase, lay people often interpret this to mean "improve" or make better. So "enhanced greenhouse effect" sounds like something positive. Many climate change communications scholars say a good metaphor can be effective. Nerlich and colleagues suggest the ever-popular metaphor of "loaded dice" is a good

response to the question of how climate change is affecting various weather events. With more greenhouse gases, we are loading the dice to come up with extreme weather events.

Despite important progress in better understanding the science and politics around climate change, persistent communication challenges remain. These include a superficial public understanding of climate science, a lack of clear communication about how to transition from awareness and concern to action, how to communicate in deeply politicized and polarized environments, and how to deal with the growing sense of hopelessness. Finally, communications are challenged by carbon-based interests or climate deniers that seek to undermine scientific literacy on the matter.

REFERENCES

Moser, S. 2009. "Communicating Climate Change: History, Challenges, Process and Future Direction." *WIREsClimateChange*. https://doi.org/10.1002/wcc.11.

Moser, S. 2016. "Reflections on Climate Change Communication Research and Practice in the Second Decade of the 21st Century: What More Is There to Say?" *WIREsClimateChange*. https://doi.org/10.1002/wcc.403.

Nerlich, B., N. Koteyko, and B. Brown. 2010. "Theory and Language of Climate Change Communication." *Wiley Interdisciplinary Reviews: Climate Change* 1. 10.1002/wcc.2.

the singular name for the nation, many now use the combined term, Aotearoa/New Zealand.

Naming the Seas

Three things make naming the seas a much more distinctive element than the naming of land features. First, large bodies of water are shared spaces, often surrounding or surrounded by different national territories. The east coast of China faces the west coast of the United States; England's south coast is France's northern coast. Thus, there is no simple hegemony over naming rights as in the case of land surfaces, which can make this a contentious issue.

We are more concerned with things closer than farther away; it is easier to agree with other countries when things are far away. Thus, both Germany and the United Kingdom refer to the Indian Ocean (*Indischer ozean* in German); it is far away from either country and both share the same toponymic history of the area. However, closer to home, the Baltic Sea in English is referred to in German as the East Sea (*Ostsee*). The English name probably draws on Russian (*boloto* = marsh) or Lithuanian (*baltas* = white), while the German refers to its location relative to Germany.

For very large bodies of water with numerous landmasses and hence a variety of different nation-states involved, the Indigenous names can be so varied and so numerous that colonial names become the standard. Take the case of the large body of water we call the Pacific. It probably had a rich variety of names as Indigenous people named it in their own languages. After the sixteenth century, it was opened up to European colonial trade and mercantile interests. In English it was named with reference to Europe and was originally called the South Sea. When Magellan crossed the ocean in 1520/21, he encountered no storms and named it Mar Pacifico. As the dominant global power of the time, the Spanish name displaced all the Indigenous names and thus persists to this day. It persisted both because of the continuing Spanish legacy in the region and because it was an easy solution to the complexity of many competing Indigenous names. Colonial namings often replace a myriad of Indigenous names. The larger the number of divergent Indigenous names, the greater the force of singular colonial namings.

Second, territories try to exert greater influence on the naming of seas closer to them. National imaginaries cast a stronger shadow over more proximate territories. States rarely argue over the names of seas far from their shores. But seas close to their coastline are part of the national imaginary and the nation's sense of itself.

Third, conflict is more pronounced when there are only a few Indigenous names in a relatively small territory. We can compare the Pacific Ocean, a vast sea body with many different Indigenous communities and nation-states, with smaller bodies of water surrounded by fewer nation-states. The East Sea/Sea of Japan is a relatively small body of water surrounded by only three countries, Korea (North and South), Japan, and Russia, for whom it was at the edge of empire. The principal proximate interests are thus only Japan and Korea. The Japanese government prefers to use the name Sea of Japan, while the South Korean government actively pushes for a change to a double naming system of East Sea/Sea of Japan to acknowledge the fact that the sea was known to Koreans as the East Sea for centuries. That a colonial relationship exists between the two countries makes the naming controversy all the more intense.

Two principles are important in naming seas: ensuring global intelligibility and a postcolonial sensitivity. We live in a postcolonial world, and the names of geographic features should reflect this fact.

Toponymic Colonialism

We can identify a **toponymic colonialism** whereby colonial powers rename the landscape, replacing local words in an act of naming that is also an act of appropriation and control. Here are two examples.

First, consider the case of central Australia, long the ancestral home of the Arrernte people. In the late nineteenth century, white British Australians extended control into the region. The names of local features were replaced with English names. Table 12.2 shows the changes. The term "gap" is literally a gap in the long MacDonnell Ranges, a dramatic feature of parallel red quartzite ridges that reach up to just over 5,000 feet at their highest point and stretch for 250 miles across the stark, eroded desert landscape. A white explorer, John McDouall Stuart, replaced the Arrernte description with the name of the governor of South Australia, Richard MacDonnell. The dramatic rock outcrops in the region were named Ayers Rock and the

TABLE 12.2 Toponymic Colonialism in Central Australia

ARRERNTE NAME	ENGLISH NAME
Mparntwe	Alice Springs
Lhere Mparntwe	Todd River
Akeyulerre	Billy Goat Hill
Anthwerrke	Emily Gap
Atnelkentyarliweke	Anzac Hill
Ntaripe	Heavitree Gap
Ntyarlkarle Tyaneme	Unnamed ridge
Uluru	Ayers Rock
Kata Tjuta	The Olgas

PHOTO 12.1 Part of the Kata Tjuta range in central Australia.

landscape, erasing the colonial heritage and replacing it with more local names.

After Vietnam achieved independence from France in 1954, the French-named streets were Vietnamized. Rue Catinat, named by the French authorities after a warship that consolidated their power in the region, was renamed Freedom Street in Vietnamese. Name changes also occur with a change in regime. After the Communists took control of all of Vietnam in 1975, Saigon became Ho Chi Minh City and Freedom Street was changed to Uprising Street in Vietnamese. However, in Vietnam and across the world, there is a commercial bilingualism as street signage displays things in English for foreign travelers (Photo 12.2).

Olgas to commemorate white explorers and their wives. However, the Indigenous terms Uluru and Kata Tjuta are now the preferred names and to use these terms is to show respect for Indigenous culture (Photo 12.1).

Another example, this time from Asia: Korea became a Japanese protectorate in 1905 and came under formal Japanese control in 1910. Japanese colonial domination took many forms; among them, they suppressed the Korean language, especially in name usage. In 1939, Koreans were encouraged to change their surnames to Japanese forms. This linguistic colonialism was also evident in maps. Japanese place names appeared in maps of Korea from 1900 onward. Maps of Seoul showed the name Keijo or Kyongsong. From 1914, a new Japanese naming system was applied to maps. New administrative place names were adopted after annexation that added Japanese generic names such as Tong, Jeong, and Jeongmok or directly copied Japanese names such as Bonjeong, Gilyajeong, Hwang Gumjeong, and Taepyong Tong. Roads and natural features were renamed in Japanese style. As well as Japanese inscriptions, there were Korean erasures. In Seoul, for example, 186 districts with Japanese names replaced 661 administrative districts with Korean names. Larger regional features were also renamed. The sea between Japan and Korea had multiple names through the centuries, predominantly named by the Koreans as East Sea. After annexation, as part of the new wider Japanese Empire, it was named the Sea of Japan. The contemporary struggle over the name of the sea is reflective of decolonization efforts.

The process of **decolonization** also involves a renaming and the act of reclaiming. As colonies became independent, one of their first acts was to rename the

PHOTO 12.2 Commercial street sign in Hoi An, Vietnam.

Although countries mark their independence by renaming places, streets, and cities in their own language, they also connect with global consumers by using English.

There are some holdovers. Georgetown remains a popular name for towns across the world, including an affluent neighborhood and a university in Washington, DC, a distant reminder in an independent country of when King George III ruled over the American colonies.

Contact Languages

Trade and especially the spread of commercial and political empires brought different languages into contact. New language forms emerge in the interaction of people with different languages. Various forms of **contact language** can be identified. A **pidgin** is a simplified language for communication between different languages. Pidgins developed as the British Empire extended its trade and political links around the world. West African Pidgin English, for example, was the contact language along the West African coast. It was also known as Coast English. Remnants of the language can be found in the more mountainous parts of Jamaica.

Creoles are formed in the contact between dominant and local languages and are adopted permanently by a group of people and passed on to their children. Various creoles can be identified. Consider two cases from the Caribbean—a melting pot of African, American, and European languages.

Papiamento is one of the two official languages of the island of Aruba; the other is Dutch. Papiamento is a creole language that draws on Amerindian, different African languages, Portuguese, Dutch, Spanish, and English. The pidgin stock of Papiamento is the simplified language that developed in the interaction between Portuguese slavers and African slaves and slavers. Later additions to the language embody a pre-Columbian linguistic heritage, Spanish control of the island for one hundred years, and, since 1636, Dutch control. With its rich mix of languages from three continents, Papiamento embodies the complex history of the region.

The Garifuna are a distinct ethnic and language group emerging from the slave trade. West Africans, shipwrecked off the coast of Saint Vincent, intermarried with indigenous Caribs who had been shipped to the island by the French. In 1763, the British gained control over Saint Vincent and again the Caribs resisted. The British defeated them in 1796 and deported around 5,000 people, separating the more African-looking Caribs from the Amerindian-looking ones and banishing them to an island off the coast of South America. Only around 2,500 survived the voyage. The people soon moved off the island and settled along the Caribbean coast. They are now known as Garifuna and number close to 600,000. They are found in Honduras and Guatemala and, with subsequent migrations, also in New York City and Los Angeles. Their language derives from the Amerindian languages of Arawak and Carib, as well as English and French, and combines Arawak grammar with African, English, and Spanish loan words.

Rather than being unitary, creoles are more accurately considered as a continuum where people can register different dialects according to the setting. All languages, not just creoles, have a fluid quality. Language use varies across space and place.

Language and Place

Map 12.2 is deceptive. It shows dominant languages as singular phenomena, covering entire spaces in a homogeneous categorization. In reality, of course, language is far more complex and interesting. There are even differences in the way we speak the same language. Although Table 12.1 uses the term "Chinese" as a singular category, different types of Chinese are spoken, including Mandarin (1.052 billion as a first and second language), Wu (77 million), and Yue (71 million). The category of English covers a rich variety of different types of spoken English, from Scotland to Australia and from Jamaica to the United States.

Let us look at some of the variation in more detail.

The Regional Geography of Language

Spoken language varies by place. Linguists identify the variation by mapping the enunciation of the same word or plotting variant descriptions of the same phenomenon. For example, small lobsters are known as crawfish in the lower Mississippi delta but as crayfish in New England and crawdads in the middle Mississippi Region. Similarly, pop, soda, and Coke describe the same thing in different places in the United States.

From this work it is possible to generate **dialect maps** that show how a single spoken language varies over place. In countries with centuries of settlement such as China, there are very distinct regional dialects even within Mandarin or Cantonese Chinese. British dialects have evolved over close to 500 years. You can take a fascinating dialect tour of the United Kingdom by visiting this website: https://www.youtube.com/watch?v=-8mzWkuOxz8. The accents are enunciated as the cursor moves over a map of the country. It is fascinating to see how speech changes over space, giving places a distinct dialect. Dialects are the localization of language. In Scotland, for example, the accent of Glasgow has much harder consonants than the softer, singsong dialect of the Highlands. John's accent is from the rural central region of Scotland. When he was a young boy and visited the big cities of Edinburgh and Glasgow, his

PHOTO 12.3 A sign in the Shetland dialect of Scotland.

expression in the national media. More recently, however, there has been conscious acceptance of regional dialects and different forms of spoken English. Now, when you listen now to the BBC, you hear a wider array of accents.

Even in the United States, which is considered more linguistically homogenous, at least twenty-four different dialects have been identified among English speakers (Map 12.4). The region depicted as the Southwest, for example, contains elements of the Mexican dialect of Spanish. Along the Carolina and Georgia coastline, Gullah is identified in as regional dialect. It is in fact a creole language. More than a quarter million people speak Gullah, mainly descendants of enslaved people brought from Africa who developed a language based on African roots and the Coast English of West Africa blended with local English. It is language that reminds us of the slave trade that connected the United States with Africa.

There are also localized differences in how people use the same language. One study compared speech patterns of New Yorkers with those of non–New Yorkers. The former tended to exaggerate intonations and quickly uttered questions. The language of the New Yorkers was defined as high involvement while that of the non–New Yorkers was characterized as high considerateness.

Map 12.4 assumes homogeneity by region. In fact, there are variations within regions. We can identify speech communities composed of people who share a common language. One such community is composed of about 200,000 people in Brooklyn and the South Bronx who speak Garifuna, a language we have already described as a creole language from the Caribbean. Speech communities survive best with large numbers of speakers, dense internal social networks, and limited external connections. Speech communities with a small number of speakers are vulnerable. With few or weakened internal connections, as people die or move away, speech communities lose their linguistic oxygen. When members of a speech community have more contact with external institutions, people, and practices, the localized language can be overwhelmed and eventually displaced. Some physical environments, such as dense inner-city neighborhoods, can solidify speech communities. So Garifuna, with significant numbers of speakers in dense urban neighborhoods and strong internal connections, survives even as members of the community, and especially their children, interact with the wider community.

Local language usage can vary from the official policies. In Israel, the two official languages are Hebrew and Arabic, with English widely spoken and understood. In 2014 there were demands from some Israeli politicians to make Hebrew the sole official language, in effect hardening the notion of Israel as a Jewish state. A study of the language of signs in three types of neighborhoods in Israel highlighted marked differences. In predominantly Jewish neighborhoods, signs were either in Hebrew or in Hebrew and English. In non-Israeli Palestinian areas, dominated by

accent marked him out as someone from the rural world. Photo 12.3 shows a sign in the written form of the spoken dialect of the Shetland Islands in the very far north of Scotland. Can you guess what it says?

In the United Kingdom, dialects are often compared with standard speech, effectively the speech of the elites of London and Southeast England. The term "Queen's English," for example, describes the standard form and hints at its elite status. Dialects are not innocent of social connotations. Indeed, what is considered a dialect and what is considered the standard form is a political act of categorizing people in a social hierarchy based on how they speak. The definition of dialect and standard speech codifies social differences and political power.

When John was a young boy growing up in Scotland, he listened to the national radio or television and never heard anyone who sounded like him, his family, or any of his friends and neighbors. A more formalized English was spoken, called "Received Pronunciation." Scots' accents were restricted to the periphery and the marginal and were denied

Arab speakers who were not citizens of Israel, most signs were either in Arabic and English or Arabic. In Israeli Palestinian neighborhoods, where most people were Arab citizens of Israel, the highest percentage of signs were in Hebrew and Arabic.

Language, Place, and Social Differences

Map 12.4 is also deceptive. It implies variation only by location. Language variation occurs not only by location but also by social difference, such as gender, age, and social class. Indeed, social differences are often embodied in different ways of speaking.

Place and socioeconomic status combine to produce social geographies of language. The higher socioeconomic groups tend to speak a similar national language despite their various locations, while further down the socioeconomic hierarchy, the national language can vary more by distinct regional dialects. Figure 12.1 shows the pattern for the United Kingdom. Regional variations in speech patterns are most marked at the lower levels of the hierarchy.

Speech patterns also reflect and embody socioeconomic differences, even within the same city. A pioneering study by the social linguist William Labov identified the speech

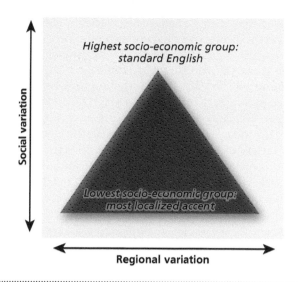

FIGURE 12.1 Social and regional accent variation.

patterns of different socioeconomic groups in New York City. Language varied by pronunciation and in the range and type of words used. The work of sociolinguist Deborah Tannen shows how speech patterns also vary between men and women. Holding other things constant, women use

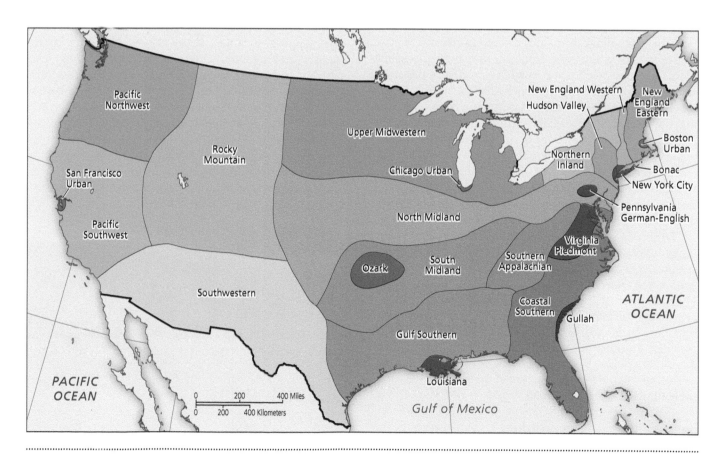

MAP 12.4 Dialect map of the United States.

less slang than men. Language also varies by age, with distinct youth cultures often having their own forms of speech.

We have a complex geography of language because people speak the same language differently depending on where and who they are. Despite the linguistic homogeneity implied by language maps such as Map 12.2 and the official language policies of national governments, the linguistic reality is that most countries are populated by people speaking a variety of languages. People use different languages and even the same language in different ways in different contexts. A variety of distinct language domains can be identified, such as work, school, and home. Adding to this diversity are differences in language use between public and private and by age, socioeconomic status, and gender.

12.3 The Place of Language

Place has an important role in language use. Language is a flexible communication device that can be tweaked and changed according to the context. We can speak differently in different places to different people in different social settings. Even people who speak only one language have a wide repertoire of language use, a continuum of different words, phrases, and intonations that can vary by place and domain. A **language domain** is defined in relation to place, role relationship, and topic. Language domains include such places as home, workplace, neighborhood, and the street. Language is a performance dependent on the role we are playing (parent, teacher, student, coworker, friend, stranger) and the domain. We speak differently with our friends compared to our colleagues. For example, John's accent hardens when he telephones or visits with family in Scotland. It softens when he speaks with his family in California. A distinction can be made between a formal domain, such as church sermons and official speeches, and a more informal domain of everyday conversation and email between friends.

Diglossia refers to the existence of two or more varieties of language in a speech community. In the German-speaking parts of Switzerland, two varieties of German are used, often referred to as High German and Swiss German. High German is spoken by people from Germany; there are many German immigrants working in Switzerland, but local Swiss will often speak to each other in Swiss German. In much of the Arabic-speaking world, classical Arabic is spoken by the elites and in more formal surroundings, while local and national dialects are the informal language of the home and street.

In some cases different languages are used in different domains. The term **code-switching** is applied when people use different languages or dialects according to context. Thus, US-born children of Mexican parents in Los Angeles may speak Spanish at home but English at school. And the forms of Spanish and English they speak will vary. High schoolers may have a different speech code with their teachers compared to their English-speaking friends, and the Spanish they use with their parents may vary from the Spanish they use with friends.

We can imagine a rich variety of instances of different code-switching. Turkish parents in Germany may speak Turkish to each other, but some blend of Turkish and German with their children. They may speak Turkish when they are at home but German in the public realm.

The spatial division between public and private, formal and informal, home and outside is often a site of code-switching and changes in the way that language is employed.

Language Shifts

People move. They take their language with them, but if the move is to a different linguistic region, their language and especially that of their children may shift. Marked **language shift** occurs as children are socialized into the dominant language.

We can follow the shift by looking at language use by different groups in Australian cities. The first generation noted in Table 12.3 is the percentage of people born outside Australia who speak English at home. Notice the low figure for people from Chile, Hong Kong, and Lebanon. The second generation is the people born in Australia who speak English at home. Notice the marked shift toward English speaking for all children, especially for those with Austrian and French parents. Where there are substantial speech communities (there are lot more people in Australian cities who speak Arabic and Chinese than French and German), then the original language remains stronger. And if there are substantial cultural differences—wider between Australia and the world of Arabic- and Chinese-speaking immigrants compared to the average immigrant from Austria and France—then the language shift is less marked.

TABLE 12.3 Language Shift in Australia		
ORIGIN (LANGUAGE)	FIRST GENERATION SPEAKING ENGLISH AT HOME	SECOND GENERATION SPEAKING ENGLISH AT HOME
Austria (German)	48	89
Chile (Spanish)	9	38
France (French)	37	77
Hong Kong (Chinese)	9	35
Lebanon (Arabic)	5	20

12.3 The Linguistic Landscape

Our environment is filled with language signs. A number of different types of **linguistic landscapes** can be found. First, there is the monopolistic landscape. Photo 12.4 shows signs on the side of a building in Seoul. They are almost entirely in the local language. By contrast, Photo 12.8 shows a streetscape in Shanghai. Notice how the linguistic landscape is now bilingual, with signs in Chinese as well as English. The metro system of the city is entirely bilingual, with all signs in English and the automatic voiceover also in English as well as Chinese. In towns and cities across the world, global English words and phrases are now part of the local linguistic landscape.

Then there are trilingual signs, such as that shown in Photo 12.5, taken in Vancouver's airport. The city-state of Singapore has a bilingual policy. All students are taught English at school but can also learn a second language: Chinese, Malay, or Tamil. The four languages are commonly used in Singapore, and only two of them use Roman script. Photo 12.5, of a warning sign next to a government building, depicts all four languages to make sure everyone gets the message to stay out.

PHOTO 12.4 Streetscape in Seoul.

REFERENCE

Ben-Rafael, E., E. Shohamy, M. Hasan Amara, and N. Trumper-Hecht. 2006. "Linguistic Landscape as Symbolic Construction of the Public Space: The Case of Israel." *International Journal of Multilingualism* 3:7–30.

PHOTO 12.5 Sign in Singapore.

12.4 The Political Geography of Language

Language and National Identity

National identity is created, maintained, and reproduced in language. Graham Robb, in his wonderfully descriptive account of the historical geography of France, outlines the process whereby the national language of French finally replaced a myriad of language and dialects. France encompasses at least five major languages—Catalan, Franco-Provençal, Gallo-Italian, Langue d'Oc, and Langue d'Oïl (now considered standard French)—with at least ninety distinct dialects. At the time of the French Revolution, French as we know it in the early twenty-first century was spoken only in Paris and the surrounding area. Creating a national language was a major goal of successive republican governments. Even as late as the mid-nineteenth century, most of the people living south of the Loire did not speak French, and in Brittany most spoke Breton. The national education system fostered the exclusive use of French, so that French became the national language, a process heightened by the space-time convergence that Robb describes as the "contraction of space and the gravitational pull of Paris." But even today, local variants persist, a testimony to the continuing connection between language and locality. Almost 2 million people speak Occitan, 1.5 million Alsatian, and 500,000 Breton. Local dialects persist, although the higher up the socioeconomic hierarchy, the more common the everyday usage of "proper" French (see Figure 12.1).

There is a complex relationship between language and national identity. A national language is more often than not a deliberate creation. In 1783, Noah Webster (1758–1843) wrote of the need to "establish a national language as well as a national government." His *American Dictionary*, published in 1828, was written to create a distinctly American English different from the English of the colonial power.

We have more recent examples. In 1947, the British Empire in India was partitioned into independent countries. Urdu was the dominant language of North India before partition. It is filled with Arabic, Persian, and Turkish words, a legacy of the ebb and flow of empires. Hindustani was almost identical, but written in Sanskrit. Urdu was in Arabic script. After 1950, the Indian government created a national language of Hindi from Hindustani with all the Arabic, Persian, and Turkish words replaced with fabrications from Sanskrit. The result was an invented national language expunged of any Muslim influence.

Hebrew as a language has an ancient history and a more recent emergence. It was the language of the early Israelites, but disappeared as an everyday language around 300 CE, surviving only as the **liturgical language** of the Jews. It was revived in the nineteenth century as part of early Zionism and is now one of the two recognized languages of Israel, the other being Arabic. Almost 90 percent of Israeli Jews are proficient in Hebrew, including 70 percent of Israeli Arabs.

12.4   Equality, Disability Justice, and Sign Language

This chapter has discussed language as predominantly an oral form of communication; however, it is important to realize that visual languages are also forms of communication.

Fabio de Brito and Rosangela Gavioli Prieto studied the history of the Deaf social movement in Brazil to achieve legal recognition of "Libras" (Língua Brasileira dos Sinais), or Brazilian Sign Language. In Brazil, the Deaf social movement was a product of a larger social movement that began in the early 1980s when the military dictatorship was on the verge of dissolution, and there was a renewed sense of optimism about political openness and redemocratization. In this moment, new movements emerged focusing on social minorities, including women, slum dwellers, and the LGBTQ+ community, as well as people with disabilities, including the blind and Deaf.

For a long time, deaf people have been seen by many hearing people only through the lens of negative stereotypes of deafness, where they are portrayed as depressed, anxious, uncommunicative, socially isolated, pitiable, and suffering from the effects of their disability. In Brazil, the Deaf social movement aimed to subvert this dominant narrative, and in a series of protests, activists took to the streets to show they were confident and proud of their Deaf identity.

For many Deaf activists, the main battle has focused on being understood by society and by the Brazilian state as members of a Deaf community with its own culture and language. To protect this culture and preserve their Deaf identity,

activists worked to ensure that their education guarantees their right to fulfill their potential in sign language. The Brazilian Deaf social movement demanded sign language be legally recognized by the government and used in education and government services. In 2002, Brazil recognized Libras as the first language of the Deaf in Brazil. Today, Brazilian Sign Language is its own language, with its own grammar, syntax, and morphology, and has dialects across Brazil reflecting regional and sociocultural differences.

This example is part of a wider disability justice movement underway in many countries around the world. By 2024, forty-one countries recognized sign language as an official language and there was legal recognition of national sign languages in seventy-one countries, from Albania to Zimbabwe, but not the United Kingdom or the United States (it should be noted that the US federal government does not recognize any language, spoken or signed, as an official language).

REFERENCES

De Brito, F. B., and R. G. Prieto. 2018. "We Did It Ourselves: The Deaf Social Movement and the Quest for the Legal Recognition of the Libras Sign Language in Brazil." *Disability Studies Quarterly* 38:4. https://dsq-sds.org/article/view/6241/5140.

World Federation of the Deaf. 2022. "The Legal Definition of National Sign Languages." https://wfdeaf.org/news/the-legal-recognition-of-national-sign-languages/.

Language policies vary from standardization onto a national language to a diffusion of standard languages into the vernacular. The official sign shown in Photo 12.3, for example, would have been unthinkable in the Shetland Islands in 1960, when the emphasis was on standardization. In new nations, standardization is often the way to ensure national cohesion. In a more postmodern world and more settled nation, vernacularization avoids bland uniformity and heightens the sense of place and regional identity.

PHOTO 12.6 Sign at the Vancouver airport.

The Written Word

While the spoken word varies, the written word is more standardized. Throughout the nineteenth and much of the twentieth century, the emphasis was on the construction of both written and spoken national languages. On closer inspection, many "national" languages arose from the promotion of one language at the expense of others. The promotion was aided by national education systems, print cultures of books and newspapers creating a national readership, and the marginalization and suppression of other languages within the national territorial space.

Language Policies

In reality, a variety of languages are used in most countries. Sometime this is reinforced and encouraged; other times it is suppressed in the pursuit of single language use. We can identity four types of state language policy. *Domination/exclusion* refers to the continuing preference given to the dominant language and the suppression of minority languages. For example, the Hawaiian language was banned in public schools from 1896 to 1986. In the case of *assimilation*, the state encourages minority language speakers to learn the dominant language. This can also involve making immigrant children, who may speak a different language at home, acculturate to the dominant language through educational practices. Under a *pluralist* policy, the state promotes tolerance of minority languages and the practice of multilingualism. Last, in a *confederation*, languages are identified with specific regions of a country. The different categories are not mutually exclusive. The United States and individual states fall between the assimilationist and pluralistic approaches.

Some countries have an explicit policy of multilingualism with a number of official languages. Canada is a bilingual society, with both French and English used in official records and documentation. Photo 12.6 shows signs at Vancouver International Airport. Notice that while the two official languages are used, so is Chinese, because of a recent history of Chinese immigration and the many Chinese tourists and visitors to the city. Sri Lanka also has two official languages, Sinhalese and Tamil, although English is widely used, a legacy of colonialism and the dominant language when the country achieved independence (Photo 12.7). New Zealand has three official languages: English, Maori, and New Zealand Sign Language. Switzerland has four: German, French, Italian, and Rumantsch. South Africa has one of the largest numbers of official languages, eleven: Afrikaans, English, Ndebele, Northern Soto, Soto, Swazi, Tsonga, Tswana, Venda, Xhosa, and Zulu.

The official languages do not cover all the languages spoken within a country. In Canada, for example, there are a large number of Indigenous languages as well as new immigrant languages; at least eighty-three other languages are spoken, and more than 1 million people speak Chinese.

Even in officially monolingual states, a multiplicity of languages is used on a daily basis. Although English is the

PHOTO 12.7 Language use in Sri Lanka.

official language of Australia, at least 388 different languages are spoken in the country, including many Indigenous languages and the speech communities of recent immigrants. In Mexico, Spanish is the official language, but of the total population of around 120 million, between 10 and 20 million speak an Indigenous language. Indigenous speakers are disadvantaged in an officially unilingual society. While Amharic is the only official language of Ethiopia, it is spoken by only around 30 percent of the population. Around 34 percent speak Oromo, and in the marketplaces of Addis Ababa, merchants freely discourse in Amharic, English, and Tigrinya. At least 188 languages are spoken in the United States, 44 in the United Kingdom, 42 in Germany, and 39 in France, all polyglot societies with a large number of immigrants from around the world.

Official language policy can change abruptly with political ruptures. In Spain, at the fall of the Franco dictatorship in 1975, the promotion of Castilian Spanish was replaced by a greater multilingualism. In the region of Catalonia around Barcelona, public signs now display Catalan as well as Spanish words. The Basques were also given greater freedom to speak and write in Basque. During the apartheid era in South Africa, Afrikaans and English were the official languages; Afrikaans, a creole of English and Dutch, was the language of the white supremacists, but was also spoken by Black and colored South Africans. After the fall of the regime, English became the preferred official language, while the state proclaimed eleven official languages. However, speaking the dominant language of English confers advantages in the labor market and in civic organizations.

Sometime the change occurs without political rupture. Around the world there is increasing recognition that the language rights of minorities should be protected. As an example, in 1977 the BBC launched a Welsh radio station.

Language Loss, Replacement, and Rebirth

We are losing linguistic diversity. Approximately 6,000 languages are spoken today, but by century's end it is estimated that only 600 are sure to survive. We are losing a language every four months. In a world dominated by fewer languages, we gain global communication but lose diversity, linguistic subtlety, and something that is almost impossible to resuscitate. Linguistic variety, like ecological diversity, is a sign of health and vitality. There is a strong correlation between areas of biodiversity and linguistic diversity at numerous scales of analysis. A loss of biodiversity is associated with a decline in linguistic variety. The loss of both impoverishes our world.

Languages are not just reflectors of the external world; they embody it. How we describe the world is crucially dependent on where we are and how we speak and write. Daniel Nettle and Suzanne Romaine drew attention in 2000 to the extinction of language in their evocatively named book *Vanishing Voices*, and K. D. Harrison made a similar point with his 2007 book *When Languages Die*.

The consequences of the loss of language are profound. Indigenous languages of the Central Desert region of Australia, for example, are the living bond between people and the country given to them by their creators. The loss of the language cuts the cord with the cosmological understanding of the world, destroys the empathetic rendering of the environment, and undermines the cultural framework that gives social meaning and religious purpose. Without language, sacred space is rendered secular and meaningless.

At the point of contact, original languages of places are modified by new languages. In some cases they are replaced. Language replacement can occur when small speech communities are simply overwhelmed by the dominant language. They can also occur through the formal policies and naming practices of the dominant linguistic group.

Thousands of different languages were spoken in the Americas before the Columbian Encounter. Consider the case of the United States, where English became the dominant language and replaced a variety of local languages deeply rooted in place. Small tribal languages were particularly vulnerable because of the small number of native speakers and the overwhelming power and influence of the nonnative language. In some cases the replacement was long part of official educational policy.

In the United States, only about 170 Indigenous precontact languages are still spoken. Navajo is one of the largest speech communities, spoken by 250,000 people, but fluency is declining. While 25,000 people describe themselves as Mohawk, only 15 percent can speak the language. More than 74 languages in the United States are on the brink of extinction, including Quechan, whose speakers number no more than 2,500. They live in Arizona, and while elders can speak the language, it is losing its appeal to the younger generation. Only around 100 people know the language well. With such a small base, one generation's indifference can doom the language. A further 58 languages have fewer than 1,000 speakers and 25 have only between 1,000 and 10,000 speakers. With such small numbers, language revival is precarious.

Minority languages survive if they have large numbers of speakers. In Central America, for example, Mayan is still spoken by more than 7 million people. In South America, 8 million people or more speak Quechuan, the language of the Incas. In the Andes, the Incan Empire spread along an extensive road system, diffusing Quechua throughout a wider region. Today, between 8 and 10 million people speak it as their first language in the Andean belt of Bolivia, Peru, Ecuador, Colombia, and Argentina (Map 12.5).

In some cases, formal recognition of a language can improve the chance of survival. Quechua, for example, is one of the official languages of Bolivia, Ecuador, and Peru. Its recognition is part of the growing political power of Indigenous people. The 2006 election of Evo Morales in

AREAS WHERE A QUECHUAN LANGUAGE:

- is spoken
- is an official language
- is a regional language

MAP 12.5 Quechuan language.

Bolivia, for example, marked the first time that an Indigenous person became president of that country; there had been seventy-nine presidents before him.

It is no surprise that Mayan and Quechua survive; they were the languages of sprawling empires that spread a common language across a wide area. Map 12.5 is in effect a language map that corresponds to the reach of the Incan Empire along the roads of its imperial system. By contrast, the languages of isolated groups in Amazonia, with their small population bases, are more vulnerable to extinction. However, if they retain their internal coherence and have little contact with the outside, then, despite their small population base, they may survive.

Despite (or perhaps because of) the replacement and death of languages, there is evidence that people are aware of the need for linguistic revival and protection. The Nooksack language, for example, effectively died out from 1977 to 2001 with the death of the last native speakers. However, a cultural rebirth, including the place name inventory that we discussed at the beginning of this chapter, as well as people using recordings of native speakers, taking language classes, and recognizing that their language is an important part of their cultural legacy, has given Nooksack new life.

12.5 Globalization of Language

The globalization of certain languages is a recent trend. Spanish, for example, was one of the first European languages to be exported as part of the expanding Spanish Empire. Spanish became the language of the elites throughout Central and South America. The emigration of Chinese speakers spread the language along the migration routes to diasporic communities across the world. But the most global language in terms of its widespread diffusion and adoption as a second and third language is English, now the hegemonic language of the contemporary world. Waves of globalization centering in the United Kingdom and United States, the extensive impact of the British Empire during the colonial period, and the dominance of the American economy, culture, technology, and politics in the contemporary world have all led to the dominance of English. It is now the language of global interaction. Most scientific, technological, and academic information in the world is expressed in English, and over 80 percent of all information stored in electronic retrieval systems is in English. It has become the language of international entertainment, popular music, movies, and advertising.

A core–periphery model of "global Englishes" is suggested by Braj Kachru: an inner ring of native speakers, an outer ring of institutionalized local varieties—often a function of extended periods under British colonization—and an expanding circle of countries that use English as part of international communication (Figure 12.2).

FIGURE 12.2 Circles of English use.

The adoption of English has varied. In Jamaica, English became part of a creole language, while in Singapore, English joined Chinese and Malay, and later Hindi, as part of a multilingual society.

Over seventy-five countries officially recognize English as a primary or secondary language. They include Botswana and Cameroon, India and Liberia, Rwanda and Sudan. The number of people who speak English as a first or second language is estimated at 573 million. A further 670 million may have native-like fluency in the language. Even nonspeakers are exposed to the language on a daily basis through advertising, government functions, or interpersonal communication. David Crystal describes them as members of an expanding circle who have been exposed, via the dynamic processes of globalization, to the usage of English on many informal levels. Almost 1.6 billion people, over a fifth of the world's population, exhibit some reasonable competence in the language.

Growing global English competency has accelerated the rate of **cultural globalization** by facilitating the movement of ideas and information. The expanding use of English around the world, however, is not simply the result of diffusion outward from a dominant center; for many people worldwide, its adoption is conscious. People in non-English-speaking countries are trying to learn English to participate more fully in international activities. English is required to be competitive in global markets. Many countries around the world have adopted English as a second language and emphasize it as an important subject in their schools. In the early days of the People's Republic of China, English was labeled as an "imperialist language" and suppressed. After the break with the Soviet Union, teaching Russian was abandoned and teaching English was revived and relabeled as "the instrument for struggles in the international stage." In the contemporary People's Republic of China, use of English has greatly increased, fueled by the increasing number of students educated in English-speaking countries, especially the United States, and rising demand for English-language skills among professionals and workers in the booming international trade sector. English-language teaching is now an important industry. Even street signs now use English as well as Chinese (Photo 12.8).

For individuals in many countries, English skills are an invaluable asset in the job market. English is a form of

PHOTO 12.8 Street signs in Shanghai.

cultural capital for national governments seeking to produce a globally competitive workforce and individuals wishing to achieve more employment opportunities in a globalizing world.

Not only has English expanded to become the dominant worldwide medium of communicative practice, but also it has undergone considerable reinvention by nonnative speaking communities, many of whom are speaking English in their own ways. Like many other cultural forms, English is being both **deterritorialized**, as it spreads from its cultural hearth, and **reterritorialized**, as nonnative speakers reshape the language to best suit their purposes. Instead of adopting a standard form of English, the language is being reterritorialized as a global medium of communication. And standard English is being changed in the process. The subtitle of David Crystal's survey of global English is "One Language, Many Voices."

Robert McCrum writes of "Globish" as the new spoken international language, primarily used in economic transactions. It joins the list of other hybrid languages, such as Spanglish (Spanish–English). One colleague from Copenhagen assures us that she speaks fluent Danglish (Danish–English).

The complex history of English as a language of both the colonizer and the colonized is summarized in the recent history of India, which was an important part of the British Empire. To govern this vast and populous land, the colonial authorities created an English-speaking Indian intellectual elite to help run the empire efficiently. The English-speaking elite became the leaders of the independence movement. In 1950, the constitution of the newly independent India was written in English. And the languages of the colonized were incorporated into English. Words like *nirvana*, *bungalow*, and *jungle* are now part of standard English.

The Written Language

English is both spoken and written. Whereas spoken English is a much more flexible form of communication—it is legitimate to speak of global Englishes, including Spanglish—written English is more rule-bound, programmatic, and slow to change. While we can think of a reterritorialization of spoken English, a reterritorialization of written English is less apparent. The use of English as a written global form of communication is particularly visible in the communities of knowledge. English has become the language of global intellectual discourse and the dominant language of intellectual communities involved in the production, reproduction, and circulation of knowledge. English dominates the **global epistemic community**.

Cited References

Crystal, D. 2003. *English as a Global Language*. 2nd ed. Cambridge: Cambridge University Press.

Crystal, D. 2010. *Evolving English: One Language, Many Voices*. London: British Library.

Harrison, K. D. 2007. *When Languages Die: The Extinction of the World's Languages and the Erosion of Human Knowledge*. New York: Oxford University Press.

Kachru, B. 1992. *The Other Tongue: English across Cultures*. Urbana: University of Illinois Press.

Krupnik, I. 2011. "How Many Eskimo Words for Ice? Collecting Inuit Sea Ice Terminologies in the International Polar Year 2007–2008." *The Canadian Geographer* 55:56–68.

Labov, W. 1972. *Sociolinguistic Patterns*. Philadelphia: University of Pennsylvania Press.

McCrum, R. 2010. *Globish*. New York: Norton.

Moser, S. 2009. "Communicating Climate Change: History, Challenges, Process and Future Direction." *WIREs Climate Change*. https://doi.org/10.1002/wcc.11.

Moser, S. 2016. "Reflections on Climate Change Communication Research and Practice in the Second Decade of the 21st Century: What More Is There to Say?" *WIREs Climate Change*. https://doi.org/10.1002/wcc.403.

Nettle, D., and S. Romaine. 2000. *Vanishing Voices: The Extinction of the World's Languages*. New York: Oxford University Press.

Richardson, A., and B. Galloway. 2011. *Nooksack Place Names: Geography, Culture, and Language*. Vancouver: University of British Columbia Press.

Robb, G. 2007. *The Discovery of France: A Historical Geography from the Revolution to the First World War*. New York: Norton.

Select Guide to Further Reading

Basso, K. 1996. *Wisdom Sits in Places: Landscape and Language among the Western Apache*. Albuquerque: University of New Mexico Press.

Eira, I. M. G., C. Jaedicke, O. H. Magga, N. G. Maynard, D. Vikhamar-Schuler, and S. D. Mathiesen. 2013. "Traditional Sámi Snow Terminology and Physical Snow

Classification—Two Ways of Knowing." *Cold Regions Science and Technology* 85:117–130.

Lupyan, G., and R. Dale. 2016. "Why Are There Different Languages? The Role of Adaptation in Linguistic Diversity." *Trends in Cognitive Sciences* 20:649–660.

Robson, D. 2012. "Chilly Words: How Eskimos Really Say 'Snow.'" *New Scientist* 216:2896–2897.

Short, J. R., A. Boniche, Y. Kim, and P. Li. 2001. "Cultural Globalization, Global English and Geography Journals." *Professional Geographer* 53:1–11.

Thurman, J. 2015. "A Loss for Words: Can a Dying Language Be Saved?" *The New Yorker*, March 30, 2015, 32–39.

Urban, M. 2021. "The Geography and Development of Language Isolates." *Royal Society Open Science* 8:202–232.

The Global Geography of Culture

13

Culture is what makes us social animals. Cultural forms are rarely fixed; they are best considered as circulating flows of ideas and practices that are always in the process of being embedded in place, as well as diffusing over space. In this chapter we will elucidate these ideas of fixity and flow with reference to architecture and tourism and show how culture is commodified as well as celebrated.

Culture is a tricky concept. It is used to cover things in the material world as well as systems of belief, ideas, and practices. The word has historically implied the notion of cultivation, of tending to something. This is an important point, because it reminds us that culture is not an a priori set of characteristics; it is something that is produced and cultivated as a response to specific circumstances. Culture is always provisional, elastic, and in a continual process of being made and remade.

Until the late eighteenth century, the word "culture" was often used interchangeably with "civilization," referencing a belief in the Enlightenment ideology of human mastery over nature. Yet at the end of the eighteenth century, thinkers such as the German philosopher Johann Gottfried Herder (1744–1803) wrote of the need to conceive of a plurality of cultures. There were entire cultural worlds below the privileged life of the court and beyond the refined manners of the elite. Local peoples also had culture. This view was reinforced in the nineteenth century through a self-conscious creation of "national" and "local" cultures. For example, in the nineteenth century, the Czech National Revival

LEARNING OBJECTIVES

13.1 Define the concept of the cultural region and give examples.

13.2 Describe how culture spatially diffuses.

13.3 Explain the sources of cultural production.

13.4 Explain how cultural products become deterritorialized or reterritorialized and how cultural practices are transplanted or transformed.

13.5 Explain the concept of "glocalization."

13.6 Explain the limitations with the myth of homogeneity.

was a broad-based cultural and political movement in which a Czech identity was crafted in music, arts, and literature. Czech composers such as Bedřich Smetana and Antonín Dvořák drew on folk motifs in their chamber music and operas. It was an attempt to forge a distinctive national culture at a time when the ruling elite spoke and wrote in German, and globalization, according to some, was sweeping away nationally distinct and uniquely local cultures. The defense of culture involves the invention of culture. Wherever and whenever a "culture" is perceived as weakened by contact with a foreign other, local cultures are not only preserved and defended, but also created and expanded. Whenever a wave of globalization sweeps around the world, there is a corresponding revival of the local and the national. Resistances to globalization often take the form of more explicit concern and promotion of national and local cultures as a way for groups to forge, redefine, or re-establish a place in the world. Culture is both a resistance to and an accommodation with wider social forces.

13.1 Cultural Regions

A cultural or culture region is defined as an area of similar culture. Different types of cultural region can be identified on the basis of language, religion, ethnicity, or shared history. The historical geographer Don Meinig, for example, identified a Mormon culture region consisting of a core in the Wasatch Mountains, a domain over much of Utah and southeastern Idaho, and a wider sphere extending from eastern Oregon to Mexico.

A culture region is different from other areas because of human behaviors and/or perceptions. For example, the American South is an area of distinct speech, history, and sense of itself compared to the rest of the United States. The exact boundaries are more fluid than fixed. With cultural regions, we need to adopt Meinig's notion of core and surrounding peripheries.

The existence of a cultural region is in part a self-identification by the majority of people in the region as well as by external perceptions. Map 13.1 shows the level of agreement about the American South by respondents in Michigan and Indiana. Notice how 91 and 96 percent of all respondents agree on a very narrow demarcation. Most would agree on the inclusion of Charleston in South Carolina (Photo 13.1).

The culture in cultural regions is sometimes contested rather than given. Not everyone in the US South celebrates the traditional definition of the South. Its complex history as a place of segregation and rebellion against federal authority makes for complicated cultural celebrations. Cultural regions are not homogeneous. Not everyone in the Bible Belt is Christian or even religious. Cultural regions are complex, part created, sometimes imposed, often contested, and never easily defined, especially outside a very narrow core area.

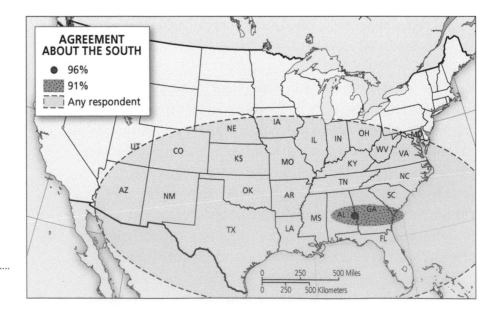

MAP 13.1 Levels of agreement about the location of the US South.

PHOTO 13.1 Downtown Charleston, South Carolina.

13.1 Confronting Cultural Symbols of Inequality

For some in the American "South," the Confederate flag symbolizes a memorial to the Confederate soldiers and an expression of their identity as Southerners. But for many others, it has become an enduring symbol of racism and opposition to civil rights. Adele N. Norris reflects on the Confederate flag as a symbol that continues to represent violence, fear, and intimidation among Black people:

I was a senior at Mississippi State University when voters, in a state-wide election, chose to keep the flag in 2001. My professor at the time held a mock election in one of my agricultural economics classes. Immediately, my classmates began declaring, "It's our heritage. Who's it hurting?" As the only Black student in the class, I became acutely aware that my feelings toward the flag didn't count. The mock election resulted in one vote in favor of changing the flag. I remembered the relief that apparently overcame the class and my professor afterward,

although the "heritage" that my classmates were in fear of losing was never articulated or discussed: Mississippi's relationship with the Confederate flag.

The Confederate emblem is directly linked to the Confederate cause during the Civil War. After the war, the flag came to symbolize the preservation of the "old ways" where Confederate veterans and supporters engaged in acts of terror in their efforts to re-establish the South. The flag was waved and displayed in activities that ranged from assaults, lynchings, and the general terrorizing of Black southerners ... the Confederate banner was used to strike fear and intimidation among Black people.

REFERENCE

Norris, A. 2020. "Mississippi Removes the Confederate Emblem from State Flag: Confronting Heritage." *Social Science Matters.* https://www.palgrave.com/gp/blogs/social-sciences/norris.

13.2 Spatial Diffusion

While traditional cultural geography focused on the identification of cultural regions, much of contemporary cultural geography is concerned with culture as flow, how culture is dynamic and shifting, and is both influenced by and influences national and global processes. Before we

discuss this, however, we need to understand the concept of spatial diffusion.

Spatial diffusion is the movement of new ideas, things, and practices through time and space. The adoption of innovations follows predictable patterns. Let us take, for example, the adoption of Christianity from its center in the Middle East. It spread along lines of communication, the trade routes that linked the cities around the

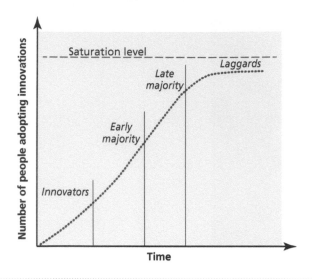

FIGURE 13.1 The spatial diffusion model for a new innovation.

Mediterranean. From there it spread quickly through towns, into the countryside, and around the world.

Not everyone adopted the new religion. Figure 13.1 shows the typical pattern for a new innovation. Innovators are those people who quickly pick up and adopt new trends and ideas. They are the first person you know with the newest version of the iPhone and the latest app. Then there are the early majority who come in the wake of the innovators, then the late majority, and finally the laggards. Lisa is an early majority adopter, but John is most definitely a laggard in the adoption of new technologies. He is always the last person with a cellphone or an Uber app.

Diffusions also vary across the urban hierarchy. Figure 13.2 could be used to model the early adoption of Christianity. It occurred first in the towns and was later diffused through the countryside. The slow diffusion into the

rural areas leaves us with the linguistic legacy of "pagan," derived from an early Latin word for villager.

Ideas and practices are abandoned as well as adopted. The rejection of a former innovation is referred to as a **paracme**. There are three situations. In situation (a), there is symmetry between a diffusion and a paracme. Situation (b) occurs when there is long, slow adoption and quick rejection, as in the case of smoking in the United States, which developed slowly over a century and then plummeted with health scares. In situation (c), there is quick diffusion and a slow paracme, as with the case of the rapid adoption of railways in the nineteenth century, followed by a long, slow decline.

13.3 Culture as Flow

Traditional cultural geography highlighted the connection to the local. Emphasis was on vernacular building forms, local ways of doing things, and the construction of specific landscapes. More recently, the field has been enlivened by an awareness that culture is shaped not only by people's connection to the local, but also by their interaction with the national and the global, which in turn can influence the national and global.

Consider just one example, chocolate. It developed first in the Mayan region of Central America, a region that also saw the first cultivation of corn and tomatoes. The local people made an unsweetened chocolate drink from locally grown cacao seeds. The Aztecs refined the process by pouring the chocolate liquid through blowers. They annexed the region around 1400 and soon levied taxes paid in cacao. Later, the Spanish adopted chocolate as a favorite food, adding sugar and cinnamon. The concoction was then imported into Europe, where it became popular. The demand grew so high that industrial production levels were needed and plantations were established in West Africa. In 2022, more than two-thirds of the world's cocoa was produced in West Africa, but the chocolate is manufactured in Belgium, Switzerland, and Hershey, Pennsylvania. So, a food grown from seeds in Mesoamerica is refined by the Aztecs and the Spanish, imported into Europe, and creates huge demand that leads to cocoa plantations in Africa (where many countries rely on illegal child labor for cocoa harvesting) and manufacturing plants around the world. When you bite into a chocolate bar, you are taking part in complex cultural flows that have spanned the globe for at least 500 years.

There are three dynamically linked sources of production of culture: the connection with the local, the relationship with the national, and the interaction with the global. In this chapter, we will focus on the connections between the global and the local.

Cultures are never static, fixed in time and place. A subtler view of cultures sees them as resulting from ongoing interactions between the local, the global, and the

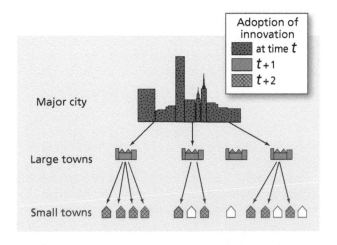

FIGURE 13.2 Diffusion of an idea through the urban hierarchy.

13.2 The Diffusion of Diseases

The spread of disease is an example of both diffusion and space-time convergence. Space-time convergence, the shrinking of the time taken to cover distance, influences the diffusion of disease. The travel time between India and Fiji during the time of sailing ships was longer than the life span of the measles virus. The virus was thus not imported when Indian laborers were shipped to the islands as cheap labor for the plantations. People infected with measles when they left India had enough time to recover from the disease before they docked in Fiji. But travel times during the age of steamships decreased to within the incubation period of measles. Measles contracted in India was then imported into Fiji, with devastating results. In 1875, one-third of Fijians died of the measles, a striking example of a historical mass mortality resulting from the diffusion of an infectious disease. Similarly, innovations in transportation such as the railway in the United States made it easier for cholera epidemics to sweep across the country in more frequent outbreaks.

In our globalized world, there is a greater possibility of global pandemics. The influenza pandemic of 1918 killed between 3 and 5 percent of the world's population, with total death estimates ranging from 50 to 100 million.

The collapse of distance brought about by recent transport improvements, such as jet travel, has further increased the ease and frequency of the transfer of communicable diseases. The swine flu epidemic of 2009, similar to the 1918 pandemic in that it was an H1N1 influenza virus, infected one-fifth of the population and more than half of all children in nineteen countries; it killed more than 200,000 people around the world. Although very contagious, it was not very lethal, killing fewer than 2 of every 10,000 infected.

A novel respiratory disease was identified in the city of Wuhan, China, in December 2019. It developed sometime between October and November 2019 when a bat coronavirus, perhaps combined with a pangolin virus, entered the human population, probably through people who were processing bat carcasses for traditional Chinese medicine. The virus was given the name of severe acute respiratory system coronavirus-2 (SARS-CoV-2). The disease it causes, a respiratory illness that can be fatal, was named COVID-19. Within months, perhaps weeks, of the first outbreak, the disease seeded and spread around the world. Globalization has "smoothed" the world, making disease transmissions, especially one spread by person-to-person contact, all the more rapid and virulent. Globalization is an accelerant capable of turning outbreaks into epidemics and epidemics into pandemics. By the summer of 2022, more than 538 million people had been infected and 6.32 million had died. In the United States, 86 million cases had been recorded and over 1.1 million people had died.

Scientists mapped the spread of the disease. Map 13.2 shows the rapid spatial diffusion of the disease in early 2020 in Italy. From a hotspot in northern Italy, it quickly spread across the country.

REFERENCES

Cliff, A., and P. Haggett. 2004. "Time, Travel and Infection." *British Medical Bulletin* 69:87–99.

Cliff, A. D., P. Haggett, and M. S. Raynor. 2004. *World Atlas of Epidemic Diseases*. London: Hodder Arnold.

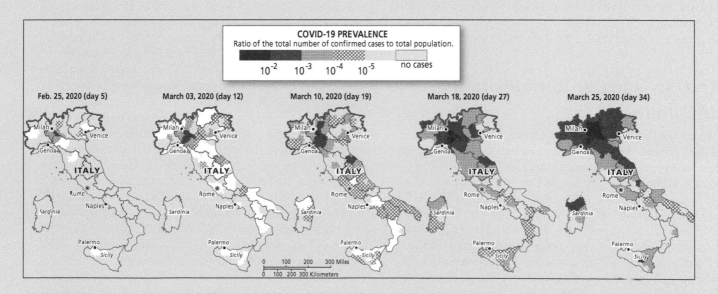

MAP 13.2 The early spread of COVID-19 in Italy.

13.2 continued

Gatto, M., Bertuzzo, E., Mari, L., Miccoli, S., Carraro, L., Casagrandi, R., and Rinaldo, A. 2020. "Spread and Dynamics of the COVID-19 Epidemic in Italy." *Proceedings of the National Academy of Sciences* 117:10484–10491.

Kerkhove, M. D., Hirve, S., Koukounari, A., et al. 2013. "Estimating Age-Specific Cumulative Incidence for the 2009 Influenza Pandemic: A Meta-Analysis of A(H1N1) pdm09 Serological Studies from 19 Countries." *Influenza and Other Respiratory Viruses* 7:872–886. http://onlinelibrary.wiley.com/doi/10.1111/irv.12074/abstrac.

Penman, B. S., S. Gupta, and D. Shanks. 2017. "Rapid Mortality Transition of Pacific Islands in the 19th Century." *Epidemiological Infections* 145:1–11. https://doi.org/10.1017/S0950268816001989.

national. The processes that shape local and national cultures are not one-way interactions but rather are dynamic and multifaceted, so that hybrids of the "newly arrived" and the "previously there" are constantly reconfigured and remobilized in and through global flows of people, ideas, and beliefs. Culture is not just passive audiences watching imported television programs or eating fast food from global chains. Culture is an active process; one definition of culture is "tending, husbanding resources," with the same root as "cultivation." A culture is not just consuming things; it is the production of a worldview, an aesthetic sensibility, the active creation of a history and geography, the working out of a place in the world. Culture is the active use of global flows of ideas and practices by local and national groups adapting, adopting, or rejecting these flows in their own specific worldviews, which in turn become part of further flows of culture used and adapted, rejected and resisted, by others in turn.

Global flows of culture are continually shaping and reshaping the world. We can identify flows of people, technologies, capital, media images, and the political configurations of such ideas as freedom, welfare, rights, sovereignty, representation, and democracy. Flows circulate globally and become re-expressed through and in local contexts.

The (Re- and De-)Territorialization of Culture

When we think of culture, we often think of distinctive songs, dances, foods—all those things that comprise collective memories and group identities. In much of our understanding, these are all tied to place: Mexican food, Indian music, Italian design, the Armenian language. The term "**territorialization** of culture" signifies this connection between culture and specific places. However, it is also important to consider the process of the deterritorialization of culture, in which cultures are lifted from particular places, and the reterritorialization of culture, whereby cultural practices are transplanted and transformed.

Migrant communities throughout the world are important networks of information flows, capital flows, and economic exchanges. Just as chains of migration are routed through families and friends, so business deals are conducted among family members dispersed throughout the world. Economic globalization is intimately linked to this particular form of global population dispersal. Businesses are culturally embedded. A flattening world is linked to the deepening economic interactions of far-flung communities, with shared family ties, ethnicity, and race also functioning as conduits of international trade and commerce.

International migration shapes the culture of both destination and origin areas. When migrant communities are established, they bring their culture with them. Sometimes it is frozen in the time of departure and lasts longer than it does at home. Yet cultures always have enough plasticity to adapt to the new and the foreign. Both home and away, origin and destination are constantly changing in response to these continual interactions and flows.

Globalization, it has often been argued, undermines local identities. However, an alternative case can be made that globalization has both deterritorialized and strengthened local cultures. In the nineteenth century, there was a massive immigration to the United States from many European countries. Italians, Germans, Poles, and Irish all moved in the millions to the United States. Their connection with home was always limited, separated as they were by rudimentary communication links. Letters could take weeks. New identities were created. People from different parts of what is now Italy became "Italians" in the United States: a kind of Italian nationality was achieved in the United States even before there was an Italian nation-state. New identities were shaped. Irish Americans took on the mythic elements of Irish nationalism. Folk memories did not die; indeed, in many cases they became both strengthened and stuck in time. Irish Americans, unlike the Irish in Ireland, could scarcely "remember" anything other than the famine. In more recent years, however, immigrants to the United States and other destination countries have been able to stay in closer connection with their families back home. Cheaper travel, ease of transmitting money, and instant communications all allow diasporic communities to remain in contact, to influence their new surroundings, and to be subtly transformed in the process of being diasporic. All those easier communication systems of a globalized world allow groups to be more in touch with their home areas.

The widespread global diffusion of flows of people and ideas has broken the simple, unproblematic connection between culture and place. On closer inspection, the connection between culture and place is always complex and problematic. "Mexican" food, for example, is a mixture of Spanish and Indigenous cuisines that evolved in different parts of the world in different ways for almost 500 years. When Spaniards came to the New World and enslaved Africans were shipped across the ocean, they combined with Indigenous people to create a New World cuisine that was creolized and hybridized right from the beginning. There is no unchanging Mexican cuisine; it is and always was a changing hybrid.

Cultures are always in the process of deterritorialization and reterritorialization, particularly marked during waves of globalization and reinforced by rapid space-time convergence. We can consider the reterritorialization of culture as revealed by architecture.

Architecture

For the great German polymath Johann Wolfgang von Goethe (1749–1832), architecture is "silent music." Two of the dominant melodies of this music are the vernacular and the global. **Vernacular architecture** uses local resources and traditions in the design and construction of buildings. The materials draw on local sources, whether they are wood or clay, ice or rock, and architectural design is a calibrated response to local conditions of weather and topography. Classic examples include the Inuit igloo, built of local ice, or the Swedish Lutheran church, built of local birch.

There are also **transnational architectural styles** that transcend particular places. Consider the classical architecture of Greece and Rome, so admired that they have been copied and recopied all over the world (Photo 13.2). **Neoclassical**

PHOTO 13.2 The Parthenon, Athens. Work began on this building in 447 BCE. It was dedicated to the goddess Athena.

architecture, the style that echoes the classic architecture of Rome and Greece, became popular in the mid-eighteenth century and soon was used in countries and cities worldwide, especially as colonial powers expressed their power though building neoclassical buildings in their colonies and overseas territories. Here, we come to a complex connection between place and culture. Neoclassicism meant different things in different places. In the United States, it was associated with democratic values and republican virtues (Photo 13.3).

PHOTO 13.3 The neoclassical design of the Lincoln Memorial, Washington, DC. Construction commenced in 1914.

In Nazi Germany, it was associated with the attempt to build a thousand-year Reich based on racist beliefs, with the use of "timeless" classical imagery reinforcing the political agenda of Nazi ideology.

From the 1950s to the 1980s, architectural designs around the world were associated with modernist designs. **Modernist architecture** was concerned with form following function, truth to material, and a revulsion against bourgeois decoration. It preferred clean-cut lines and soon became serialized into tall, flat-topped buildings constructed in towns and cities all over the world—so specifically meant for global usage that it was designated "the International Style." And it truly was international: Communist headquarters in eastern Europe, corporate offices in New York, government buildings in Brasilia, and public housing projects in London and Chicago all used it (Photo 13.4). Despite the differences in place, the buildings all looked the same. In the West, international modernist architecture soon reached dominance and then was so ubiquitous as to become boring. The modernist mantra of "less is more" turned into "less is a bore."

Since the 1980s, there has been a reaction to modernism, usually called postmodernism. It is a conscious distancing from the austere, flat-topped, straight line of modernist design; it includes historical referencing of the local place,

PHOTO 13.4 Modernist buildings along Sixth Avenue, New York.

ornamentation, and playfulness with material and design. Office buildings with Georgian facades, museums with Renaissance decoration, and neoclassical houses have become popular. Variants include **deconstructivism**, which involves the use of nonrectilinear surfaces (Photo 13.5). In the highly competitive world of global cities, not to have a postmodernist building is to be out of date, too far from the cusp of the future.

This hyperbrief history of architectural styles is just a small indication of the way that buildings and cities reflect transnational forms and styles.

13.4 The Production and Commodification of Culture

The global production of culture takes many forms. Here we will consider the production of space, the tourist gaze, ideas of East and West, and the commodification of culture.

PHOTO 13.5 A postmodern building in New York City, designed by Frank Gehry and completed in 2007.

The Production of Global Space

The production of global space has two distinct meanings: the creation of an understanding of the world's geography and the representation of this geography in cartographic forms. A traditional view sees both things simply as a Eurocentric process in which European powers incorporated the world into a global discourse and created new, more modern forms of cartographic representation. It is the story of how Europe explored, discovered, and mapped the world, displacing older, indigenous cartographies with more scientific mappings. Cartographic modernity in this rendering is exported from Europe to the rest of the world. There is now an exciting body of work that points to the creation of new knowledge in the process of interactions and encounters between colonial and colonized, the imperial center and the periphery, the modern and the indigenous. Knowledge was not so much imposed from one region of the world but created and circulated in a complex series of interactions in many different regions of the world. Modernity was created in the space of encounters; continually imported, transformed, and modified; and then re-exported in a continuous and ongoing global circulation of ideas and practices.

Kapil Raj describes the integration in the circulation of knowledge between South Asia and Europe from 1650 to 1900. Raj undermines the traditional view that science was exported from the European core to the colonial periphery. He shows that there was an intercultural knowledge encounter that produced new knowledge. In particular, the case of the geographical exploration of British India in the late eighteenth and early nineteenth centuries provides a good illustration of the way in which British and Indian practitioners and skills met around specific projects, how they were reshaped, and how the modern map and its uses coemerged in India and Britain through the colonial encounter.

The Tourist Gaze

Tourism is an important industry. There are almost 1 billion international tourist visits each year (prior to COVID-19) to a vast array of tourist landscapes: Florida beaches, resorts along Spain's Mediterranean coastline, the small Caribbean islands of Dominica and Barbados, and even remoter islands such as the Maldives and Fiji. It is a significant economic sector in large countries. Countries with the largest tourist arrivals are France, the United States, and China, and the countries with the largest revenues from tourism are the United States, Spain, France, and China.

Tourism used to be the preserve of wealthy English aristocrats taking the Grand Tour, visiting classical Mediterranean countries, often purchasing artwork to add culture and class to their residences back home. With space-time convergence, mass tourism became possible and brought more of the world into easier and quicker tourist access. The tourist market is now finely graded, from cheap mass tourism to exclusive and expensive tourist experiences. The cruise ship industry is finely graded by price, quality, and destinations. Cheaper cruises tend to visit sunny recreational sites in the Caribbean, while more upmarket cruises visit a wider range of destinations in luxury ships (Photo 13.6).

As tourist sites have become more popular, there is constant demand for a more selective experience, away from the crowds. There are a variety of tourisms, including beach holidays, ecotourism, heritage tourism, culinary tourism, and even slum tourism, in places such as the favelas of Brazil (Box 13.3). In the wake of Hurricane Katrina, disaster tourists visited New Orleans to see the aftermath of the storm.

John Urry coined the term, the **tourist gaze**. When people go on holiday, they look on different landscapes, settings, and peoples. This gaze is socially organized and systematized. He highlights how the tourist gaze—"gazes" would be more accurate—has changed for different groups

PHOTO 13.6 Luxury cruise ship visits Ålesund, Norway.

and places over the years. There are stock images promoting Paris as romantic, Southeast Asia as exotic from a Euro-centric perspective, English villages as representing an unchanging organic order. The gaze is not an innocent seeing, but a perception guided and shaped and freighted with subtle and not-so-subtle messages about history and geography. Cultural geographer Patrick McGreevy explores the changing meaning of visiting Niagara Falls.

13.3 Sustainable Tourism?

One sustainability trend on the rise has been ecotourism. The International Ecotourism Society, founded in 1990, defined it as responsible travel that conserves the environment, sustains the well-being of the local people, and includes interpretation and education. In theory, ecotourism can provide economic benefits to communities living around protected areas by encouraging them to protect the biodiversity in their own interest. The Society lists the following principles of ecotourism:

Minimize physical, social, behavioral, and psychological impacts;

Build environmental and cultural awareness and respect;

Provide positive experiences for both visitors and hosts;

Provide direct financial benefits for conservation;

Generate financial benefits for both local people and private industry;

Deliver memorable interpretative experiences to visitors that help raise sensitivity to host countries' political, environmental, and social climates;

Design, construct, and operate low-impact facilities;

Recognize the rights and spiritual beliefs of the Indigenous People in the community and work in partnership with them to create empowerment.

Costa Rica has long been considered an exemplar ecotourism country. It is almost one-fourth rainforest—a major draw—and boasts volcanoes and beaches and rich biodiversity. Over the past three decades, the ecotourism industry in Costa Rica has expanded extensively, making Costa Rica one of the world's most popular ecotourism destinations.

The development of ecotourism in Costa Rica resulted from a forward-thinking 1998 Biodiversity Law. The law helped establish entrepreneurship training programs tailored to the needs of each community. The programs teach business development with a focus on environmental and social responsibility and are hosted by various environmental organizations, including the Nature Conservancy and Conservation International.

One study by C. A. Hunt and colleagues on the social impact of ecotourism on local communities focused on the Osa Peninsula, a region in southwest Costa Rica. They found that the tourism industry tends to hire more local people than other sectors in the Costa Rican economy. Ecotourism can also provide jobs with higher salaries and better opportunities for advancement than other jobs in the region, especially for young people, who often have lower skills and less experience than the labor force as a whole, and women with children, who need a more flexible working schedule to balance child care. Moreover, workers employed in ecotourism are less likely to engage in illegal logging or the extraction of nontimber products, further reducing deforestation. This highlights that effective ecotourism can have multiple positive benefits for a community and its surrounding environment.

However, ecotourism is not always successful or sustainable. Revenues from ecotourism do not always reach local communities; and there have been reported negative impacts from ecotourism such as increased pollution, solid waste generation, degradation of forest, trail erosion, and disturbance to plants and animals. Political ecology scholars have found that ecotourism can replicate power relationships that further marginalize local communities.

Priyanka and Aditya Ghosh examined Sundarban Biosphere Reserve, home of the highly endangered royal Bengal tiger in one of the largest mangrove forests in India, touted as catering to both biodiversity conservation and socioeconomic development of local communities living around the protected area. They found the Sundarban ecotourism project failed to offer any benefits to the poorest and most marginal communities that surround the reserve. On the contrary, it offered disproportionately larger returns to the ecotourism companies and hotels that control lodging, food, and transportation and are not owned locally. Additionally, locals blamed tourists for increasing pollution and harming the health of the ecosystem. Their study finds that ecotourism cannot be considered a magic bullet for biodiversity conservation and local socioeconomic development.

REFERENCES

Ghosh, P., and A. Ghosh. 2019. "Is Ecotourism a Panacea? Political Ecology Perspectives from the Sundarban Biosphere Reserve, India. *GeoJournal* 84:345–366. https://doi.org/10.1007/s10708-018-9862-7.

Hunt, C. A., W. H. Durham, L. Driscoll, and M. Honey. 2015. "Can Ecotourism Deliver Real Economic, Social, and Environmental Benefits? A Study of the Osa Peninsula, Costa Rica." *Journal of Sustainable Tourism* 23:339–357. https://doi.org/10.1080/09669582.2014.965176.

He shows how the perception of the place is bound up with themes of remoteness, death, nature, and the future. The experience itself has changed over the years from an almost sacred pilgrimage, part of a romanticism of nature, into a more commodified tourist package. This commodification extends to sacred sites and monuments and memorials.

Consider the example of how sites of grief have become commodified. Many monuments and memorials about war or loss include visitor centers, gift shops, or kiosks, selling an assortment of "souvenirs" such as books, refrigerator magnets, shot glasses, and t-shirts, none of which interprets or enhances the memorial experience. With a tourist gaze, today's visitors often focus more on taking selfies for Instagram, or buying souvenirs, rather than thinking through the complexities of the experience, especially those around loss of life. The memorial has become commodified, something to be consumed, something that ultimately trivializes the message of the memorial.

Orientalism, Ornamentalism, and Occidentalism

There is a hierarchy to the production of geographical knowledge. The world is always described by certain people and from certain places. It is a legacy of the British Empire that to this day Iran and Iraq are described as the Middle East. They are only the Middle East if you are describing them from London. Cultural critic Edward Said described **orientalism** as the writing and description of these places by those in the West. The term is not innocent of political meaning and the operation of military power. To be described and portrayed is to be discursively captured.

Riffing on this idea, historian David Cannadine describes an **ornamentalism** that he defines as the importance of class and status as much as race and ethnicity, especially for the British Empire. He stresses the construction of affinities between different elites in the encounter between East and West. Cannadine's assertion is that class was as much of a key factor—if not more of one—as race and ethnicity in the social distinctions of the imperial experience.

There is also an **occidentalism** as well as an orientalism. While orientalism is a much-studied phenomenon, occidentalism receives rather less attention. The West is also understood and imagined by its enemies. Ian Buruma and Avishai Margalit trace the roots of anti-Western ideas. These two authors point to the important role of the sinful Western city in this ideology. Purifying the evil city is a major trope. They argue that the United States and especially its cities are often projected as a "rootless, cosmopolitan, superficial, trivial, materialistic, racially mixed, fashion-addicted civilization."

13.5 The Commodification of Culture

Globalization has not disrupted the long connection between culture and place, but it has added some new wrinkles. The commodification of selected cultural forms has transformed some local cultures into globalized forms. Hollywood movies, for example, are now more international, dealing with broader themes than simply US concerns. Indian restaurants serve food that is only ever found in Indian restaurants around the world rather than replicating dishes from India. Sushi restaurants are found not only in Japan but also in cities all over the world and have become an important signifier of "foodie coolness."

In some cases, cultural globalization is aided by the nature of national cultures. Big-budget Hollywood movies tend to draw on US themes to provide narrative drive and meaning. But what exactly is this America that is being represented? It is an America that has to appeal to a heterogeneous experience even within the United States. For things to sell well in the United States, they already need to have a level of smoothing. Much of the warp and weft of life in specific parts of the United States is filtered through the need to reach a more general national market. A case can be made that American culture has been so successful in achieving global penetration because it has already been partly globalized just to reach a wide US audience. The country is so diverse and heterogeneous that US cultural products have to be smoothed out to succeed, and that primes them for global dispersal.

National and local cultures in turn also influence global cultural forms. As more and more US films are sold abroad, US filmmakers must make allowances for foreign markets. The particularities of US life are replaced in expensive Hollywood movies by images, plots, and characters that are instantly recognizable by overseas viewers. A more recent trend has been to make movies that feature a major US actor and a popular foreign actor, in the hopes that it will be a box-office hit with both domestic and international audiences. For example, the 2016 *Great Wall* was a Hollywood–Chinese co-production that starred Matt Damon, Jing Tian, and Andy Lau. The film grossed nearly $335 million.

The commodification of culture does lead to a certain similarity in the cultural mix available to audiences around the world. But the more astute companies tailor their cultural products for specific audiences. Akio Morita, the legendary head of Sony, first coined the term **glocalization**, which represented a business strategy to create worldwide operations that were attuned to local markets and conditions. Two waves of Coca-Cola globalization can be identified. The first and earliest involved the marketing and sale of Coca-Cola. The second and more recent involves the sale of Coca-Cola alongside the sale and marketing of "local" brands, often owned by Coca-Cola. In Brazil, soft drinks are often marketed by Coca-Cola as rainforest drinks with "forest" tastes and plant sources.

Culture is not divorced from commerce. Fredric Jameson notes that in the latest stage of capitalism, symbolic meanings and associations increasingly determine the economic value of goods. Imposing symbolism, meanings, values, and emotions onto goods blurs the boundary between the image of those products and their concrete reality. Economics and culture are now intrinsically linked. There is even a range of cultural industries that includes music, dance, theater, literature, the visual arts, crafts, and many newer forms of practice, such as video art, performance art, computer art, and multimedia art. Economic geographer Allen Scott introduces a broader list of "cultural-products sectors," which includes high fashion, furniture design, news media, jewelry, advertising, and architecture.

13.6 The Myth of Homogeneity

Debates on cultural globalization often polarize into whether the recent surge of cultural flows and global consciousness has increased or decreased sameness between people and places around the world. One popular argument is that the world is becoming more alike as greater numbers drink Coca-Cola, eat at McDonald's, and watch Disney movies. It is a compelling image that captures some, if not all, of the complexity. The fact that people across the globe are watching CNN and MTV, McDonald's restaurants are opening throughout the world, and Hollywood films dominate the world film market are taken as indisputable evidence of the homogenization and Americanization of the world (Photo 13.7). In this view, ideas and products flow from the core (the Global North) into the rest of the world. The **homogenization thesis** assumes that the same things are consumed in the same ways.

An alternative argument is that, while particular television programs, sport spectacles, network news, advertisements, and films may rapidly encircle the globe, this does not mean that the responses of those viewing and listening will be uniform. Goods, ideas, and symbols may be diffused globally, but they are consumed within national and local cultures. Ideas, symbols, and goods that circulate around the world are consumed in national contexts and in local circumstances. Similar goods mean different things in different places. There is enough variation to suggest that there is little prospect of a unified global culture; rather, there are a variety of global cultures. Specific cultural backgrounds are not just empty containers for the receipt of global

PHOTO 13.7 Fast food in Shanghai.

messages; they are critical to how messages are received and consumed. People in the contemporary world have become increasingly familiar with the presence of different cultures, rather than being sucked into a single cultural orbit. To be sure, more people around the world can draw on a similar range of flows, but there is still a wide variety of cultural formations. In fact, cultural globalization has led to a more explicit concern with more local cultures as certain groups seek to redefine or re-establish their place in the world. Consider the multidirectional flow of music. Many local genres of music have penetrated international markets and influenced musicians around the world. One example would be reggae, which originated in Jamaica in the 1960s. Its popularity then influenced and inspired other genres, including Western punk and rap. The recent explosion of K-pop bands such as B.I.G is another example.

We can replace the popular image of local cultures under assault from a global culture with a more complex picture of local cultures being generated and recreated in response to globalization. The sense of a pervasive globalization has not so much overwhelmed local cultures as helped create them. The search for authenticity in "pure" local cultures (which are in fact mixtures) has helped in the creation of world music, ethnic cuisine, and Indigenous culture. Religious fundamentalism is less a return to a pure theocratic ideology and more the self-conscious recreation of religious beliefs in the face of secularization and globalization. Cultural authenticity is less an excavation of the pure and more a contemporary representation of the contrived.

Cited References

Buruma, I., and A. Margalit. 2004. *Occidentalism: The West in the Eyes of Its Enemies.* New York: Penguin. (Quote is from page 6).

Cannadine, D. 2001. *Ornamentalism: How the British Saw Their Empire.* Harmondsworth, UK: Allen Lane.

International Ecotourism Society. https://ecotourism.org/.

Jameson, F. 1991. *Postmodernism, or, The Cultural Logic of Late Capitalism.* Durham, NC: Duke University Press.

McGreevy, P. 1994. *Imagining Niagara: The Meaning and Making of Niagara.* Amherst: University of Massachusetts Press.

Meinig, D. W. 1965. "The Mormon Culture Region: Strategies and Patterns in the Geography of the American West, 1847–1954." *Annals of the Association of American Geographers* 55:191–220.

Norris, A. 2020. "Mississippi Removes the Confederate Emblem from State Flag: Confronting Heritage." *Social Science Matters* (blog). https://www.palgrave.com/gp/blogs/social-sciences/norris.

Raj, K. 2007. *Relocating Modern Science: Circulation and the Construction of Knowledge in South Asia and Europe, 1650–1900.* New York: Palgrave.

Said, E. 1979. *Orientalism.* New York: Vintage.

Scott, A. 2017. *The Constitution of the City.* Cham: Palgrave.

Urry, J. 2002. *The Tourist Gaze.* 2nd ed. London: Sage.

Select Guide to Further Reading

Anderson, B. 2020. "Cultural Geography III: The Concept of 'Culture.'" *Progress in Human Geography* 44:608–617.

Knox, P. 2011. *Cities and Design.* New York: Routledge.

Lees, L. 2001. "Towards a Critical Geography of Architecture: The Case of an Ersatz Colosseum." *Cultural Geographies* 8:51–86.

Pieterse, J. N. 2019. *Globalization and Culture: Global Mélange.* Lanham, MD: Rowman & Littlefield.

Short, J. R. 2012. *Korea: A Cartographic History.* Chicago: University of Chicago Press.

Zelinsky, W. 1992. *The Cultural Geography of the United States.* Englewood Cliffs, NJ: Prentice Hall.

PART 5

The Political Organization of Space

There are many relationships between space and politics. There are the various organizations of political space, such as empires, states, and administrative units. Politics also revolves around issues of space and place, such as when neighboring states have boundary disputes or when a city government tries to locate a new incinerator. Finally, many political processes take place in space, as in the geography of elections. In this part we will look at some of these themes. Chapter 14 will focus on the political geography of world order, with particular emphasis on the rise and fall of empires and identifying the elements of a new world order. Chapter 15 will consider the state and how it operates in a national and global context.

OUTLINE

14

World Orders

The very earliest form of human society was small groups of extended families that, over time, merged into larger social units, eventually creating such diverse socio-spatial entities as empires, city-states, and nation-states. In the first part of this chapter we will look at empires as a spatial unit of political organization and the rise and fall of **imperialism**, which played an important role in shaping global human geography; then we will focus on contemporary global political geography.

14.1 The Geography of Empire

Some of the earliest empires were city-states in Mesopotamia. The city of Uruk emerged around 6,000 years ago in the southern part of Mesopotamia. Settled agriculture, sophisticated irrigation, and a marked division of labor propelled Uruk's growth. The city broadened its influence, referred to as the **Uruk expansion**, through a wider net of control in the upper reaches of Mesopotamia, between the Euphrates and Tigris Rivers. Surrounding peoples were brought into the cultural and economic control of Uruk. The city's elites developed more expensive tastes, and the search for more exotic goods and materials to meet this demand drove the expansion. Eventually, expansion reached its peak and then turned into decline. In a process that is repeated over and over again, the empire collapsed, local traditions re-emerged in regions formerly controlled by Uruk, and in the main city, invaders razed the monumental buildings of the temple.

14.2 Early Empires

There are a number of themes in this empire's brief story that figure in later imperial experiences: empires emerge, extend their reach, and then collapse from external pressures and internal conflicts. Across time and over space,

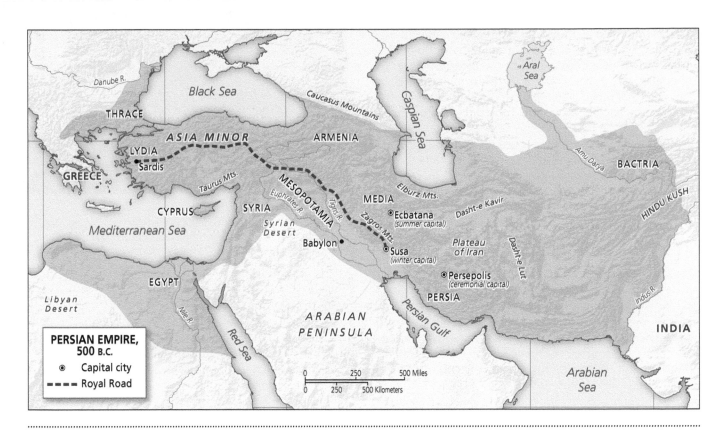

MAP 14.1 The Persian Empire.

the process is repeated. What is different is the increasing territorial extent of the empires and the enduring consequences of imperial incorporation. Subsequent imperial history is of powers becoming larger and their territorial reach more extensive as space-convergence improvements extend the effective imperial range. In the Middle East, after Uruk came the Hittites, the Assyrians, and the Persians. The Persian Empire peaked from around 559 BCE to 331 BCE (Map 14.1), its spatial expansion aided by the greater use of more mobile cavalry and better roads. At its greatest extent it covered a vast area, from the Indus valley in the east to Greece in the west and from Central Asia in the north to southern Egypt on its southern reaches.

We can make some generalizations about the geography of empire. Through time, empires grow in size as they are more able to overcome the tyranny of distance. While the Uruk expansion was limited to the Tigris and Euphrates River basins, the Persian Empire stretched from the Mediterranean Sea to the Indus River. Better transport and improved road networks enable empires to extend their spatial reach. Yet empires, no matter how vast, have territorial limits to their power. As they expand farther from the core, their ability to wield effective and overpowering force tends to weaken. There is a distance-decay effect in their military power and political reach. Far from the empire's center, borders are vulnerable. To defend them, early empires relied on

force, whose effectiveness declined markedly with distance from the imperial center. The farther an early empire stretched, the more difficult it was to control the ever-more-distant periphery. There were limits on early imperial expansion because of the marked distance-decay effect of military reach and political control.

14.3 Global Integration

The historical geography of the world is filled with the rise and fall of empires and the spatial spread of imperial influences. Alexander the Great reversed the flow of Persian influence by extending Greek influence throughout Eurasia, taking over much of the remnants of the Persian Empire. The Roman Empire turned the Mediterranean into a Roman lake. These and other empires, such as the Mali in West Africa and the Mayan, Aztec, and Inca in the Americas, created a more homogenous political surface that aided cultural mixing, economic ties, and religious dispersals, which diffused creeds and beliefs across space. Empires pulled peoples and regions together in shared experiences of (often forced and brutal) spatial convergence. Empires promoted global integration, often at the cost of local cultures and economies.

The more recent the empire, the greater its enduring consequences. English is spoken around the world because of the wide sphere of influence of the British Empire. The experience of empire is one of the most important global phenomena of the past 500 years. We will discuss three spatial elements: global integration, **imperial overstretch**, and imperial disintegration.

The global integration of space is the defining feature of early **modernity**. In the period 1400 to 1800, empire building across the globe (including the Chinese, Ottomans, Mughals, and Safavids as well as the Spanish and British) created international markets and global exchange networks. These global empires encouraged, forced, and facilitated the movement of people, the spread of new technologies, the diffusion of cultures, and the transmission of religion and scientific practices. The result was a tighter integration of global space. Increasing contact between different cultures led to cross-cultural borrowings as well cultural destructions. Portugal's trade with Africa, Asia, and South America in the sixteenth and seventeenth centuries resulted in a complex aesthetic blending and technological borrowing: West African wooden and ivory sculpture that incorporated Portuguese figures, precious objects produced in Goa using Christian iconography for the Portuguese market, Japanese paintings that depicted Portuguese merchants, and the introduction of European firearms and maps into Japan.

Whether in intellectual discourses and practices, such as cartography and mapmaking, or in sports, language, religion, and forms of government, modern empires have integrated much of the world: Spanish-speaking South America, cricket playing in India and rugby in Fiji, Catholicism in Goa, and British-inspired representative government in Australia and Canada are just some of the many consequences of the globally integrating effects of modern empires.

Empire was not undertaken just for the opportunity of cross-cultural integration; there was a hard material core. Empires provided access to resources. The push to empire in the modern world was largely driven by economics. Spain and Portugal laid claims to vast territories in the New World, Africa, and Asia to provide gold, silver, spices, and raw materials for the home countries. The British-based East India Company, and subsequently the British Crown, created a commercial empire on the Indian subcontinent because it provided a rich source of raw materials for British industry as well as a captive market for goods produced in Britain. The Dutch East India Company, and subsequently the Dutch state, annexed land in South America and islands, often at the expense of local people in the Caribbean (Photo 14.1).

It is possible to identify a core–periphery spatial structure to the global political economy that emerged from this imperial relationship. The core consisted of the imperial

PHOTO 14.1 The Dutch Empire in the New World. Willemstad, Curaçao, retains its Dutch architectural heritage, but has been integrated with traditional Caribbean colors. The Dutch controlled the island from 1634. It was a center of the slave trade.

centers in Europe, with a vast periphery of regions throughout the world under their control. The success of the core was predicated on the exploitation of the periphery.

From the sixteenth to the twentieth century, much of the world was incorporated by these core powers. Incorporation was achieved through a variety of means, including colonization, informal control, and direct imperial annexation. The British Empire, for example, colonized North America and Australia, had informal control in South America, and annexed much of the Indian subcontinent. Empire could be costly—much better to have economic control without political responsibilities. British economic interests were vigorously pursued in South America without formal annexation. Latin America was the real success story of the British commercial empire, because there was market penetration without enacting the responsibilities of formal empire. It was a cheap form of economic integration for the imperialists. Direct imperial annexation was, in a sense, a sign of failure to achieve economic ends without direct political intervention. Formal annexation became more common in the late nineteenth century because of increasing imperial rivalries between core countries. Formal imperialism also resulted when the ruling elites of peripheral countries refused to go along with business arrangements.

Colonization involved a rewriting of space to show who was in control and who was controlled. Timothy Mitchell identifies three broad strategies for the framing of colonial states: producing a plan of urban segmentation that includes racial segregation, creating a fixed distinction between inside and outside, and constructing central spaces of observation to keep an eye on things and to show the presence of colonial power. Under British rule in Accra, now the capital of Ghana, new areas of European settlement were created that were sharply demarcated from the "native" areas. Laws barred "natives" from living in areas reserved for whites. Further north, Tunis came under French rule in 1881. The French built neoclassical buildings and created straight boulevards and symmetrical squares as a direct response to the tight and convoluted street pattern of the old Arab town: Gallic "rationalism" counterpoised to Islamic "chaos."

Local Elites

A key concept is that of the ruling elites, the groups that have power in a country or region. Collaborative elites reproduce the imperial center. This was the case of Anglo-Americans in North America before the American Revolution and the British colonial settlers in Australia and New Zealand. Collaborative elites can also encourage imperial incorporation. Throughout much of Africa and India, for example, the British authorities used indirect rule, working through local rulers who were rewarded for their efforts. Local leaders and existing hierarchies were employed to secure colonial rule. If and when the collaborative elite lost legitimacy and their grip on power, then the colonial powers often had to intervene.

Core–Periphery Spatial Structure

The spatial structure of core–periphery relation, in which the periphery is used as a source of raw materials and as a market for goods produced in the core, laid the basis for subsequent **economic development**. The core industrialized while the periphery remained a cheap resource base and a secure market for manufactured goods. The Industrial Revolution in core countries was built on the

14.1 The Caribbean as Imperial Shatter Zone

The shifting balance of imperial victory and defeat is evident in the changing ownership of islands in the Caribbean. From 1492, the Spanish expanded their imperial control to the large islands of the Caribbean and most of the mainland in the search for gold, precious minerals, and land. It was difficult to retain effective control over the scatter of islands in the Caribbean; they were too many and too widely scattered to police effectively. From 1600 to 1800, other European powers moved into the Caribbean to establish sugar plantations. They were able to annex various groups of islands and the more remote slices of the mainland. The Dutch, for example, gained control of slivers of land in South America as well as of the smaller islands of Aruba, Bonaire, and Curaçao. The French gained control of various islands, some of which still remain part of France. In 1657, the Danish West India Company established the town of St. Thomas, in what is now the US Virgin Islands, as the center of a sugar plantation economy based on slave labor.

The need for labor to work the mines and plantations and the decline of the Indigenous population led to the slave trade, in which almost 20 million people were captured in Africa and shipped to the New World. The Caribbean was transformed in the process as millions of African slaves became a vital part of the economic geography of the region, an important cultural element, and a nascent political force. The sugar colonies were the scenes of constant eruptions as slaves continually revolted against the appalling living and working conditions. Effective control was always difficult to maintain. These islands were far from the core countries. And after emancipation, the former slaves sought both economic freedom and political independence.

The competition and shifting alliances between European powers were often embodied in changes of territorial control. Consider the island of Dominica, which was under French control from 1632 to 1761, then British control, then French control again from 1778 to 1783, and then British again until it achieved independence in 1978. Most of the small island nations of the Caribbean have similarly complex histories. Trinidad was under Spanish control from 1532 to 1797, when it became a British territory until achieving independence in 1962. St. Lucia was held nine different times by France and six different times by the British, including the last colonial period of 1803 to 1979, when it finally became independent. The French legacy can be heard in the language of the locals.

The imperial legacy lives on. France still controls St. Bart's, Martinique, and Guadeloupe, the latter two being departments of France (equivalent to states in the United States). Montserrat remains a British territory, as do the British Virgin Islands. The Dutch have effective control over Aruba, Curaçao, and Bonaire and share control of the small island of St. Maarten/St. Martin with the French. In 1898, the United States invaded Puerto Rico, and in 1904 it acquired land along what was to be the Panama Canal. In 1917, the United States bought the Virgin Islands from Denmark, less an economic move (because the sugar economy had slumped) and more part of a geopolitical strategy of maintaining a more visible military position in the region and protecting its interests in the Panama Canal.

foundation of this **unequal exchange** between the core and the periphery. The poverty of much of the developing world is a direct result of the core–periphery structure.

Age of Imperialism

The late nineteenth century also saw an intensified period of imperialism. From around 1880 to 1918, there was a scramble for territory as old and new European powers struggled to incorporate more of the periphery. A map of Africa in 1880 shows the limited range of European influence on the continent. But by 1914, most of the continent had been annexed by European powers, large and small (Maps 14.2 and 14.3). The age of imperialism resulted from growing competition between core powers, a belief that national economic development was made easier by possessing colonial territory, and the feeling that each significant power had to do something to avoid being left out of the "spoils." Even tiny Belgium laid claim to a vast territory in the middle of the African continent. All these

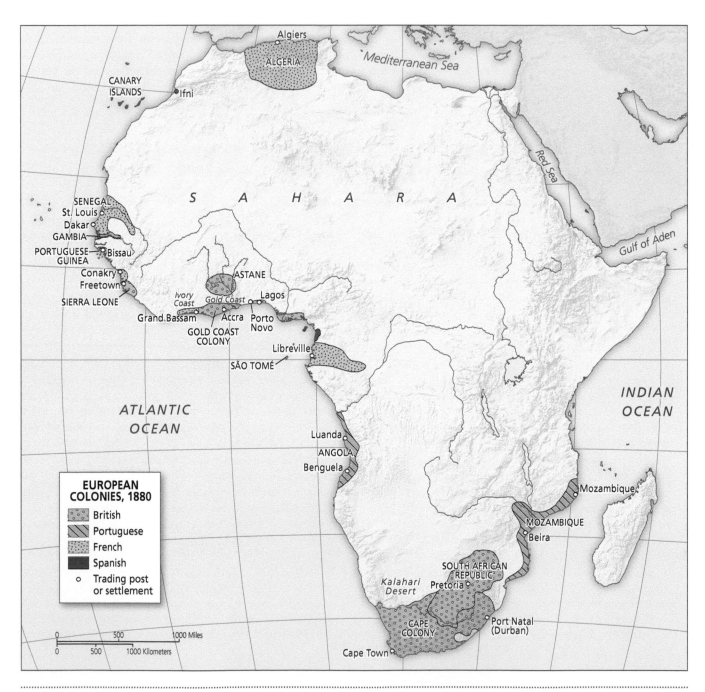

MAP 14.2 Africa, 1880.

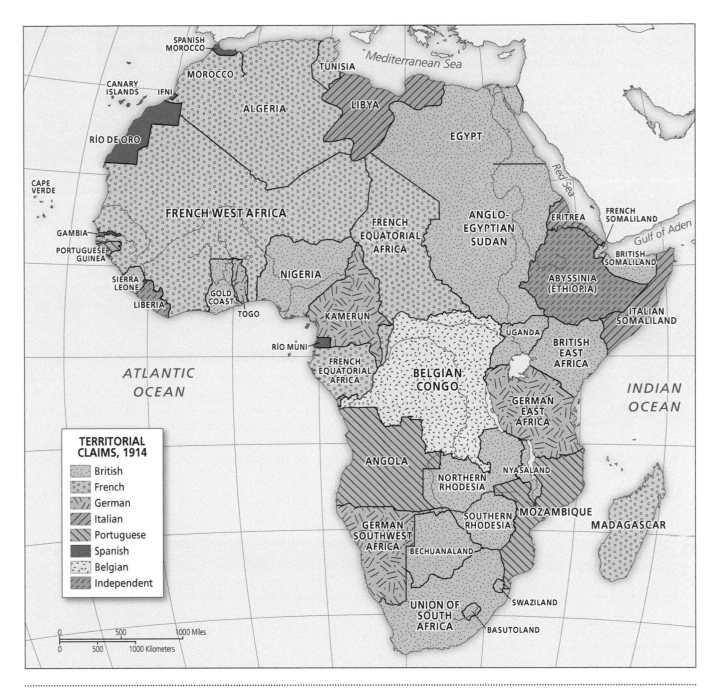

MAP 14.3 Africa, 1914.

developments exacted brutal and debilitating costs on local people throughout the region.

By the second half of the twentieth century, the struggle for global supremacy was being fought between the United States and the Union of Soviet Socialist Republics (USSR). Each had a long history of territorial annexation. The United States expanded from a narrow sliver along the eastern seaboard through treaties, purchase, and conquest to become a continental power. In the late nineteenth century, it also began an overseas expansion,

purchasing Alaska from the Russians in 1867 and picking up territories from a declining Spanish Empire, such as Puerto Rico and the Philippines. Hawaii was annexed in 1898. We have an interesting record of this imperial expansion in a popular children's book of the time, *The Navy Alphabet*, first printed in 1900 and written by Frank Baum, who also wrote *The Wizard of Oz*. An illustration that accompanies the text clearly shows the imperial nature of the United States, with explicit use of the term "colonies" (Photo 14.2).

Across the OCEAN vast and blue
Our warships plow the billows through
To guard our Colonies afar
And carry aid in time of war
They sail to fair Manila from
The sunny little Isle of Guam,
And then from Porto Rican blooms
To where Hawaii grandly looms.

PHOTO 14.2 Teaching the US Empire, 1900.

The Russian Empire grew from a tiny area centered on Moscow to become a vast continental power stretching from the Baltic to the Pacific and from the Arctic to the subtropics. In 1462, Muscovy was a small, isolated region. From this tiny enclave a Russian empire was created through annexation of territory. There was an early move east, drawn in part by the lure of the fur trade. By 1639, the Russians claimed land all the way to the Pacific. The empire was also extended to the west and south. In the north and west, territory was obtained in war with Sweden and Poland. In 1703, St. Petersburg was built by Peter the Great as a major port and as a "window on the west." In the south, the Russians annexed territory from the Tatars and other Turkic peoples, and in 1794 Odessa became the principal Russian port for trade through the Black Sea and the Mediterranean. A small, landlocked territory had expanded to become a major continental power. When the Bolsheviks grabbed power in 1917, the Russian Empire became the Soviet Empire.

The Cold War

The struggle between the United States and the USSR, called the Cold War, went through several phases. The chilliest period was from 1947 to 1964, as the USSR laid effective claim to most of Eastern Europe by installing and maintaining Communist regimes. In 1947, the Truman Doctrine announced that the United States would adopt a

14.2 Imperialism, Race, and Ethnicity

Race and ethnicity continue to be the most significant reason for discrimination and treatment. They can determine an individual's ability to be successful. Race is socially constructed and linked with physical characteristics such as skin color. Ethnicity is linked with an identity mostly on the basis of language and shared culture (religion, history, geography/territory). Both race and ethnicity are complex and problematic concepts linked to the power of empires and colonization.

As European empires were expanding, the idea of race originated among anthropologists and philosophers in the eighteenth century, who used geographic location and skin color to place people into different racial groupings.

Cedric Robinson introduces the idea of racial capitalism. He argues that the modern system of capitalism could not have proceeded without colonial expansion, the slave trade, and plantation slavery. He also writes that capitalism was "racial" not because of some conspiracy to divide

workers or justify slavery and dispossession, but because racialism had already permeated Western feudal society. The first European proletarians were racial subjects (Irish, Jews, Roma or Gypsies, Slavs, etc.) and they were victims of dispossession (enclosure), colonialism, and slavery within Europe. Robinson's work sheds light on how deeply implicated capitalism is in racial subjugation in many countries in the world.

Like race, ethnicity is socially constructed. An interesting example can be found in the country of Rwanda. Under Belgian colonial rule from the 1880s to the 1950s, Belgians regarded the Tutsi minority as racially superior and gave them preferential access to privilege and to positions of authority. In contrast, the Hutu majority were considered inferior. Belgians first identified physical characteristics as defining features: Tutsis were characterized as taller and more intelligent, while Hutus were smaller and more ignorant. As Deborah Mayersen has noted, Rwanda's colonizers ranked each "race" hierarchically, and over time, this racialized hierarchy was institutionalized within Belgian colonial policies and internalized by the Rwandan population. Belgium was able to "divide and conquer" within a country dominated by Black Africans who were not all that different from each other.

Although race and ethnicity may be socially constructed, and largely abstract, this does not negate their potent real-world influence. It is not simply that we have racial categories, but that we have constructed racial hierarchies that create layers of inequalities that persist even to this day.

REFERENCES

Mayerson, D. 2014. *On the Path to Genocide: Armenia and Rwanda Re-examined.* New York: Berghahn.

Robinson, C. 1983. *Black Marxism: The Making of the Black Radical Tradition.* Chapel Hill: University of North Carolina Press.

global posture against the USSR. Both countries established bases all over the world, formed alliances, and built up their military capabilities. Direct conflict did not occur, and a period of détente (less overt conflict) from 1964 to 1979 (when the USSR invaded Afghanistan) marked a warming of the interactions between the two superpowers. The military buildups and geopolitical maneuverings around the world continued, however, like two chess players with the world as their board, moving pieces to gain advantages. Local elites remained in power with imperial backing if they tied their existence to their backers' imperatives. Throughout much of the Cold War, for example, local elites in countries around the world, such as the Philippines under Ferdinand Marcos, maintained their grip on power by reminding the United States of their anti-Communist credentials or, in the case of Cuba, informing the Soviet Union of their anticapitalist policies.

A new Cold War emerged in 1980 with a significant rise in military buildup. As each side responded to the other's military increases, there was steady escalation, and the dialogue of détente was replaced by the rhetoric of confrontation. The gradual decline of the USSR and its eventual collapse in 1991 meant the end of the bipolar structure of global geopolitics. The United States remains the world's only superpower with a global reach.

The US Empire

The United States is not an empire in the old sense of colonial possessions. However, the US Empire is a sprawling archipelago of military bases around the world, a vast military–industrial–security complex backing a pervasive worldview dominated by visions of global **hegemony**. There are now at least 700 bases with half a million US personnel spread around the world in at least 130 countries. The United States is the most powerful military state, with the ability to assert global reach. The contemporary US Empire is not like previous empires that annexed territory and peoples into a singular political orbit. The earlier fragments of US Empire, such as Hawaii and Puerto Rico, have long since been incorporated. The US Empire is less about annexing territory and subjugating peoples and more about orchestrating a global order to extend and maintain global economic connectivity.

14.3 The Clash of Civilizations?

In 1993 the political scientist Samuel Huntington wrote an essay with the provocative title "The Clash of Civilizations." He argued that people were divided along cultural lines drawn by religion, history, and geography. He identified nine different types of cultures, including Western, Islamic, Orthodox, Latin American, and Confucian. It is an odd mixture of the religious, such as Islamic and Confucian, and the purely geographical, such as Latin American and African. The most contentious conflict was between the West, characterized as societies founded on principles of pluralism, democracy, and individualism, and the Islamic, characterized as less tolerant of difference and authoritarian, with a primary commitment to religion and not to the nation-state.

Several criticisms are made of this argument. First, the categories are very large, homogeneous spatial blocs, whereas the real world is more fractured, with religious and cultural groups more widely dispersed and the spatial units more often heterogeneous than homogeneous. Flows and heterogeneity rather than closed boundaries and homogeneity mark a globalizing world. Second, the historical record of contact between Islam and the West

14.3 continued

reveals shared human values, commercial trading, and cultural exchange, interaction, and penetration. Third, as the **Arab Spring** of 2011 revealed, the characteristics that Huntington asserted for the Islamic world were not eternal verities but the product of specific regimes and particular times. The demand for democratic inclusiveness is also a very important strain in parts of the Islamic world. Huntington's ideas were given a wider play in the wake of 9/11. The idea of a clash of civilizations between the West and the Islamic world is also central to the ideologies of Islamic terrorist groups such as al-Qaeda. But despite the ongoing terrorist threat, the contemporary world is a more complex place than the broad cultural categories proposed by Huntington can encompass and is more marked by civilization exchange and dialogue than outright clashes.

REFERENCE

Huntington, S. P. 1993. "The Clash of Civilizations." *Foreign Affairs* 72:22–49.

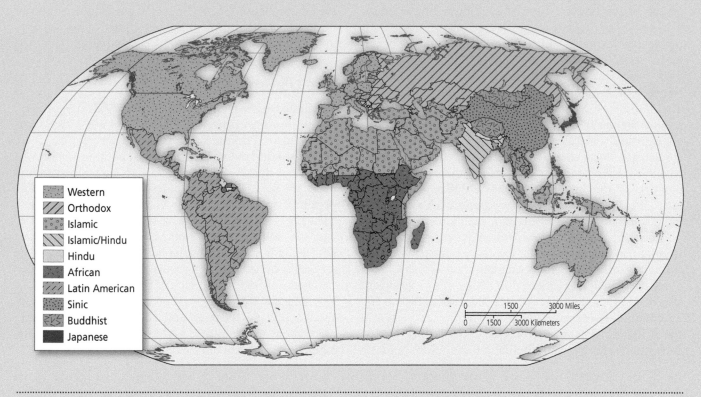

Legend:
- Western
- Orthodox
- Islamic
- Islamic/Hindu
- Hindu
- African
- Latin American
- Sinic
- Buddhist
- Japanese

MAP 14.4 Huntington's clash of civilizations.

14.4 The Fall of Empires: Imperial Overstretch and Disintegration

The rise and fall of empires is an enduring metaphor for the cyclical nature of success and failure, as well as a constant reminder of the spatial limits of territorial expansion. What lies behind the fall of global empires? Paul Kennedy (1987) identifies "imperial overstretch," the tendency for empires to expand beyond their ability to maintain economic dominance and military power. The tendency to initiate and expand military commitments undercuts the economic ability to pay for them. In other words, the hubris of empire tends to lead to entanglements that gnaw at its economic foundations. The collapse of the Soviet Union, for example, was in part caused by the inability of the state to maintain its expensive military posture as its economy was disintegrating. The United States is currently facing a fiscal crisis as it tries to pay for expensive social programs as well as the escalating cost of empire, especially the costs of overseas wars and "interventions."

There are numerous limits to empire. The first is information overload. The bigger the empire, the more the information that needs to be analyzed and turned into convincing narratives. The second and most important is the fiscal limit, as imperial overstretch bumps up against the ability to pay for all the commitments. The possibilities for involvement are infinite, while the ability to pay for them is finite, especially at times of economic distress and declining relative economic power. Third, there are the strategic limits. As strategies and forces are continually restructured according to the demands of the more recent conflict, there is always the possibility that they will be unable to deal with new conflicts. For example, in 2001, the US military was very well able to fight against another superpower, but unable to conduct an asymmetric war against small bands of transnational terrorists or undertake the nation building necessary after invading Afghanistan and Iraq. Big military powers have long reaction times that in a quick-changing world make them vulnerable to outdated assumptions, failed strategies, and incorrect tactics. Fourth, there is also the possibility of a **legitimation crisis**, which, in this particular context, can be defined as a major loss of public support for military involvements. These are more pronounced in more democratic societies, because the public can more easily voice its disapproval. Quick and successful troop deployments are easy to justify. The stunning success of the Gulf War in 1991 was, for many in the United States, a source of national pride, this single campaign erasing many of the wounds of Vietnam as the US military was applauded and honored rather than shunned and vilified. Fast-forward twelve years: as the invasion of Iraq soon turned into a quagmire, with the mounting carnage shown nightly on television, public enthusiasm for the war faltered and waned. When empire involves seemingly ceaseless, costly campaigns, public opinion comes into play. The visible costs of empire can provoke a crisis of legitimation if wars and military engagements are too onerous for too long. Empires come and go, but their going is rarely easy.

Decolonization

After the end of the Second World War, there was a wave of decolonization as imperial powers either left or were effectively removed from their colonial possessions. From 1945 to 1954, the wave was felt most intensely in Asia. The Dutch fought off independence struggles in Indonesia, but the persistent resistance finally led to the country becoming independent in 1949. The French had control over what was called French Indochina, consisting of what is now Cambodia, Laos, and Vietnam, but the independence struggle against French colonialism was interpreted by the United States as Communist insurgency. An anticolonial nationalist struggle was entangled by a superpower's geopolitical strategy as the Vietnamese fight for independence became the Vietnam War.

The decolonization of Africa came in the second wave, with independence for Sudan (1956), Ghana (1957), and Kenya (1963). In some cases, the struggle was intense. British authority in Kenya was contested especially by the members of the Kikuyu tribe, who mounted a guerilla campaign in pursuit of their goal of independence. The British responded by not only waging a counterinsurgency but also setting up a system of detention camps where up to 1.5 million people were rounded up and imprisoned. In a detailed study, historian Caroline Elkins came to the conclusion that "there was in late colonial Kenya a murderous campaign to eliminate Kikuyu people, a campaign that left tens of thousands, perhaps hundreds of thousands, dead."

With few "national" borders before colonization, many of the independence movements in Africa existed within the boundaries drawn up in arbitrary divisions of spoils by imperial cartographers. The result was that the national boundaries of postindependence Africa emerged from this cartographic legacy. These national boundaries, drawn initially for colonial convenience and compromise, cut across centuries-old tribal and ethnic differences that in some cases re-emerged in the wake of decolonization. Compare Map 14.3 with the contemporary political map of Africa in Map 14.5 to see the enduring legacy of colonial borders and boundaries on present-day African nation-states. A major problem for the newly independent states was how to create national unity and national identity from such a variegated population. In some cases, the new nations are continually split apart by the resurfacing of tribal and ethnic differences.

Collapse of the Soviet Empire

The fall of the Soviet Union in 1991 is just the latest in a long line of imperial collapses. When popular uprisings took hold in East Germany in 1989, the Soviet leadership did not send in the troops—something they had done in Hungary in 1956 and in Czechoslovakia in 1968, where they put down the popular uprising known as the Prague Spring. The system was economically and politically bankrupt, with its more astute leaders able to see the limits of the Soviet Empire. The unpopularity of the Soviet-dominated Communist system in Eastern Europe fueled public discontent. The people, when given an opportunity, showed a profound lack of support for the regime and a strong desire to be rid of the old leadership and scrap the Communist system. Events moved quickly. In June 1989, the popular political movement in Poland, Solidarity, won the first free election held for two generations. In September, the Hungarians set a date

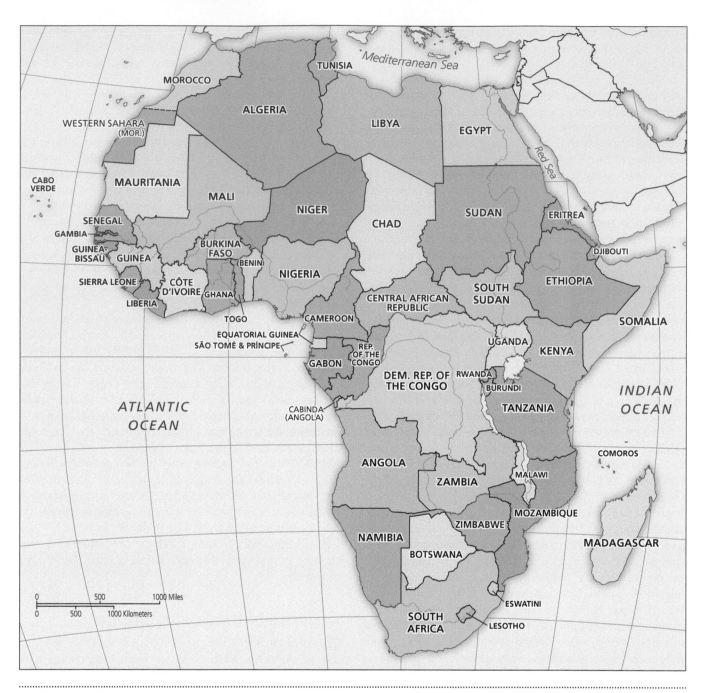

MAP 14.5 Political map of contemporary Africa.

for free elections. On November 9, the Berlin Wall ceased to be a meaningful barrier as East Germans scrambled freely over it, and Berliners, from both East and West, delighted in demolishing the symbol of partition and Communist control. Later that year, the Communist leadership in Bulgaria, Czechoslovakia, East Germany, and Romania was overthrown by popular uprisings. The corrupt system collapsed as more and more people voiced their disapproval. It was one of the rare examples of imperial decline brought about by popular uprising and bottom-up resistance rather than top-down regime change.

New countries emerged from the breakup of the Soviet Empire. Some of them had mature identities, such as the Baltic republics of Estonia, Latvia, and Lithuania, which had been swallowed whole by Soviet expansion in the 1940s. Others were broken off from larger national units, such as East Germany, a break that was healed with reunification with West Germany. Some were multiethnic agglomerations, such as Yugoslavia, which quickly began to break up over regional and ethnic tensions after 1989 (and would eventually become seven independent states, including Slovenia, Croatia, Serbia, and Montenegro). In yet others, new states emerged in the postimperial context. Political geographer Nick Megoran tells the story of Kyrgyzstan and Uzbekistan. Before 1924, these countries did not exist. The whole area was a complex, multiethnic place. It was annexed by the expanding Russian Empire in 1876, and in 1924 the Soviet Empire created the administrative structure of separate Kyrgyz and Uzbek units. This division helped forge separate identities, although not separate economic units. The collapse of the Soviet system in 1989 gave space for national emergence (see Map 14.6). In 1991, the two former regions of the USSR became separate states. Map 14.7 outlines the borders of the new state of Kyrgyzstan. An uncertain, arbitrary, and hazy distinction now became codified into two separate states sharing an international boundary. Tensions between the two states erupted in 1998 and resulted in the closing of the border. The invented borders of the Soviet era moved from acts of imagination to lines of national conflict.

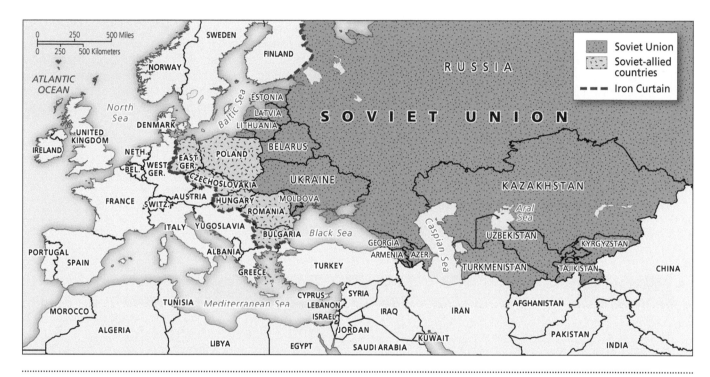

MAP 14.6 The Soviet breakup.

MAP 14.7 Kyrgyzstan: Soviet imperial disintegration leads to new international boundaries in Central Asia.

14.4 The Environmental Legacy of the Soviet Union

The collapse of the Soviet Empire in 1991 also revealed the significant extent of pollution under Communist rule. It became clear that the USSR prioritized economic development at the expense of the environment. The Soviet Union's emphasis on industrial and production and military strength polluted both air and water, degraded agricultural lands, and poisoned the land with toxic waste and radioactive fallout.

For almost three-quarters of a century, however, the full picture about the state of the environment was obscured by a lack of transparency. Soviet ideology proclaimed that socialism (as practiced by the USSR) was inherently better for the environment than Western-style capitalism, so any complaints about the environment could only reflect disloyalty. And disloyalty was punished by a trip to gulag. As a totalitarian state, there was very little information shared with its citizens, many of whom had to live with the health consequences.

The collapse of the Soviet Union uncovered a host of environmental horrors. Peterson notes that air quality in industrial centers such as Nizhnii, Tagil, and Bratsk was so severe that drivers frequently had to turn on their headlights during the day to safely negotiate city streets. The Aral Sea shrunk to one-third its size as the water that feeds the lake was diverted by Soviet civil engineers to support the region's cotton monoculture. The Cold War and the nuclear arms race meant that there were numerous instances where high-level radioactive waste in top-secret nuclear weapons production labs was dumped or leaked into local rivers.

But perhaps nothing epitomizes the legacy of a toxic environment like the disaster at Chernobyl in Ukraine. The 1986 explosion at the nuclear power plant affected the lives of hundreds of thousands of people. The tragedy underscored the lengths to which the Soviet government went to cover up the incident. Initially, the Soviets denied there had been a problem, but after a few days, they finally admitted that there had been an accident at Chernobyl. No one in the Soviet government informed the hundreds of emergency response workers of the true risks, and many of them suffered radiation poisoning. The initial denial meant the nearby town of Pripyat was not evacuated until thirty-six hours afterward. Even today, the long-term health effects are uncertain, as is the impact on ecosystems.

REFERENCE

Peterson, D. J. 2019. *Troubled Lands: The Legacy of Soviet Environmental Destruction*. New York: Routledge.

14.5 Critical Geopolitics

Geopolitics is the study of the power relations between empires and states. Early proponents include Rear Admiral A. T. Mahan of the US Navy, who wrote in 1890 about the influence of sea power, and the Englishman Sir Halford Mackinder, who provided a geopolitical strategy for the British Empire. Mackinder's 1919 book *Democratic Ideals and Reality* is concerned with geopolitical strategies in the immediate aftermath of the First World War.

More recent contributions include George Kennan, writing during the Cold War of the need for a US policy of containing the USSR. Even more recently, popular writers such as Robert Kaplan have argued that a nation's position on the world map is a primary determinant of conflict. The world is viewed as a series of states competing over space, with emphasis on the notion of geographic pivots.

Since the 1990s, a more critical form of geopolitics has emerged that looks at the social construction of political spaces, exposes the material interests involved in the narratives used to explain these spaces, and explores the spatial construction of social identity. It also interrogates the ideological underpinning of standard geopolitics. Gerry Kearns re-examines the imperial context of Mackinder's life, works, and geopolitical ideas. There is the study of popular geopolitics that looks at the role of popular culture and mass media in structuring national identities and popular geographical understandings of the world. There is also a reworking of the spatial nature of war and conflict. Derek Gregory, for example, looks at the geographical dimension of contemporary war through an examination of three global borderlands: Afghanistan–Pakistan, the United States–Mexico, and cyberspace.

Geopolitics—traditional, popular, and critical—explores the meaning, contestation, construction, and maintenance of the political organization of space.

REFERENCES

Gregory, D. 2011. "The Everywhere War." *Geographical Journal* 177: 238–250.

Kearns, G. 2009. *Geopolitics and Empire: The Legacy of Halford Mackinder*. Oxford: Oxford University Press.

Short, J. R. 2021. *Geopolitics: Making Sense of a Changing World*. Lanham, MD: Rowman & Littlefield.

14.5 A New World Order

In the past thirty years, at least four significant trends can be noted. The first is that the United States remains the world's global superpower. It has the largest economy, the most powerful military, and the global reach, with bases around the world, that enables troops to be placed anywhere in the world within seventy-two hours. The US military hegemony comes at a cost. The United States is dragged into wars and conflicts around the world, even when its immediate material interests are not at stake. It is blamed for many of the world's ills, leading to a strange asymmetry. While many people in the United States see the country as a force of good, many in other countries see it as a regressive force that props up undemocratic regimes.

A second trend is the rise of security groupings and mutual defense organizations such as the United Nations peacekeeping forces, the North Atlantic Treaty Organization (NATO), and the African Union, which consists of all fifty-four countries on the continent except Morocco. The African Union provides troops in certain instances. African Union troops were involved in recent military-policing interventions in Darfur (7,000 troops) and Somalia (8,000 troops) and more recently served in Burundi, Liberia, Côte d'Ivoire, and the Democratic Republic of Congo.

Another trend is the rise—or in some cases the re-emergence—of alternative power sources to the United States in the world. And here we can make a distinction between these and organizations such as the European Union, which is an alternative economic center of power to the United States that still shares similar global strategic goals. For critics in the United States, some countries and organizations such as Japan and the European Union get to achieve the economic success of a peaceful world underwritten by the military expenditure of the United States. Then there are countries that are both economic and military rivals. The two most obvious cases are Russia and China. Russia emerged as the rump of the former USSR. After a very rough transition from a Communist system to something closer to a market system, Russia began to benefit from the sale of its vast oil and natural gas reserves. As world prices of both increased in the 2000s, government coffers swelled, which allowed an increase in military spending and national assertiveness.

The Rise of China

The most significant entry onto the global stage since 2000 is China, which now has the second largest world economy. Before the trade liberalization that began in the late 1970s, China was a poor country and a bystander in the everyday operation of global society. China's rise is marked by a greater engagement in global organizations such as the World Trade Organization in 2001, a more active role in international affairs that parallels its growing economic connections with countries around the world, and a steady buildup in military capabilities. China's huge demand for resources from across the world has led to significant foreign investments in South America and Africa, including land purchases and building ports, roads, and railways to

14.6 The Russian Invasion of Ukraine

Under President Putin, Russia has reimagined itself as a world power eager to play a larger role in the world order, from annexing Crimea in 2014 to getting involved in the military campaign in Syria to cyberhacking adventures around the world. Russia's invasion of Ukraine in 2022 marked a radical change in the world order. Fearful of Ukraine joining NATO and eager to reclaim Russia's imperial holdings, Putin ordered an unprovoked invasion of Ukraine. When he was unable to achieve a quick victory, the war became a bloody struggle as Russian troops sought to extend de facto control over eastern and southern Ukraine. Ukraine did not have any mutual security pact with other countries, so had to fight off Russia alone. However, the US and European countries quickly organized a punishing economic boycott of Russia and lent considerable military assistance to Ukraine. Russia is now regarded as a pariah state. It still wields power because of its rich oil and gas deposits, but Western countries are distancing themselves from economic relations with Russia as they search for alternative sources of energy. This may cause a fiscal crisis for the Russian state, which may undercut its future projection of power. The invasion was a strategic blunder for Russia. Its poor military performance undercut its reputation, its aggression led to countries such as Sweden and Finland being eager to join NATO, and its actions solidified a Western alliance that was teetering on the edge of irrelevancy. Putin made a huge strategic error because the invasion of Ukraine weakened rather than strengthened Russia. The invasion was a tragedy for Ukraine as towns and cities were bombed, thousands were killed, and millions had to relocate or flee the country.

secure access to commodities. Africa is such a large recipient of Chinese money, trade, and migrants that some writers refer to it as China's second continent.

A recent case of China's growing role in the global economy, beyond simply importing vast amounts of raw materials and exporting manufactured goods, is the emergence of the Asian Infrastructure Investment Bank (AIIB). When the US Congress refused to support an increased role for China in the International Monetary Fund, China established the AIIB, which, despite US protestations, has now attracted a large number of partners, including traditional US allies such as Australia, Brazil, Saudi Arabia, and the United Kingdom. The AIIB has pledged to provide $40 billion worth of funds for infrastructure improvement, much of it targeted at a Belt and Road Initiative, a new Silk Road of improved communication and transport across Central Asia to link China with Europe. The initiative is a global infrastructure project, announced in 2013, that involves Chinese investment in over 150 countries to improve Chinese global trade links. For China it is a way of utilizing its vast foreign exchange, currently valued at a staggering $4 trillion, and of signifying its growing international prestige. The Belt and Road Initiative and the AIIB signal China's emergence as an active stakeholder in global affairs.

Significant brakes on the rise of China include the reliance on high and constant levels of economic growth, which may be difficult to achieve as the economy transitions from a low-wage, low-value to a high-wage, high-value economy; low gross domestic product per person;

and potential social unrest as internal criticisms are perceived as existential threats to established political power. Its military buildup has been rapid and impressive, but its reach is more regional rather than global. We will examine the issue of the South China Sea in the next chapter. Moreover, other countries are reacting and responding to China's rise. The territorial disputes that are occurring between China and her neighbors are in part a function of China's rise, but also because of the resistance of its neighbors to territorial claims. The American "pivot to the Pacific" announced in 2010 was in large part a response to China's growing regional assertiveness. US foreign policy is being recrafted to address the perceived threat of China.

Non-state Actors

Another significant element in the contemporary political geography of the world is the rise of non-state actors. These include social movements, some of them temporary, such as the Arab Spring, which swept across the Middle East in 2011 as young people in particular took to the streets to protest corruption, repression, inequality, and limited economic opportunities. There are also transnational terrorist groups, the most prominent being al-Qaeda and more recently ISIS (Photo 14.3).

Al-Qaeda is a global terrorist organization with significant offshoots including al-Qaeda in Islamic Maghreb, al-Qaeda in Syria, and al-Qaeda in the Arabian peninsula. It seeks to promote an Islamic fundamentalism that sanctions war against non-Muslims and non–Sunni Muslims.

14.7 China: Rising Power, Rising Emissions

China's rise as an economic and political power has environmental consequences because the high levels of economic growth since 1980 have been accompanied by increasing emissions of greenhouse gases. Since 2006, China has been emitting more carbon dioxide (CO_2) than any other country—however, it is also by far the most populous country (Figure 14.1). If we examine total CO_2 emission per capita, the United States is still the largest CO_2 emitter.

The top ten largest emitter countries account for more than 67 percent of the global total. The global disparity in carbon footprints is profound. Yet it is also important to understand that countries such as China and the United States emit on behalf of other countries, when factoring in the dependence

on Chinese and American exports to the global economy. Recent research estimates that about 25 percent of emissions from China are a result of the transportation of goods by rail or cargo ships going to other markets. While it is important to understand the geography of emissions, the interdependent nature of global economy makes this a more nuanced issue. Countries with low emissions may depend on purchases of goods from the big economies of the global superpowers of the United States and China.

As China asserts its growing political power, will it become more of a leader in addressing climate change, or will internal pressure to continue rapid economic growth come at the expense of greenhouse gas emissions?

Total vs. Per Capita CO$_2$ Emissions (2019)

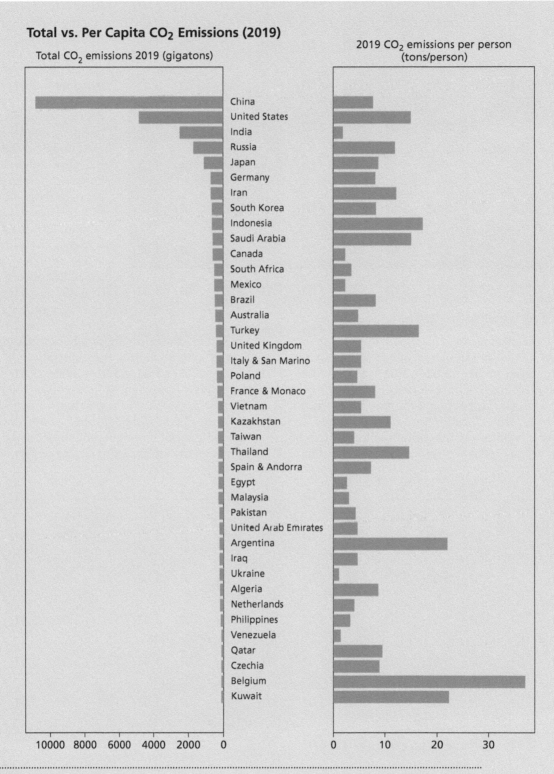

FIGURE 14.1 Carbon footprints.

PHOTO 14.3 ISIS militants parade through Northern Syria in 2014.

It is marked by violent terrorist acts. ISIS (the Islamic State of Iraq), also known as Daesh, is a Sunni Muslim movement that emerged in postinvasion Iraq from Sunni terrorist groups affiliated with al-Qaeda involved in the Sunni–Shia civil war. It became a military force when it linked up with former Baathist military personnel displaced by the postinvasion Shiite dominance in Iraq. By 2013, it had become a terrorist group motivated by Sunni Islamic fundamentalism, including violence against Shia Muslims and Christians, with a muscular military staffed by former Iraq Army personnel. It became self-funded by capturing oil depots, taxing residents and businesses under its control, extortion, and smuggling. In the failed states of Syria, undergoing civil war, and Iraq, beset by religious tensions, ISIS flourished, and by 2014 it had captured territory across both countries with active affiliates, also sometimes described as provinces of ISIS, as shown in Map 14.8. ISIS gained some measure of support from alienated and bored Muslim youth who felt marginalized in Europe, North America, and Australia. Of its 31,000 fighters in 2015 in Iraq and Syria, almost 12,000 were foreign nationals from eighty-one countries. Fundamentalism appeals to a youthful population without jobs, hope, or any sense of belonging. ISIS also has sophisticated marketing using social messaging and the Internet. Its images of pornographic violence also attract certain individuals.

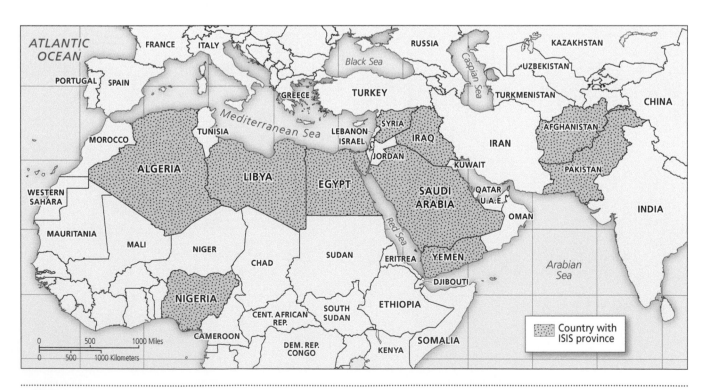

MAP 14.8 Countries with ISIS provinces.

ISIS has extended its reach beyond Syria and Iraq through the declaration of separate provinces in the Afghanistan–Pakistan border region, Libya, Egypt, Yemen, Saudi Arabia, Russia (North Caucasus), and Nigeria. In some cases, this involves encroaching on al-Qaeda, as in the Afghanistan–Pakistan border area, Russia, and Yemen. In other cases it involves the incorporation of existing insurgent terrorist groups such as Boko Haram in West Africa.

Cited References

Elkins, C. 2005. *Imperial Reckoning: The Untold Story of Britain's Gulag in Kenya.* New York: Holt. (Quote is from page xvi).

Kennedy, P. 1987. *The Rise and Fall of the Great Powers.* New York: Random House.

Megoran, N. 2012. "Rethinking the Study of International Boundaries: A Biography of the Kyrgyzstan–Uzbekistan Boundary." *Annals of Association of American Geographers* 102:464–481.

Mitchell, T. 1988. *Colonizing Egypt.* Berkeley: University of California Press.

Select Guide to Further Reading

Alcock, S. E. 2001. *Empires.* Cambridge: Cambridge University Press.

Applebaum, A. 2012. *Iron Curtain: The Crushing of Eastern Europe, 1944–1956.* New York: Doubleday.

Brzezinski, Z. 2012. "Balancing the East, Upgrading the West: U.S. Grand Strategy in an Age of Upheaval." *Foreign Affairs* 91:97–104.

Burbank, J., and F. Cooper. 2010. *Empires in World History: Power and the Politics of Difference.* Princeton, NJ: Princeton University Press.

Butlin, R. 2009. *Geographies of Empire: European Empires and Colonies c. 1880–1960.* Cambridge: Cambridge University Press.

Darwin, J. 2010. *After Tamerlane: The Rise and Fall of Global Empires, 1400–2000.* London: Bloomsbury Press.

Elkins, C. 2022. *Legacy of Violence: A History of the British Empire.* New York: Alfred A. Knopf.

French, H. 2015. *China's Second Continent: How a Million Migrants Are Building a New Empire in Africa.* New York: Vintage.

Levenson, J. A., ed. 2007. *Encompassing the Globe: Portugal and the World in the 16th and 17th Centuries.* Washington, DC: Sackler Gallery, Smithsonian Institution.

Marples, D. R. 2016. *The Collapse of the Soviet Union, 1985–1991.* London: Routledge.

Pagden, A. 2008. *Worlds at War: The 2,500-Year Struggle between East and West.* New York: Random House.

Parker, G. 1998. *The Grand Strategy of Philip II.* New Haven, CT: Yale University Press.

Parsons, T. H. 2010. *The Rule of Empires: Those Who Built Them, Those Who Endured Them, and Why They Always Fail.* Oxford: Oxford University Press.

Short, J. R. 2021. *Geopolitics: Making Sense of a Changing World.* Lanham, MD: Rowman & Littlefield.

Short, J. R. 2021. *Stress Testing the USA.* 2nd ed. New York: Palgrave Macmillan.

Tolan, J. V., G. Veinstein, and H. Laurens. 2012. *Europe and the Islamic World.* Princeton, NJ: Princeton University Press.

Veltmeyer, H., ed. 2017. *Globalization and Antiglobalization: Dynamics of Change in the New World Order.* London: Taylor & Francis.

Zhao, S. 2015. "Rethinking the Chinese World Order: The Imperial Cycle and the Rise of China." *Journal of Contemporary China* 24:961–982.

15

The State

The **state** is an important building block of the political organization of space. In this chapter we will consider the state as a spatial entity. We also explore challenges to state stability and issues around territoriality. Finally, we examine the geography of elections at different scales of analysis.

15.1 The Geography of States

A state is a self-governing political unit. It has monopoly control over a territory and boundaries with other states. For it to be legitimate, it must be recognized by other states and most of its population. There are currently almost 200 states in the world that cover the land surface of the Earth and make claims on the surrounding and adjacent seas and oceans. States are spatial entities.

The land surface of the world is divided up into states, separate units of political authority (Map 15.1). There has been a steady rise in the number of states. In 1900, there were 57 independent countries; there were 70 by 1930 and 195 by 2024. There are now 193 member states in the United Nations. The increase is a result of decolonization, whereby former colonies such as Kenya, India, and Vietnam achieved independence; and breakups, as formerly large empires and states fractured into separate national units. The breakup of the Austro-Hungarian Empire after the First World War led to the creation of a large number of smaller units, including all or parts of Austria, Bosnia, Croatia, the Czech Republic, Hungary, Italy, Montenegro, Romania, Serbia, Slovakia, and Ukraine. Some of these states were reassembled as part of larger units, such as the Soviet Union or Yugoslavia. The fall of the Soviet Union in 1991 in turn led to the creation of 15 new states. The breakup of Yugoslavia after 1989 eventually resulted in 6 new independent countries. The fracturing continues: in 2008 Kosovo seceded unilaterally from Serbia, and in 2011 South Sudan seceded from Sudan.

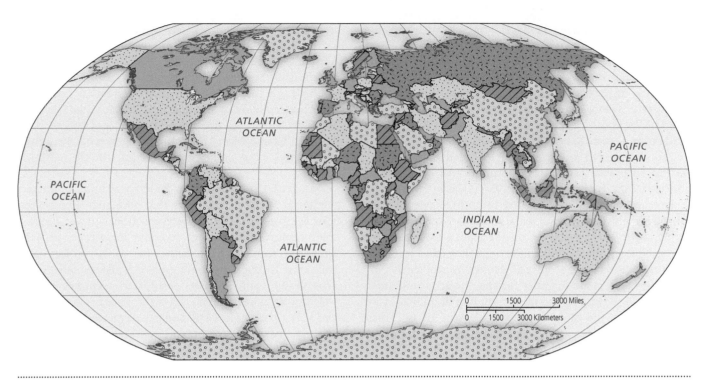

MAP 15.1 Political map of the world.

The Size of the State

At a time of increasing global integration, it is a seeming paradox that the creation of new states continues apace. But global integration allows smaller states to break off from difficult "marriages" as they find it easier to negotiate their place in the world, even as a small, diminished country.

The Size of the State

States vary in their spatial characteristics. There are large and small states. Size can be measured by territory and by population. Table 15.1 lists the top ten countries by area and population. Some countries, such as Australia and Canada, have a large territory but relatively small population. Some are small in area but large in population, such as Bangladesh. Others are big in both respects. Besides the United States, four other countries make both top tens: Brazil, Russia, India, and China. These big four are often grouped together along with South Africa—termed the **BRICS**, after the first letter of each country's name—because they all combine large size with emerging economies. Their joint population of 2.8 billion constitutes just over 40 percent of the total world population. Indonesia, with 279 million people, is a possible future member of the BRICS. Although its economy is perhaps poised to become more dynamic, in part fueled by a demographic dividend, Indonesia's problems of poor infrastructure, corruption, underemployment, and poverty may limit its future growth. Despite the hurdles faced by each of these

countries, they have the necessary size and rate of growth so that their combined potential represents a possible shift in geopolitical and geo-economic power away from the core of rich countries. Their potential is in part based on their territorial sizes, which include large amounts of natural resources (as in Brazil and Russia) and great potential for growth through the demographic dividend

TABLE 15.1	The Top Ten "Big" Countries
BY AREA	**BY POPULATION**
Russia	China
Canada	India
China	United States
United States	Indonesia
Brazil	Pakistan
Australia	Nigeria
India	Brazil
Argentina	Bangladesh
Kazakhstan	Russia
Algeria	Mexico

(especially pronounced in Brazil, India, and Indonesia). Size does matter.

One way that smaller countries try to offset their limited size is through organizations with other states. For instance, **economic unions** try to extend the size of the market. The European Union, for example, grew from the perceived need of European countries to avoid further conflict in the wake of two world wars, but also to provide a large enough economic entity to compete with the United States. It began as the European Economic Community of six countries formed by the Treaty of Rome in 1957. It was renamed as the European Union and expanded to include twenty-seven countries with a single market. The recent fiscal crisis, highlighted by the debt problems of Greece but also the very real potential for similar crises in Portugal, Italy, and Spain, reveals that unions do not solve all economic problems.

States also come together for specific economic interests. We already noted the important role that OPEC plays in regulating the global supply and price of oil. There are also looser economic organizations like the North American Free Trade Agreement between Canada, Mexico, and the United States.

The Location of States

States also vary by location. One significant feature is access to sea transport, which in general tends to be far cheaper than overland or air transport. For landlocked states to access global markets, they have to ship goods overland through another state, which increases costs and tariffs (Map 15.2). There is no simple relationship between level of economic development and being landlocked. On a variety of measures, Switzerland is one of the richest countries in the world, with many inhabitants enjoying high per capita income and a full range of quality public services. However, for many poor countries such as Bolivia, Chad, Mali, Mongolia, Niger, and Paraguay, landlocked status makes economic growth all that more difficult.

As spatial units, states can be lucky or unlucky in their territorial endowment. States that have rich oil reserves, for example, can wield effective political power because their resource endowment provides huge revenues. The Saudi government's oil revenues pay for its projection of soft power through the promotion of fundamentalist Islam around the world. In some cases, rich resources provide welcome revenues. In 2011, when the price of a barrel of oil was $111, revenues from oil and gas accounted for one-half of Russia's federal budget, and raw materials constituted 80 percent of exports. While a generous resource endowment may provide short- and medium-term riches, it can, over the longer term, lead to problems. Easy reliance on nature's bounty may restrict the innovation necessary for long-term economic growth. This is the so-called **Dutch disease**, based on that country's experience with oil revenues that created a huge trade surplus and large currency reserves, which led to overvalued domestic industries that were uncompetitive in global markets. Resource prices are

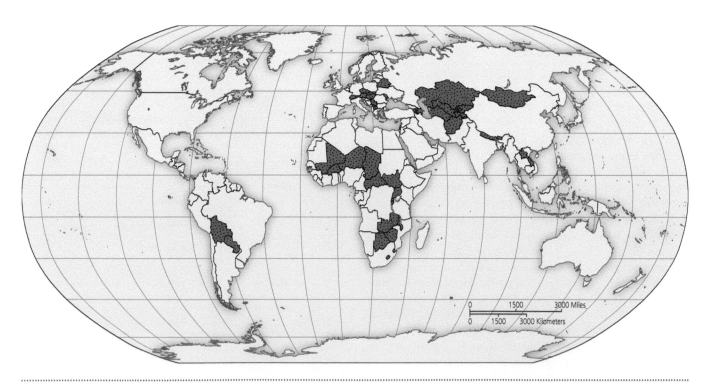

MAP 15.2 Landlocked states.

also very volatile, so that revenues may vary from one year to the next. In high-price years, governments may spend to shore up popular support, but that support may vanish when prices fall and spending has to be curtailed. By 2016, the price of a barrel of oil had fallen to $33. The revenues of the Russian government were effectively reduced by 70 percent in just five years. Commodity price volatility is one manifestation of the **resource curse**, sometimes known as the paradox of plenty. While valuable resources can create short-term booms, they may depress long-term economic growth. In less democratic countries, wealth from resources is easily diverted, stolen, and misspent.

Nations, States, and Minorities

A distinction can be made between a state and **nation**. A state is a political organization that controls a particular territory. A nation is a community of people with a common identity, shared cultural values, and a commitment and attachment to a particular area. In some cases, there is congruence between nation and state. In the case of the small landlocked country of Swaziland, the nation and the state are almost identical, with the great majority of the population ethnically Swazi. Such instances are rare, especially with international migration creating more complex patterns.

The incongruence between nation and state takes two main forms. First, some nations lack a state. For example, the traditional homeland of the Kurds is in the Near East. After the fall of the Ottoman Empire, just after the First World War, their homeland was divided by Western powers that drew boundaries to suit their geopolitical interests rather than to honor national differences. The Kurdish nation was divided among separate states, including Turkey, Syria, Iran, and Iraq, and Kurds became vulnerable minorities within them (see Map 15.3). In some cases, the Kurds have been successful in creating regions of relative autonomy. After the US invasion of Iraq, the Kurdish region in northern Iraq achieved a large measure of autonomy. However, the Kurdish region in eastern Turkey is still a scene of ongoing conflict between the Turkish government and Kurdish separatists (Photo 15.1).

There are also states with more than one nation. In Belgium, for example, the centuries-old rift between the Dutch-speaking north and the French-speaking south,

despite their shared Catholic religion and rising living standards, remains always just below the surface and always capable of breaking out into explicit calls for separation. In some cases, rifts widen to civil war. Sudan was a colonial possession of the United Kingdom. It consisted of two regions: a northern part centered on Khartoum in which most people are African Arab and Muslim, and a southern one where most people are Black Africans and either animist or Christian. When the country achieved independence in 1956, the two parts were joined in one state. It was never a happy union. Power was retained in the north, and the south languished. Decades of civil war caused casualties and created famine. More than 2 million people died, and millions more were displaced. The south was able to secede only after years of war and starvation. The state of South Sudan came into existence on July 9, 2011.

The population of a state is rarely homogenous. Differences in religion, ethnicity, and language, if they have distinct spatial expression, can create tensions in the internal coherence of the state. Minority groups are often targeted when economic difficulties provoke social unrest. In Indonesia, for example, there is a significant ethnic Chinese population, located in specific neighborhoods of cities and towns. Of a total population of 240 million, more than 2 million are estimated to be ethnic Chinese. Their experience has not been a happy one. There is a long history of anti-Chinese feeling in Indonesia. In 1740, economic unrest led to demonstrations and mass killings. Ethnic Chinese could travel only with special permits. After independence from the Dutch in 1949, the ruling junta regularly provoked anti-Chinese feeling. In 1998, anti-Chinese

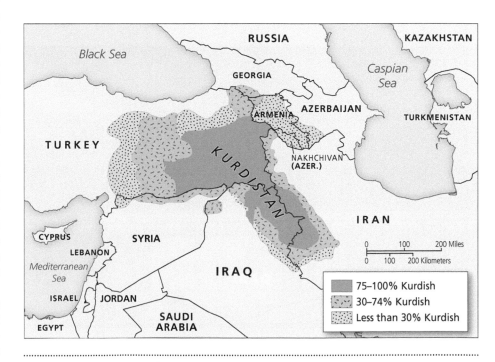

MAP 15.3 Distribution of Kurds in the Middle East.

PHOTO 15.1 Kurds celebrating the festival of Newroz ("New Day") in Van, Turkey.

latest era of struggles between the Catholic and Protestant communities lasted for almost forty years, from 1970 until a power-sharing administration began in 2007.

The protection of minority rights is a fundamental part of the Charter of the United Nations and is now an important element of the global discourse of international politics. However, genocide in Bosnia, Darfur, and Rwanda indicates that the human rights of minorities can still be violated. When a state is committed to harming minority groups within its borders, the international community must overcome distance, time, and political inertia to stop that state from killing its own citizens. International involvement is slow, costly, and difficult to organize.

rioting in the capital city of Jakarta erupted in violence and damage to Chinese Indonesian communities. Minorities can often become scapegoats for wider economic issues. When economic conditions decline, the Other is often blamed.

In many states, there is some measure of tolerance for people who are different from one another. In others, however, the differences remain, often exacerbated and embodied in years of struggle. In Northern Ireland, for example, the

National Environmental Ideologies

National identity is bound up with attachment to territory and its creative representation (Photo 15.2). National environmental ideologies build on myths of wilderness, countryside, and city. These myths often have a bipolar quality. Consider wilderness. Its defeat is part of many national histories. In the United States, for example, the defeat of the wilderness in the nineteenth and early twentieth centuries was an important element in the sense of national identity, and the moving frontier gave a sense of an expansive national trajectory. Slowly, an alternative began to emerge, that of preserving the remaining wilderness. As wilderness disappears, its survival can also become a national project. The national park movement in the United States took as its guiding principle the defense of remaining wilderness areas. The wilderness and its point of contact with the civilized world, the frontier, are the background to one of the essential US texts, the Western movie. Countryside can also be idealized, especially in a rapidly industrializing and urbanizing country. An idealized representation of countryside is still at the core of English environmental ideology. The idea

PHOTO 15.2 This picture a of rustic church at Geirangerfjord embodies an idealized picture of the national landscape of Norway.

The State | **257**

of the city can range from Jerusalem, a place of social uplift, the "city on the hill" of the early Puritans in North America, to Babylon, the place of vice and debauchery populated by the foreign Other. In the United States, there is a long anti-urban tradition that extols the virtue of the small town and rural life.

Environmental ideologies are expressed in paintings, novels, and cinema. Australian landscape painting, for example, depicts the changing idea of the outback; many English novels deal with the tension of modernity in an idealized countryside; and the US Western provides mythic depictions of the frontier between the settled and the wild.

While the general categories of wilderness, countryside, and city are used in varying ways to construct and challenge national environmental ideologies, more specific features are also employed. In a fascinating study, Tricia Cusack tells the story of how five rivers became part of national identity. She tells the story of the Hudson as an icon of American nationalism; the Thames as embodying ideas of monarchy, empire, and commerce in England and Great Britain; the Seine and consumer culture in France following the War of 1870; the Volga and its role in Russian romantic nationalism; and the Shannon and its connection with Celtic Irish cultural identity. Rivers are more than just hydrogeomorphic features. They are texts that are used to generate and sometimes unify diverse narratives and competing claims on national identity. Forests, mountains, rivers, coasts, valleys, and other natural features become elements of a national landscape. National identity in the United States, for example, was shaped by contact with forests. The seemingly endless supply of trees created a throwaway culture in which things were not built to last. Only when the forest cover was threatened was a conservation movement initiated. Trees shaped American identity just as Americans transformed the forest.

A "national" landscape is crafted from and represented by the raw materials of the environment into texts of historic meaning and political significance.

The Different Power of States

Not all states are equal. There are vast differences in power and the ability to influence the global order, wealth, and measures of inequality. Along the dimension of power, we can distinguish between superpowers, major powers, and minor powers. Superpowers have the ability to influence events around the world. Successive superpowers have risen and fallen. In the late sixteenth and early seventeenth centuries, Spain was the first global power, with possessions spread across the world. By the nineteenth century, Britain emerged after years of struggle with France to emerge as the world's superpower, its influence embodied not only by its territorial annexation but also by its naval power, which enabled it to influence events all over the world. Britain's success encouraged emulation. The United States still remains the single largest superpower with the

---

15.1 Imagined Communities

The construction of national identity is an important subject of recent writings. In an argument of great subtlety, Benedict Anderson argued that nations are not so much facts of race or ethnicity; rather, they are what he terms "imagined communities." Anderson paid particular attention to the role of print capitalism in creating a national discourse. In a later elaboration, Anderson identified three institutions of power: the *census*, the *map*, and the *museum*, which together allow the state to imagine the people under its dominance, the geographic territory under its control, and the nature of historical legitimacy.

National imaginaries are rarely coherent, consistent, or stable. The dominant national imaginary in Australia has shifted from celebration as a white Australian, Anglo-Celtic colonial outpost to the more uncertain embrace of a postcolonial, multicultural society. A major element in this shift is the changing role assigned to Indigenous peoples.

There are national histories and national geographies, national characteristics and national claims to greatness. "National"

events are enacted and re-enacted; "national" stories are told and retold. National identities are also created through commemorative activities. Exhibitions, fairs, and sites of historical memory and commemoration are common vehicles for celebrations and claims of national identity. In the centennial celebrations of the United States (1876) and Australia (1888), the Indigenous peoples were excluded. Fast-forward a hundred years and ritual and symbol were mobilized once again. Both bicentennial celebrations, of 1976 and 1988, praised political liberties and economic prosperity, yet the fundamental belief in the forward march of progress had weakened over the course of the intervening century. Now cultural diversity was celebrated rather than ignored. The Indigenous peoples were given a more central role; in Australia, this emerged mainly as a result of Aboriginal protest at the concept and the practice of the celebrations.

REFERENCE

Anderson, B. 2006. *Imagined Communities: Reflections on the Origin and Spread of Nationalism*. Rev. ed. London: Verso.

ability to achieve global military reach and project its power and influence around the world. This ability is a result of the large distribution of military bases spread around the world, a large military force, and commitment to high levels of government military spending. Table 15.2 lists the ten largest military spenders in 2022. The United States is clearly dominant, responsible for almost 40 percent of the world's total military spending, more than the next sixteen biggest spenders combined. Military spending is a function of wealth and size. The richer countries, such as France and the United Kingdom, are able to devote some of their considerable national wealth to the military, while the larger countries, such as China, Russia, and India, are able to mobilize their vast resources. Saudi Arabia devotes the largest proportion of GDP to military spending. These ten countries are responsible for more than three-quarters of all military spending in the world.

Only a few countries have the military ability to project national power on the global stage. While the United States is the superpower, the remaining nine countries are major powers able to exert some national influence beyond their boundaries. The Chinese, for example, are building up their navy enough to back up substantial territorial claims in the South China Sea. Not all states are equal, with the more powerful able to exert an outsize role in world affairs. There is marked asymmetry in global affairs, with only very few countries having the military strength to impose their will across space.

A distinction can be made between soft and hard power. **Hard power** is the ability to coerce others through the superiority of military forces. **Soft power** is the ability to co-opt others to your point of view without the use of force. The United States has hard power through its military superiority, but also elements of soft power through wide dispersal of its cultural forms. Political scientist Joseph Nye, who drew the distinction, argues for the use of smart power, which combines hard and soft power in successful strategies.

States also have different levels of wealth and inequality, both among states and within. In China, for example, there are substantial differences between the poor rural interior and the wealthier, urban, industrial coastal cities. The average income figure also masks substantial variation between households within one country. The level of income inequality is an important indicator of how income is spread across the national population. The CIA produces a ranking of inequality using the **Gini index**, which measures the spread of wealth across the population. The values range from the most unequal, Lesotho at 63, to the least unequal, Slovenia at 23. The most unequal states are low- and middle-income countries such as Haiti and Chile. The most equal societies are small, affluent countries such as Sweden, Finland, Hungary, and Norway. However, a very poor country such as Bangladesh has a more equal income distribution than India, China, and Peru—it is even more equal than the United States.

The CIA collects data on national income inequality because it is an important predictor of political stability. More equal societies tend to be more peaceful, less riven by conflict. Even societies that are unequal but have been so for some time have a measure of stability. But societies where the rate of change in inequality is marked and sudden can become

TABLE 15.2 Military Spending, 2022				
	COUNTRY	SPENDING ($ BILLION)	% OF GDP	WORLD SHARE (%)
Rank	World total	2,113	2.2	100
1	United States	877	3.5	41.5
2	China	292	1.7	13.8
3	Russia	86	4.1	4.0
4	India	81	2.4	3.8
5	Saudi Arabia	75	6.6	3.5
6	United Kingdom	68	2.2	3.5
7	Germany	55	2.1	2.6
8	France	53	1.9	2.5
9	South Korea	46	1.1	2.1
10	Japan	46	1.1	2.1

Source: Adapted from Stockholm International Peace Research Institute data.

unstable as social movements emerge to resist and contest the sudden and marked concentration of wealth in just a few hands. The Arab Spring of 2011, for example, emerged in the wake of sudden increases in inequality resulting from globalization, privatization, and the adoption of neoliberal policies. Rapid growth that brings a quick increase in inequality can provide the basis for social stress and social upheaval. Even for more stable and affluent societies, the level of inequality is an important indicator of well-being. Richard Wilkinson and Kate Pickett found that in certain states of the United States and in the twenty richest countries, where there was a big gap between the incomes of rich and poor, there was more mental illness, drug and alcohol abuse, obesity, teenage pregnancy, and homicide. The more unequal a society, the shorter the life expectancy and the poorer the children's educational performance. Scandinavian countries and Japan consistently do well on a range of such social indicators; they have the smallest differences between higher and lower incomes and the best levels of overall physical and mental health. Obesity is twice as common in the United Kingdom as in the more equal societies of Sweden and Norway and six times more common in the United States than in Japan. It is not just that the lives of the poor are worse; the lives of everyone in the society are made worse. Life expectancy even among the rich is lower in more unequal societies. The United States is wealthier and spends more on healthcare than any other country, yet a baby born in Greece (at least before the current economic crisis), where average income levels are about half those of the United States, had a lower risk of infant mortality and longer life expectancy than an American baby. It is not only the wealth of society that structures health outcomes and social performances; so does the level of inequality.

Even within a country there are profound differences. National averages mask racial and ethnic differences. The infant mortality rate, measured in terms of deaths per 1,000 live births, in the United States is 5. This figure varies by race. For non-Hispanic Blacks it is 10.8 per 1,000 live births; for non-Hispanic whites it is 4.6. In other words, Black babies born in the United States are 2.3 times more likely to die than white babies. It also varies by scale. States such as West Virginia, Arkansas, Mississippi, and Alabama have the highest infant mortality rates. Cities such as Baltimore, Detroit, and Cleveland have the highest infant mortality rates, with Cleveland ranking last, at 14.1. The Black infant mortality rate in the United States is higher than the national average in Costa Rica (7), Sri Lanka (6), or Thailand (7) and closer to the national rates of much poorer countries such as Albania (9), Jamaica (11), and Libya (10).

State Stability

States vary enormously along a variety of dimensions, including military might, economic wealth, and political freedom. What is also worthy of note is the rapid rate of change, with low-income countries sometimes becoming richer and authoritarian systems becoming more democratic. To be sure, there are opposing trends, as some countries become poorer and governments slide into authoritarianism. The system of states is a dynamic jigsaw puzzle, constantly changing, sometimes slowly and at other times with dramatic speed. Perhaps we need maps of changing status to get a firmer sense of this dynamism.

The openness of a state is defined as the extent to and ease with which people, ideas, information, goods, and services flow across its borders. Very closed societies such as North Korea are very stable. But once such a closed society becomes more open, there is a risk of political instability. That is why the rulers of closed states strive so hard to keep the lid on; it keeps them in power. The position of some individual states is difficult to assess. China, for example, encourages the free movement of capital and goods but keeps a firmer control over ideas that come over the Internet. It is open to the globalization of goods for its consumers, but less open to the globalization of ideas and information for its citizens.

The political geography of states is complex and dynamic. States rise and fall, expand and contract. The United States, for example, started off as thirteen colonial states pinned up against the eastern seaboard, but through annexation, invasion, purchase, and treaties it extended to continental proportions. Historian Norman Davies relates the fascinating story of states and kingdoms in Europe that no longer exist. He recounts the rise and fall of, among others, Byzantium (330–1453), Rosenau (1826–1918), Tsernagora (1910–1918), and the Soviet Union (1924–1989).

The Crises of the State

States, despite their power and seeming permanence, are, in the long-term, fragile organisms. Their fragility results from the variety of crises that can beset them. Following on and developing the work of the German sociologist Jürgen Habermas, we can identify three contemporary and interlinked crises; fiscal, legitimation, and rationality. A **fiscal crisis** is when a state has more expenditure than revenue. Military and social spending, either singly or together, can outmatch the revenue from taxes and tariffs. In the short term, the state can borrow money, and the biggest economies can borrow more, but ultimately a fiscal crisis turns into a political crisis as debts have to be repaid and unpopular cuts in spending have to be made. Fiscal crisis is an endemic feature of poor countries but is increasingly a problem for richer countries as well: either they spend a lot on social welfare and the military (the United States), or just on social welfare (western European countries), or they have a low tax base and/or rampant tax evasion (e.g., Bulgaria, Italy). The **shadow economy**, in which people pay no taxes, constitutes 32 percent of the Bulgarian GDP, 21 percent of the Italian economy, but only 7 percent of the US economy. Fiscal crises prompt political crises.

A legitimation crisis is when the state loses its popular appeal and its ability to govern. When it loses the support of an increasing part of the population, its legitimacy is at risk. During the Arab Spring, for example, the governing elites of Egypt, Libya, and Tunisia no longer had the support of the majority of the population. In Syria, the Assad regime lost the support of the majority of the Sunni population. Declining living standards, a sense of injustice, and a feeling of profound alienation between the government and the governed prompt a legitimation crisis. The crisis can take a shallow form when the party in power is highly unpopular, or it can be deeper when the entire system or regime is unpopular and wholesale change is demanded.

A **rationality crisis** occurs when the state makes enough poor decisions that other crises emerge. The US decision to invade Iraq in 2003 was based on faulty assumptions and led to the death of almost 100,000 Iraqis, the displacement of millions more, and the loss of over 4,000 US troops. In another case, in 2008 the Irish government guaranteed the losses of Irish banks. The banks received almost $4.5 billion each. By 2010, government support for the banks constituted almost a third of the country's entire GDP. The ruling party at the time of the bank bailouts, Fianna Fáil, the largest party since 1932, lost the 2011 election in a landslide defeat with only 17 percent of the votes. These examples show the interlinked nature of fiscal, legitimation, and rationality crises.

Centripetal Forces

We can identify **centripetal forces** that unify the state's power across space and centrifugal forces that disrupt it. Centripetal forces include external aggression, which may stimulate national bonding against a common enemy; federal structures that allow the safe expression of regional and other subnational differences; and national mass media and education, which create a shared culture and language. In some cases, a measure of regional autonomy can blunt separatist claims. In Spain, for example, under the Franco regime (1936–1975), power was centered in Madrid and used to suppress both Basque and Catalan autonomy. Support for the soccer team of Barcelona was and still is a vigorous expression of Catalan identity. Since the fall of the Franco dictatorship, there has been a more pronounced commitment to power sharing. In other cases, separatist movements may lie dormant but then erupt. Scotland joined with England in 1707, and for three centuries Scottish nationalism in lowland Scotland was little more than a whisper, although many in the Highlands resisted vigorously in 1715 and 1745; recent years, however, have seen more vocal calls for separation.

Centrifugal Forces

Centrifugal forces include political and economic inequality and high levels of social and cultural diversity within a population. One reason behind a resurgence of Scottish nationalism is the large gap in living standards between Scotland and the southeast of England, the center of power and home of the elites. Despite the homogenizing abilities of the state, religious, ethnic, and language differences can persist and survive attempts at their eradication. In Spain, the Basque language is still a principal vehicle and embodiment of Basque identity and difference from Castilian Spain.

The state's responses to separatist movements within its national boundaries can range from repression to some form of accommodation. Many states began with repression. When the British state defeated the Highland rebels at the Battle of Culloden in 1746, it criminalized wearing tartan, playing bagpipes, or speaking Gaelic. The territorial expansion of both the United States and Canada involved the defeat and removal of many Indigenous nations. States can also accommodate separatist movement by granting a degree of political autonomy. Since 1998, the British state has given Scotland its own parliament with greater fiscal powers to head off the appeal of Scottish nationalism.

Territorial Integrity

A state has a monopoly of legitimate physical force. But in some instances, the rule of the state does not extend to all of its territory; when it cannot effectively control it all, a territorial crisis occurs. In some cases, the territorial divisions erupt into civil war. The US Civil War (1861–1865) was fought between northern and southern states, with the South declaring itself independent from the United States. It was a bloody war. By 1865, over 620,000 Americans had died as a direct result of the war, more than all the Americans killed in twentieth-century conflicts. On one day, September 17, 1862, at the battle of Antietam in Maryland, more than 4,000 men were killed and 17,000 more wounded.

The territorial control of the state varies along a continuum. At one extreme are states that wield undisputed monopoly power over their territory. At the other are dysfunctional states with limited territorial control. In Somalia, for example, the government's control barely extends beyond the government compound in Mogadishu. In Afghanistan, the president of the country is known as the mayor of Kabul, emphasizing his limited influence. In Colombia, a fifty-year civil war was fought between the central government and an insurgency. In its remote mountain areas during the 1950s, Communist guerillas and members of the Liberal Party started to organize against the army and the Conservative Party in a struggle over control of land and political dominance. More than 200,000 were killed. In 1964, FARC (Revolutionary Armed Forces of Colombia) came into existence and waged a forty-year war against the central government. The rebels were funded through their proceeds from the drug trade, while the government received aid from the United States. Only by 2016 did a signed agreement lead to a possible end to the decades-old conflict.

The internal territorial integrity of the state is more ensured where there is a homogenous population. But marked differences, overlaid by economic inequality and profound cultural differences, can exert pressures on the integrity of state control over national space. When states cannot control the national territory, they become dysfunctional. Piracy developed along the coast of Somalia because of the power vacuum created by a dysfunctional state. The pirates were halted only with the intervention of outside powers.

Weak states lacking territorial control are the breeding ground for non-state actors such as FARC, al-Qaeda, and ISIS. Dysfunctional states or states characterized by internal convulsions along ethnic and religious lines and an economic system that offers few opportunities for the majority of people, especially young people, are the necessary oxygen for non-state actors that wish to challenge the legitimacy and territorial integrity of states.

The Example of Brexit

In 2016, the voters of the United Kingdom were asked to vote whether to stay in or leave the European Union (EU). It was known as Brexit, short for "Britain's exit."

A majority in the United Kingdom voted to leave the EU. But a look at the geography of the vote provides a good example of the political geography of the United Kingdom (Map 15.4) and the centrifugal and centripetal forces that affect it. A major area of support for remaining in the EU was centered on London. Greater London comprises 7.5 million people, but the greater metropolitan region has a population of approximately 21 million. The reach of the city extends into most of the southeast part of the country and beyond. It is by far the most affluent part of Britain. London has emerged as a global financial center, attracting expertise and investment from across the globe and around Europe—it is hardwired into the financial circuits of the EU and global economy.

Since 1990, the principal aim of government economic policy was to protect London's position, largely ignoring the rest of the country. In the UK version of trickle-down economics, London is the golden goose whose droppings fertilize the rest of the country. In reality, there was not much trickle and the United Kingdom, like the United States, is becoming a more unequal society. The United Kingdom has the most marked regional inequality in Europe.

The rich and wealthy are concentrated in London and the southeast, where household incomes are higher than in the rest of the country. As the United Kingdom became a more unequal and divided society, the cleavage between London and the southeast and the rest of the country has become more marked.

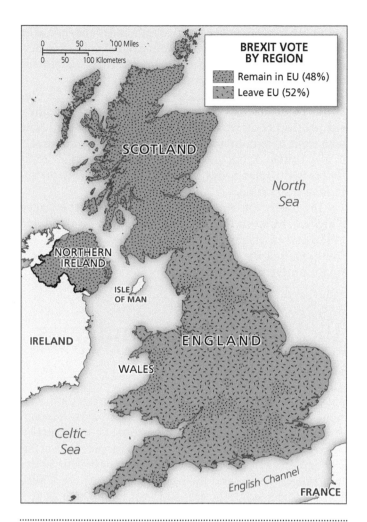

MAP 15.4 The Brexit vote.

The rest of the country in England and Wales, and especially the lower-income groups in those regions, was bypassed by the emphasis on the London money machine. But discontent could not feed into demands for political separation as it could in Scotland—there was no constitutional basis—and so the rest of England and Wales seethed at the rising inequality and neglect. The exit vote was also a vote against the dominance of London and the political establishment that captured popular resentment against the status quo.

All districts in Scotland voted to remain in the EU—a complete reversal of the 1975 referendum when Scotland voted against joining the EU while much of the rest of the country embraced the European project. What happened? The EU provided benefits to Scotland that the London-biased UK government did not. Scotland's political culture, it turns out, is closer to that of the EU than to that of London-based Tories. And years of punishing Thatcher rule soured many in Scotland against relying on the UK political system.

The Brexit vote highlights the problem of sharing a political space with divergent interests. The Scottish Parliament in Edinburgh will now move to draft legislation for another referendum on Scottish independence. The result could be different from the one in 2014, when 55 percent voted to stay in. The Brexit vote may herald the breakup of a union first created in 1707.

Northern Ireland voted to remain in the EU. Its economic position is similar to Scotland's but in a very different political context. The province is split between those who want to join up with Eire and those who want to stay in the United Kingdom. Sinn Féin, the party that seeks union with Ireland, has already proposed a vote on all-Ireland unification.

The Brexit vote revealed the deep divide in the United Kingdom between the different regions of England and Wales and especially between the affluent London and the southeast and the rest of the United Kingdom. This division is unlikely to heal soon.

15.2 Borders, Boundaries, and Frontiers

States have edges, and these boundaries create both tensions that can lead to conflict and sites for interstate cooperation. Political geographers used to make a distinction between artificial and natural borders. People made artificial borders, while natural borders were coastlines, mountain ranges, rivers, and the like. The limits of the United States are shaped in the east and west by the natural borders of oceanic coastlines, but are more artificial in the north and south. More recent work tends to view borders as social constructions, the result of bordering practices rather than the embodiment of "natural" differences.

Viewing borders as socio-spatial constructions focuses attention on their origin and evolution and interprets them as stages for the performance of national identity. Borders arise, shift, and change shape over time, their changing configuration embodying competition for territorial space. Consider the case of Hungary. During the Austro-Hungarian Empire (1867–1918), Hungary was joined with Austria in ruling over a large empire in central Europe. The breakup of its empire after the First World War reduced Austria to a small German-speaking state and started the process of shrinking Hungary's territory. In the Treaty of Trianon (1920), Hungary lost Vojvodina, Croatia, and Slavonia to the new state that became Yugoslavia and forfeited its coastline. In the north, Hungary lost what is now Slovakia to newly formed Czechoslovakia. In the east, Hungary had to cede Transylvania and the Banat to Romania. To the west, Hungary lost Burgenland to Austria. In all, Hungary forfeited two-thirds of its former area.

The borders of many countries are complex texts that tell of the country's history. Poland, located in central Europe, has a more complicated history than most. Poland's current boundary encompasses territory that once belonged to Russia, Hungary, and Germany, while formerly Polish territory can be found in Belarus, the Czech Republic, Germany, Russia, and Ukraine. One Polish city, Wrocław, was a municipality of the Habsburg Empire, of the Kingdom of Prussia, of the German Empire, and then of Poland. Its name changed in the process from Vratislavia to Breslau to Wrocław. Over the longer term, boundaries are fluid and changeable.

Boundaries are used to define and contain. The Great Wall of China, a centuries-old building project first begun 2,500 years ago, was built to keep out marauders from the steppe. During the Cold War, the boundary between East and West Germany was used to define the limits of effective power but also to limit mobility and movement. Boundaries can be used to constrain the movement of people and goods, but in a globalized, interconnected world, it is increasingly difficult to limit the flow of information. But states try. The newest "border construction" in China is the government's control over Internet traffic.

The Porosity of Borders

A distinction can be made in the porosity of boundaries. Hard, impervious boundaries are created and maintained by the state as a display of power, a performance of national identity. The increasingly fenced boundary between the United States and Mexico, for example, is a form of border theater to display US intentions against illegal immigration and cross-border drug trade. The US boundary with Canada is more porous than its boundary with Mexico. We can thus make a distinction between **hard** and **soft borders**, **porous** and **nonporous boundaries**. The boundaries between the United States and its two neighbors provide examples of hard, nonporous borders and soft, porous boundaries, although in reality they are both more porous than official pronouncements would suggest.

Along the edge of boundaries, frontier regions may develop. There are the landscapes of hard boundaries: the barbed wire, the border patrols, and the checkpoints. Between North and South Korea, the boundary is made visible on either side by a demilitarized zone of unpopulated territory filled with landmines. Map 15.5 depicts this hard, nonporous boundary. Frontier zones in such cases may develop as liminal spaces lacking investment and development.

Conflict

Boundaries can be places of conflict, especially where territorial sovereignty is disputed. Around the world, boundaries are often neither settled nor agreed on. Let us consider one example in Southeast Asia.

French cartographers created the boundary between Thailand and Cambodia in 1907, when France was the colonial power in the region. An area around the ancient Khmer

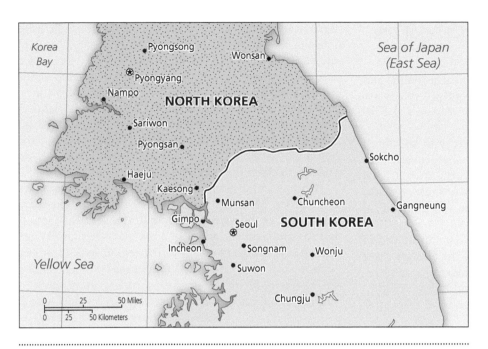

MAP 15.5 Boundary between North and South Korea.

temple in Preah Vihear was given to Cambodia. Thailand occupied this area when the French withdrew in 1953, but the International Court of Justice confirmed Cambodian sovereignty in 1962. In 2008, Cambodia submitted an application to have the temple listed as a World Heritage Site. Nationalist parties in both countries fanned the controversy to inflame public opinion for their own political agendas. In 2008, troops from both countries exchanged fire. There were sporadic exchanges until April–May 2011, when more sustained conflict occurred. In a ten-day skirmish, Cambodia claimed that Thai forces fired 50,000 shells along the border. A ceasefire was brokered on May 4, 2011. The area around the temple remains in dispute.

15.2   Conflict, Peace, and Sustainable Development

War and conflict threaten the foundation of peaceful societies and undermine sustainable development.

Every seven minutes, somewhere in the world, a child is killed by violence. Every day, 100 civilians, including women and children, are killed in armed conflicts, despite protections under international law. In 2019, the number of people fleeing war, persecution, and conflict exceeded 79.5 million, the highest level ever recorded.

Violent conflict reverses development and violates the most fundamental of human rights: the right to life. Armed violence and insecurity have a destructive impact on a country's development, affecting economic growth, and may cause longstanding grievances among communities. Armed conflict can involve the use of armed force between two or more organized armed groups, governmental or nongovernmental, within a country or between countries. Globally, 2 billion people live in countries affected by fragility, conflict, and violence. Syria, Afghanistan, Mexico, and Yemen are countries where recent conflict has caused the deaths of many.

Yemen has been steeped in crisis for decades, and the combination of famine, war, and other health emergencies has made it one of the worst humanitarian crises in the world. Afghanistan has endured four decades of conflict and nearly twenty years of direct US military engagement and

thousands of deaths. Conflict among Eritrea, Ethiopia, and Sudan has caused more than 17,000 deaths since 2020. In Ethiopia, escalating violent conflict in the Tigray region threatens to undermine reforms and destabilize not only Ethiopia, but also the entire Horn of Africa. For decades, systemic and long-standing socioeconomic development challenges kept Ethiopia classified high on fragile state and conflict watchlists. With this recent conflict, there has been a reversal of political reforms, including indefinitely postponed elections.

Many experts say international funding to fragile states must increase if we are to prevent violent conflict. More is needed to help establish stronger civil society institutions (such as legal systems and justice systems) that can create conditions for political stability, the rule of law, and lasting peace. Most of the current funding focuses on humanitarian aid in places with active wars like Syria and Yemen. Compare the amount of US funding for defense versus foreign assistance. Military funding is the second-largest item in the US federal budget after Social Security and was estimated to be $877 billion (almost 20 percent of the overall budget) in 2022. Foreign assistance, in contrast, constitutes less than 1 percent of the federal budget, of which only 11 percent is spent supporting political stability, democratic institutions, justice, and peacebuilding.

Maritime Boundaries: The Case of the South China Sea

International boundary disputes also involve islands and control over the sea. Maritime sovereignty disputes are likely to grow as the search for valuable minerals and resources extends into the sea.

China claims a wide swath of territory in the South China Sea, first claimed by the Nationalist Government in Taiwan at the end of the Second World War. It is an area of rich fishing grounds, vast oil and gas reserves, and vital sea lanes. It is of major economic and geopolitical significance.

Exclusive economic zones (EEZs), first introduced in the 1982 UN Convention on the Law of the Sea, are areas where states can claim exclusionary rights over marine resources. They have a limit of 200 nautical miles (370 kilometers) from the coastal baseline, the seaward edge of its territorial sea, but may extend beyond the 200-mile limit into the continental shelf. Where EEZs overlap, it is up to the states to delineate the maritime border. Matters are made more complex when small islands are used to define a nation's EEZ and are claimed by more than one country. A country can establish territorial seas up to 12 nautical miles from a naturally formed island and a 200-nautical-mile EEZ. Part of China's strategy in the waters of the South China Sea is to lay claim over islands, natural and created, to claim the surrounding waters.

China's claim impinges on the EEZs of Vietnam, the Philippines, Malaysia, and Brunei (Map 15.6). There are disputes over the various islands in the sea. The Paracel Islands are claimed by both China and Vietnam. When China moved a giant oil rig to the region in 2013, it was seen as a direct threat by the Vietnamese and undermined China–Vietnam relations. The Chinese have built a surface-to-air missile defense system on Woody Island, one of the largest of the Paracel Islands, and recently added a helicopter base to Duncan Island.

The Scarborough Shoal, a 60-square-mile chain of rocks and reefs, is claimed by both China and the Philippines. In 2012, China took effective control by reconstructing seven islands. Control over the islands in the shoal allows China to monitor US military activity by triangulating with Chinese bases and listening posts on the Paracel and Spratly Islands. The Philippines responded by taking the issue to the UN rather than pursuing bilateral discussion. The Philippines, unlike other states in the region, have fewer economic ties with China and close military ties with the United States.

Six claimants—China, Vietnam, Taiwan, Malaysia, Brunei, and the Philippines—make claim to the Spratly Islands, more than 740 reefs, islets, atolls, and islands. All but Brunei have built or claimed outposts among them. However, China's activity has ramped up since 2013, when it started to build on existing rock outcrops and reefs to create 2,900 acres of new land. Vietnam and the Philippines have undertaken similar ventures, but China has outpaced every other country combined. Sand is pumped up from the sea floor and

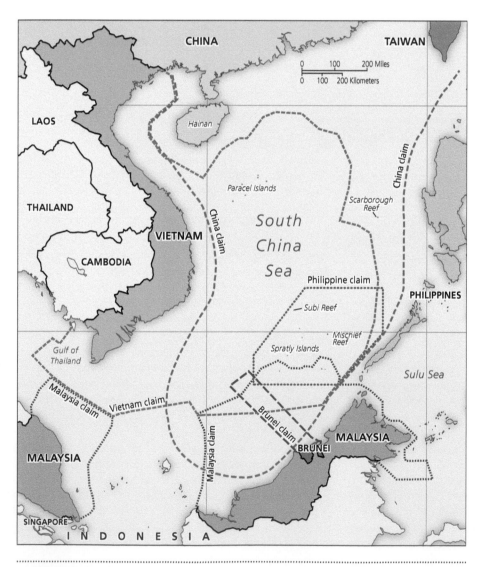

MAP 15.6 The South China Sea claims.

spread across islands, and even over the live coral of submerged reefs, which are then concreted over to become bases capable of providing docks for ships and runways for planes. Military runways have been built on Mischief Reef and Subi Reef. The Chinese have militarized the disputed islands.

On July 12, 2016, the Permanent Court of Arbitration announced that China's claim has no basis in international maritime law. China will ignore the ruling as it seeks to reach bilateral agreement with countries such as the Philippines that took the case to the court in 2013. China's claim is best interpreted as a claiming of world power status and importance. The Chinese look around the region and see the projection of US power: US bases in South Korea and Japan, a nuclear submarine fleet, a military presence in the Philippines and Australia, and a fleet of supercarriers able to project power and influence. The United States has a significant military and naval presence in the region that the Chinese see as a threat to their regional power status.

Cooperation

Boundaries can also be sites of cooperation. The opening of formerly closed boundaries and the easing of movement across once nonporous boundaries are often highly symbolic markers of regime change and interstate cooperation. The fall of the Berlin Wall in 1989 marked the end of the Cold War in Europe. When Berliners chipped away at the wall, they were demolishing not only a concrete wall but also a political regime. A hard, nonporous boundary was replaced by the free flow of national unification.

New Perspectives on Borders

The political geography of borders is now a lively area of geographical scholarship. The current emphasis is on the social construction of borders, practices of bordering, and the narration of borders as important parts of national identity. Reece Jones recounts the bordering between India and Bangladesh. He begins his interesting paper with a description of Moushumi, a domestic worker in Bangladesh who crosses the border into India to visit her son. Prior to 1947, the trip would not have been a border crossing, because both areas were part of the same space of colonial British control. After 1947, however, the partition of the continent created an international border. It took a while for the border to become bordered. A map was drawn in 1947, but the border was not properly surveyed until 1952, and only in the 1960s were border security forces deployed. The Indian government has now fenced more than three-quarters of the 4,000-kilometer border. This hard border is now "a key site for the state to establish the binaries of power that frame the world as

citizen–alien, national–foreign, here–there, and we–they" (Jones 2012, p. 691). Despite this hardening, cross-boundary flows still occur. The paper is interesting because it tells the story of the border as a place where the performances of state power are played out in conspicuous acts of security theater, but also highlights the limits to and often ragged nature of state territorial control. Borders are transgressed by smugglers and illegal immigrants and in unrecorded economic transactions. Borders are sites of performance of territorial control as well as examples of resistance, refusal, and transgression.

15.3 The Political Organization of States

In terms of political organization, three main regimes can be identified: **totalitarian regimes**, in which the government has control over wide and deep swathes of social, political, and economic life; **authoritarian regimes**, where power is concentrated but not so deeply entrenched as in totalitarian regimes; and **democracies**, where political power arises from the majority will of the people. There is some variation within these broad categories. Democracies, for example, include monarchies, such as the United Kingdom and Sweden, as well as republics, such as the United States and France. While power may ultimately reside with the people, even in full democracies its expression and implementation are deeply influenced by the distribution of wealth and power in the society.

A distinction can be made between government, state, and national territory. The government is the political expression of power. "State" is a wider term that covers the more embedded power structures, such as the army, the police, and the educational system. In some cases, a **deep state** of entrenched power operates despite changes of government. In more authoritarian regimes, government and state are more interconnected, while in more democratic states there is space between the government and the state.

There have been several waves of democracy movements. The most recent wave began in the 1990s, with the fall of communism in eastern Europe and the collapse of the Soviet Union. Countries such as Poland and Hungary replaced communism with democracy. Many of the newly independent post-Soviet states such as Latvia, Lithuania, and Ukraine transformed their economies and their political systems of government, becoming free-market oriented and democratic. Even Russia made a brief attempt at democracy before descending again into totalitarian rule under Putin. More recently, some states have taken steps to enhance or advance their democracies. Ecuador, Peru, Madagascar, and Honduras have implemented changes that improve freedom and democracy.

Democracies require government accountability and transparency; they must ensure freedom of expression and the protection of human rights. However, democracies are not easy. They are messy, difficult, and sometimes frustrating for their citizens. They require significant investment in building institutions that support democratic governance (such as the legal systems or electoral systems). They require citizens to be interested, engaged, and involved. Some of these realities may be why we have seen a decline of democracy, as some countries have regressed to authoritarian or totalitarian forms of government. Russia is one example, as are Belarus, Sudan, Afghanistan, and Myanmar. Even countries with strong democracies have been challenged by populist leaders who reject the demand of unchecked powers.

The Economist Intelligence Unit uses a **democracy index** to gauge the level of democracy using sixty indicators related to electoral process, civil liberties, functioning of government, political participation, and political culture. The index, produced each year, ranges from 10, the most democratic, to 0, the least democratic (Map 15.7). The most democratic countries include Scandinavian democracies such as Norway, Iceland, Denmark, and Sweden, while the least democratic are North Korea, Syria, and Chad. The United Kingdom stands sixteenth and the United States twenty-sixth.

The democracy index is used to classify countries into different regimes: full democracies, flawed democracies, hybrid regimes, and authoritarian regimes. The 2022 edition listed 167 countries ranked on metrics of five dimensions: electoral process and pluralism, the functioning of government, political participation, democratic political culture, and civil liberties. The United States ranked 26th in the world. At the top of the list were Norway, New Zealand, Finland, and Sweden. At the bottom were North Korea, Myanmar, and Afghanistan. There are no real surprises there, but Taiwan (8), Uruguay (13), South Korea (16), the United Kingdom (18), and Costa Rica (21) all outranked the United States. The United States had slipped over the past six years from a full democracy to a flawed democracy.

According to this classification, more than a third of the world's population lives under the duress of authoritarianism (see Table 15.3). The democracy index is a function of the variables employed. Change the mix of variables, and different results may emerge at the margins. But whatever variables are used, some countries, such as North Korea, will always come out as deeply undemocratic. At this extreme end, there are depressing categories such as **kleptocracies**, states geared to the enrichment of the tiny political elite. Some of the greediest include former Indonesian president Suharto, who reputedly stole between $15 billion and $35 billion; former president of the Philippines Ferdinand Marcos ($5 billion–$10 billion); and Mobutu, former president of Zaire, now the Democratic Republic of the Congo ($5 billion), who plundered the country and pauperized the population. Current kleptocracies include Equatorial Guinea, where Teodoro Obiang Nguema, president since 1979, has amassed a personal fortune estimated at around $600 million. He has personal control over the nation's treasury and works well

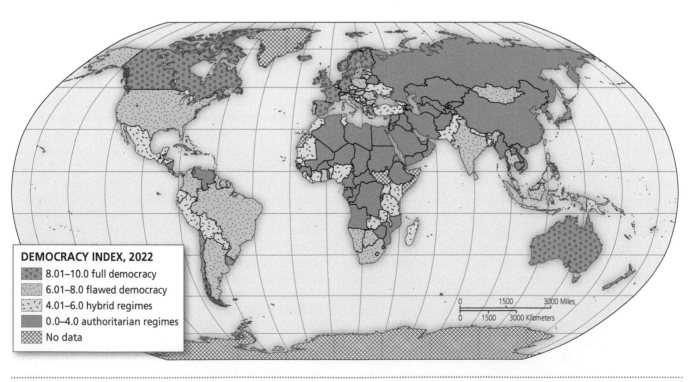

DEMOCRACY INDEX, 2022
- 8.01–10.0 full democracy
- 6.01–8.0 flawed democracy
- 4.01–6.0 hybrid regimes
- 0.0–4.0 authoritarian regimes
- No data

MAP 15.7 The democracy index.

TABLE 15.3 Political Regimes and Democracy Scores, 2022

CATEGORY	SCORE	% OF WORLD POPULATION	EXAMPLES
Full democracies	8.01–10.0	8.9	Canada, Sweden, United Kingdom
Flawed democracies	6.01–8.0	39.5	Argentina, Philippines, United States
Hybrid regimes	4.01–6.0	17.5	Bolivia, Pakistan, Turkey
Authoritarian	0.0–4.0	34.1	China, Russia, Saudi Arabia

Source: Based on Economist Intelligence Unit data.

with foreign oil companies eager to drill for oil. Although Equatorial Guinea is one of the world's poorest countries, the president lives a life of unrivaled opulence.

Classifications are snapshots in time. Popular movements and internal regime changes can alter them. Marcos was overthrown by a popular uprising, and the Philippines are now classified as a flawed democracy—not ideal, but more democratic than in the Marcos era. Indonesia is also now in the same category. Myanmar, after years of authoritarian military control, had a brief democratic moment from 2011 until 2021, when a military junta again grabbed power. Congo remains in the authoritarian category.

States vary by quality of life as well as political system. A standard economic argument is that rising income levels create a greater sense of well-being. More income, so the argument goes, creates happier people. The **Easterlin paradox**, based on the work of the economist Richard Easterlin, suggests that once basic needs are met, however, there is little difference in reported happiness with increasing incomes. In other words, the paradox is that once a certain base level of income is reached, increasing income further may not necessarily increase the level of happiness in the country. The debate continues. Including variables such as life expectancy, hours worked, and unemployment confirms the paradox. More recent papers highlight the finding that people in rich countries with more progressive taxation tend to report higher levels of happiness than those with lower and more regressive tax policies. Perhaps we need to measure **gross national happiness** as well as gross national income. What variables would you include? One of the few countries to measure happiness is Bhutan. However, their index does not capture the fears and anxieties of the ethnic Nepali minority in the south of the country, many of whom have fled to neighboring Nepal since the 1990s.

Corrupt States

Corruption affects how a state functions internally and how it interacts with the world. Corruption is the abuse of entrusted power for private gain. This includes corruption in public and business sectors, from local to international levels. It extends from petty corruption felt acutely by citizens every day to kleptocracy and high-level grand corruption damaging entire societies. Examples include bribery and embezzlement, theft, tax evasion, and profiting from illicit flows (such as drugs or illegal arms). Bribery is one of the most common forms of corruption. The World Economic Forum reports that in countries such as Yemen, Liberia, and Vietnam, at least two of three survey respondents report having paid a bribe within the past year. In Japan, the United Kingdom, and Sweden, only one in one hundred (or fewer) reports having paid a bribe. Generally, corruption is higher in nondemocratic countries, although there are exceptions, such as India. India's level of corruption has been high for many years, leading some experts to conclude that corruption has diminished India's economic growth.

Corruption is a global problem but impacts low-income countries the most. The World Economic Forum, reported in 2022 that corruption costs low- and middle-income countries $1.26 trillion. The amount of money lost because of corruption could be used to lift those who are living on less than $1.25 a day above $1.25 for at least six years. For example, in war-torn Afghanistan, of the $8 billion donated in recent years, as much as $1 billion has been lost to corruption. Countries where corruption levels are high fail to attract foreign investors, who see investment as too risky. This includes many transnational corporations, who steer clear of countries with high levels of corruption and a weak rule of law.

Although corruption is difficult to measure, the nongovernmental organization Transparency International publishes a "Corruption Perception Index," the most widely used indicator of corruption worldwide. The Corruption Perceptions Index scores 180 countries on their degree of corruption: 10 is the cleanest possible and 0 indicates endemic corruption. The five countries with the highest scores (and thus perceived as most "clean") for 2022 are Denmark, Finland, New Zealand, Norway, and Singapore. At the other extreme, the countries with the lowest scores (and highest perceived corruption) are Somalia, Syria, South Sudan, Venezuela, Yemen, and North Korea.

Corruption is often a part of daily life for many: having to pay bribes to access public services, for example, shows that small-scale corruption can have a big impact as the constant demands for bribes push people further into poverty.

Government and politics scholar Eric Uslaner writes that "corruption not only thrives under conditions of high inequality and low trust, but in turn it leads to more inequality."

Corruption is a major obstacle to economic and social development. Its political costs can include the destruction of public order and the erosion of societal trust in the institutions. In economic terms, corruption depletes wealth, contributes to further inequality, and hinders entrepreneurship.

15.4 The Geography of Elections

States maintain their popular legitimacy through the performances of elections. There is a geography to these elections. Three distinct geographies can be identified: of voting, of representation, and of electoral systems.

The Geography of Voting

This geography examines the distribution in votes cast for different parties and candidates. Consider the voting for presidential candidates in Nigeria in 2011 (see Map 15.8). The three candidates appealed to different parts of the country. One candidate won most of the northern region, another won the south, and a third candidate's support was very localized in one state. The differences reflect tribal and ethnic differences. The northern states are predominantly made up of Muslim and Hausa-Fulani tribes, while the south is Christian and dominated by Yoruba and Igbo tribal members.

There are national and regional cleavages in voting. In the 1968 US presidential election, there were three candidates. While the Republican, Richard Nixon, won most of the country, the Democrat, Hubert Humphrey, won most of the northeastern states as well as his home state of Minnesota. The third-party candidate, George Wallace,

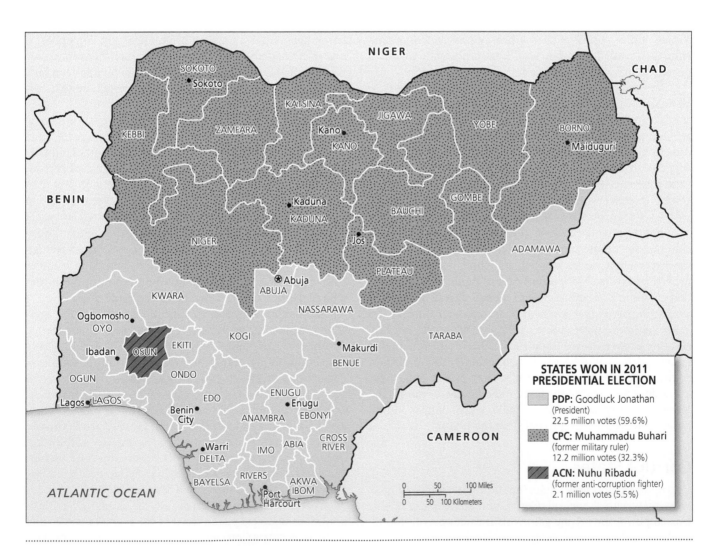

MAP 15.8 Presidential election voting in Nigeria.

who pledged to maintain racial segregation, won only states in the Deep South.

There is also an even more local effect, known as the **neighborhood effect**. This refers to the fact that neighborhood can trump traditional allegiance. Democratic-leaning voters in a predominantly Republican neighborhood will vote for Republicans and vice versa. We have detailed empirical demonstration for this neighborhood effect in voting outcomes. The precise causal factors are difficult to disentangle. Does the neighborhood cause the shift, or do people predisposed to the shift mark their intentions by moving to the neighborhood in the first place? The geography of voting is influenced by distribution of income and racial and ethnic differences, but also by local factors such as neighborhood and the effect of friends and neighbors.

A marked spatial polarization in voting has been noted in the United States. Since 1992, across the different spatial scales of 3,077 counties, 49 states, and 9 census divisions, there was an increase in the polarization of votes for one of the two main political parties in presidential elections. The vote distribution shifted from areas with a relatively even split to areas more heavily skewed toward either Democrat

or Republican. This may be the result of selective relocation as the population shifts into communities of similar political persuasion. And these communities, in turn, may influence those who subsequently move in. Whatever the exact combination of factors, the end result is that more people live in areas where neighbors amplify and reinforce rather than contest and question their political beliefs and prejudices. This spatial polarization may be one of the causes behind the increasing social polarization along party lines in the United States.

The 2016 US presidential election provides an interesting example of the geography of voting. A map of Trump's America, places where he picked up the majority of votes, consists of a vast landmass of the interior of the country. Clinton's America, on the other hand, consists of the large cities along the coasts and large inland cities and places with college towns and substantial Black and Hispanic populations (see Map 15.9). The swing to Trump, as indicated by those counties that voted for Obama in 2012 but Trump in 2016, highlights his appeal in the old industrial heartland of the upper Midwest and Northeast, as well as his success in mobilizing racial and ethnic animus (Map 15.10). Trump's wins in Michigan,

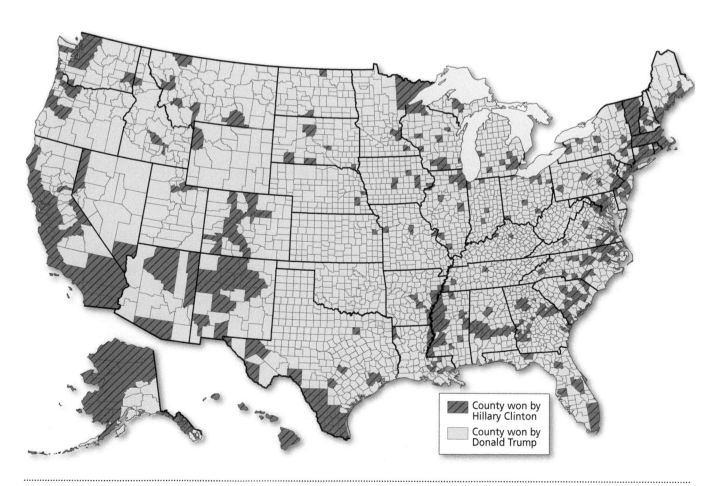

MAP 15.9 Clinton's America.

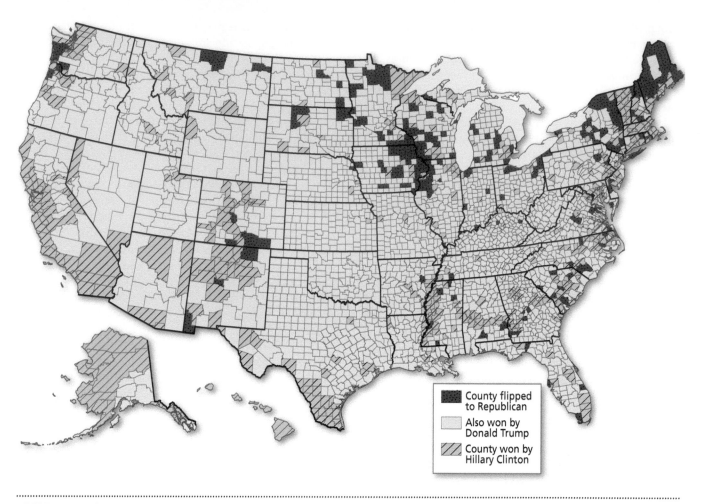

MAP 15.10 The swing to Trump.

The Geography of Representation

National territories are divided into political constituencies. The geography of this representation has an impact on the results. **Malapportionment** refers to the imbalance between the number of voters and the population in each constituency. In some cases, numerical malapportionment is accepted, indeed codified into the legislative framework. The rigging of the voting system for the US Senate so that some electoral votes count more than others is not new; it is a foundational reality, an integral part of the political architecture of the country. Under the US constitution, each state receives the same number of senators, despite differing population size, while the number of representatives afforded to a state is based on its population. It started at the beginning of the Republic, when each state was allocated two senators, despite differing population size. At

Pennsylvania, and Wisconsin extended Republican support beyond its usual base and ensured his victory in the Electoral College.

the time of the First Congress in 1789, the population of the largest and smallest state, respectively, Virginia and Delaware—and here we will only include free white males over the age of sixteen, as befits the prioritization of the time—was 110,936 and 11,783. This is roughly a 9-fold differential. By the time of the 2020 presidential election, the population of the most and least populous states, respectively, California and Wyoming, was 39.35 million and 581,348. The differential has increased to 67-fold, although the senator allocation has remained the same. Senators from small states with reliably consistent voting preferences can amass seniority that bestows enormous power beyond their demographic significance. A longtime leader of the Senate, Mitch McConnell, co-represents a state with a 2020 total population of only 4.4 million that is 89.4 percent white with only 3.5 percent foreign-born, while the US average is 71.7 percent white and 12.9 percent foreign-born. In the United States, for example, each state receives two seats in the Senate, yet in 2016, although California and North Dakota each had two senators, California had a population of 39.1 million versus North Dakota's 756,927.

This disparity was accepted by the Founding Fathers as the price to be paid to create a union of fiercely independent states. Malapportionment is a function of political power. Those who gain from the existing system rarely want to change the system.

There is also **gerrymandering**, which refers to the manipulation of voting boundaries to engineer specific political outcomes. The term originated in the activities of Elbridge Gerry, who in 1810 as governor of Massachusetts signed a bill that demarcated boundaries that favored his party (Democratic-Republican). The bizarrely shaped new district was depicted in a political cartoon as a salamander, which opposition politicians quickly described as the "gerrymander." The system was so "gerrymandered" that while the Democratic-Republicans won only 50,164 of the votes compared to the 51,766 gained by the Federalists, they won twenty-nine of the forty seats. Gerrymandering is a recurring feature of political boundary making and an ongoing reality of congressional boundaries in US states.

The US Constitution requires each state to establish new congressional districts every ten years to reflect the population changes measured by the census. This congressional redistricting is freighted with partisan political interests. In Utah, the results of the 2010 Census revealed enough population increase to justify another congressional district, increasing the number from three to four. The political geography of Utah consists of the more Democratic-leaning metro area of Salt Lake City and the Republican-dominated rural areas. The Republicans control the state legislature and thus the redistricting. If the metro area were one congressional district, it could possibly ensure a Democratic victory. To avoid this possibility, the Republican-dominated state legislature produced a redistricting plan that split the metro area across three separate districts. The Democratic vote was thus spread and diluted within more rural districts. In Democratic-controlled Maryland, the post-2010 redistricting proposals ensured that Republican support was diffused to favor Democratic voters. A new District 6 was drawn up so that the more Republican voters in rural counties were lumped together, but swamped by the more populous Democratic-voting suburban areas. District 3, the least compact congressional seat in the country, was the result of drawing boundaries to ensure a Democratic victory (Map 15.11).

The Geography of Electoral Systems

States vary in the geography of their electoral systems. The biggest distinction is between the **first-past-the-post** method, one of what are called plurality voting sys-

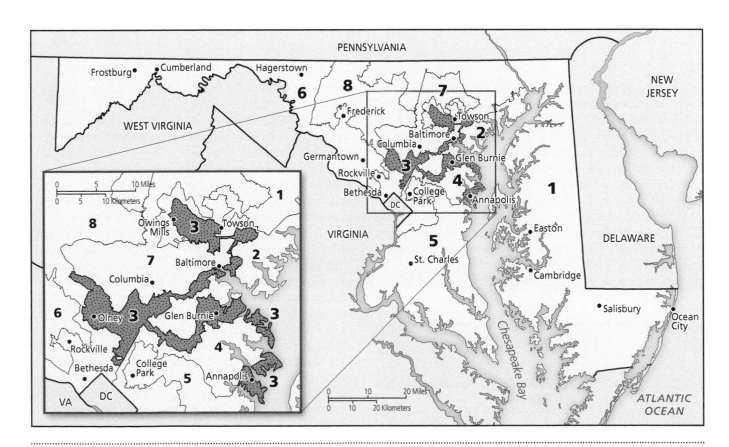

MAP 15.11 Congressional boundary redistricting in Maryland.

15.3 Racial Gerrymandering

In the United States the two most common forms of gerrymandering have been partisan gerrymandering, which attempts to secure electoral advantage for one's preferred party, and racial gerrymandering, which seeks to secure electoral advantage for one's preferred race. In 1960, the US Supreme Court ruled that racial gerrymandering was unconstitutional, finding that as states redrew legislative district lines, they often disenfranchised minorities and perpetuated systemic racism.

Although racial gerrymandering is now unconstitutional, it continues to happen. States redistrict in ways that dilute minority voting power, often under the guise of "partisan gerrymandering," which is more complicated to analyze, prove, and challenge in the courts.

In 2019, the Supreme Court considered *Rucho v. Common Cause*, a landmark case about partisan gerrymandering in North Carolina that appeared to disenfranchise many Democrats, a majority of whom were Black. Voters and other plaintiffs filed a lawsuit. A district court struck down the initial map of redistricted areas for being a partisan gerrymander. It appears that Republican state legislators then instructed the cartographers to use granular political data to draw a congressional district line that divided North Carolina Agricultural and Technical State University—the largest historically Black college in the nation—"so precisely that it all but guarantees it will be represented in Congress by two Republicans for years to come." Despite this evidence, the Supreme Court ruled 5–4 that while partisan gerrymandering may be incompatible with democratic principles, these types of political questions were beyond the reach of federal courts. In other words, it left the gerrymander in place, saying that it was up to states and Congress to pass stricter laws about gerrymandering. The Supreme Court's decision means that explicit acts of gerrymandering continue to disproportionately disenfranchise minority voters.

REFERENCES

Palandrani, J., and D. Watson 2020. "Systemic Inequality/Racial Gerrymandering, the For the People Act and Brnovich: Systemic Racism and Voting Rights in 2021." *Fordham Law Review Online* 89, Article 21. https://ir.lawnet.fordham.edu/flro/vol89/iss1/21.

Short, J. R. 2018. "4 Reasons Gerrymandering Is Getting Worse." *The Conversation*. https://theconversation.com/4-reasons-gerrymandering-is-getting-worse-105182.

Short, J. R. 2019. "After Supreme Court Decision, Gerrymandering Fix Is Up to Voters." *The Conversation*. https://theconversation.com/after-supreme-court-decision-gerrymandering-fix-is-up-to-voters-117307.

tems, that ensures that the politician with the majority of votes in a district wins that district. This dominates the electoral scene in Canada, the United Kingdom, and the United States. The system is simple and easy to understand, but it tends to solidify the political representation of large nationwide parties and marginalize third and fourth parties with wide, but not necessarily nationwide, depth of support. Table 15.4 provides the results of the 2019 general election in the United Kingdom for the three main parties in England. Notice how the Liberal Democrat Party won almost 12 percent of all votes but managed to gain less than 2 percent of the seats. Smaller parties with wide support, but not deep enough to win particular seats, are penalized by the first-past-the-post system.

A large number of countries, one example of which is Ireland, now have a system of **proportional representation** that translates votes into seats in a more equitable manner. This can be done through giving seats in proportion to the votes cast. Some countries have a mix of seats allocated by first-past-the-post and then some by proportional representation methods. Under a single transferable vote system, electors rank candidates in order of preference. If their preferred candidate does not gain enough votes, the votes are given to the second ranked candidate. There are also preferential systems. In an alternative vote system, voters have an opportunity to express their second choice, as happens in Australia, or, as in French presidential elections, in a second round of voting if no candidate has an absolute majority in the first round. Each system has a mix of advantages and disadvantages. The first-past-the-post system ensures one party has a majority, but small and minority parties are underrepresented. In the proportional and preferential system, the legislature reflects voting preferences more closely, but the large number of parties can make for fragile and unstable coalition governments.

TABLE 15.4 General Election Results in the United Kingdom, 2019

PARTY	% OF VOTES	% OF SEATS
Conservative	43.6	56.2
Labour	32.1	31.1
Liberal Democrat	11.6	1.7

Political outcomes in part depend on geography. The geographies of voting, representation, and electoral systems all play an important part in turning votes into representation. Consider again the 2016 US presidential election. One candidate, Donald Trump, gained 46.1 percent of the popular vote, a total of 62.9 million votes, while the other main candidate, Hillary Clinton, won 48.2 percent of all votes cast, 65.8 million votes. But because the US electoral system, in the form of the Electoral College, rewards those who win the most votes in each state, the winning candidate had fewer popular votes than his defeated rival.

Cited References

Cusak, T. 2010. *Riverscapes and National Identity.* Syracuse, NY: Syracuse University Press.

Davies, N. 2012. *Vanished Kingdoms: The Rise and Fall of States and Nations.* New York: Viking.

Easterlin, R. 1974. "Does Economic Growth Improve the Human Lot? Some Empirical Evidence." In *Nations and Households in Economic Growth: Essays in Honor of Moses Abramovitz,* edited by Paul A. David and Melvin W. Reder, 89–125. New York: Academic Press.

Economist Intelligence Unit. 2022. Democracy Index 2022. https://www.eiu.com/n/campaigns/democracy-index-2022/.

Habermas, J. 1975. *Legitimation Crisis.* Boston: Beacon Press.

Jones, R. 2012. "Spaces of Refusal: Rethinking Sovereign Power and Resistance at the Border." *Annals of the Association of American Geographers* 102:685–699.

Nye, J. 2004. *Soft Power: The Means to Success in World Politics.* New York: Public Affairs.

Stockholm International Peace Research Institute. 2023. *SIPRA Military Expenditure Database.* https://www.sipri.org/databases/milex.

Transparency International. 2022. "Corruptions Perceptions Index." https://www.transparency.org/en/cpi/2022.

Uslaner, E. 2011. *Corruption and Inequality CESifo DICE Report.* Leibniz Institut für Wirtschaftsforschung an der Universität München, München 9:20–24.

Wilkinson, R., and K. Pickett. 2009. *The Spirit Level: Why More Equal Societies Almost Always Do Better.* London: Allen Lane.

World Economic Forum. 2022. "*Why Fighting Corruption Is Key to Addressing the World's Most Pressing Problems.*" https://www.weforum.org/agenda/2022/12/why-fighting-corruption-is-key-to-addressing-the-world-s-most-pressing-problems/.

Select Guide to Further Reading

Bremmer, I. 2006. *The J Curve: A New Way to Understand Why Nations Rise and Fall.* New York: Simon & Schuster.

Brooks, K. 2011. "Is Indonesia Bound for the BRICs?" *Foreign Affairs,* November–December, 109–117.

Calvo, E., and J. Rodden. 2015. "The Achilles Heel of Plurality Systems: Geography and Representation in Multiparty Democracies." *American Journal of Political Science* 59:789–805.

Davies, N., and R. Moorhouse. 2003. *Microcosm: Portrait of a Central European City.* London: Pimlico.

Diener, A. C., and J. Hagen. 2012. *Borders: A Very Short Introduction.* New York: Oxford University Press.

Dittmer, J., and J. Sharp, eds. 2014. *Geopolitics: An Introduction.* London: Routledge.

Elden, S. 2013. *The Birth of Territory.* Chicago: University of Chicago Press.

Kolossov, V., and J. O'Loughlin. 1998. "New Borders for New World Orders: Territorialities at the Fin-de-siecle." *GeoJournal* 44:259–273.

Megoran, N. 2012. "Rethinking the Study of International Boundaries: A Biography of the Kyrgyzstan–Uzbekistan Boundary." *Annals of the Association of American Geographers* 102:464–481.

Popescu, G. 2012. *Bordering and Ordering the Twenty-First Century.* Lanham, MD: Rowman & Littlefield.

Richardson, T., ed. 2013. "Borders and Mobilities." Special issue, *Mobilities* 8:1–165.

Scuzzarello, S., and C. Kinnvall. 2013. "Rebordering France and Denmark: Narratives and Practices of Border Construction in Two European Countries." *Mobilities* 8: 90–106.

Short, J. R. 2014. "The Supreme Court, the Voting Rights Act and Competing National Imaginaries of the USA." *Territory, Politics, Governance* 2:94–108.

Short, J. R. 2021. *Geopolitics: Making Sense of a Changing World.* Lanham, MD: Rowman & Littlefield.

Webster, G. 2013. "Reflections on Current Criteria to Evaluate Redistricting Plans." *Political Geography* 32:3–14.

Wrong, M. 2001. *In the Footsteps of Mr. Kurtz: Living on the Brink of Disaster in Mobuto's Congo.* New York: HarperCollins.

Yayboke, E., and E. Hume. 2020. "Ending Violent Conflicts Requires Preventing Them in the First Place." Center for Strategic and International Studies. https://www.csis.org/analysis/ending-violent-conflicts-requires-preventing-them-first-place.

PART 6

The Urban Organization of Space

This section explores the urban organization of space. Chapter 16 identifies three distinct urban revolutions. Chapter 17 identifies networks of cities at the regional, national, and global levels. In Chapter 18, the focus is on the internal structure of the city and how it is shaped by capital flows, housing markets, social interactions, and political struggles.

OUTLINE

16

The Urban Transformation

A majority of the world's population now lives in urban areas. This shift constitutes a major change, because for most of humanity's occupancy of the Earth, most people lived in rural areas. In this chapter we will consider the rise of cities and the nature of this urban transformation. We can identify three distinct urban revolutions.

16.1 The First Urban Revolution

The very first cities emerged between 11,000 and 5,000 years ago. They developed alongside settled agriculture, the domestication of animals, and the creation of complex irrigation systems. A traditional argument claims that the **first urban revolution** was predicated on the agricultural revolution: it was agricultural surplus that created cities. Most commentators now reverse the direction of the causal arrow. It was cities, or at least permanent settlements, that created agriculture. An 11,000-year-old complex of megaliths in Turkey, called Göbekli Tepe, provides some insights. In an apparent act of religious devotion, the monuments were built by hunter-gatherers. The need for large numbers of people to stay in one place and cooperate to build these monuments created a demand for a stable and secure food supply: hence the invention of agriculture. It was a solution that allowed early societies to remain in place to build monuments. Cities caused the development of agriculture because they generated enough demand in one place to stimulate new sources of food production.

Progress or Regress?

The development of agriculture and the creation of cities are often depicted as an upward trajectory of increasing civilization. However, the hunting-gathering societies that preceded this revolution spent less time working than agricultural societies; they are what one commentator terms the "original affluent society." In other words, preurban, preagricultural societies had more disposable time and more freedom. Even in the early twenty-first century, the Bushmen of the Kalahari spend only around 1,000 hours a year hunting and gathering food, and that is in a harsh environment; hunter-gatherers who lived in more hospitable climes probably would have spent much less time. The rest of the time is spent entertaining, relaxing, and being with friends and families. Rice farmers in southern China, in contrast, spend around 3,000 hours a year tending to their crops.

The shift from hunting-gathering to settled agriculture involved more work. Mortality increased, and food intake declined to such a narrow range of foods that it led to an increase in anemia and vitamin deficiencies. The supply of food did not keep pace with the increasing population, so more farming land was needed, creating the constant growth that was an essential feature of the early urban empires. Agriculture and cities were a mixed blessing that, in the short term, meant declining living standards for most of the population while an elite prospered.

The urban-agricultural revolution, for the vast majority, marked a loss of freedom, greater work discipline, and more time devoted to the drudgery of work and the compulsion of social order. It was a social order that had to be imposed. As an example, there are remains of an important urban culture in the desert Southwest of the United States known as the Anasazi. This independent urban civilization was centered on Chaco Canyon, New Mexico. The traditional rendition goes like this: Between the tenth and twelfth centuries, the Anasazi culture, based on efficient agriculture, flowered into cities of vast cliff dwellings and major feats of engineering, architecture, and art. Beautiful pottery, sophisticated irrigation systems, and keen solar and astronomical observations round out a picture of an urban civilization that follows the old precept that cities equal civilization. Yet there is another interpretation of the Anasazi, a darker side suggesting that the Anasazi culture developed from the Toltec Empire, which lasted from the ninth to the twelfth centuries in central Mexico. This was an empire centered on human sacrifice and cannibalism. Thugs from the Toltec Empire moved north into what is now New Mexico and found a pliant population of docile farmers whom they terrorized into a theocratic society. Social control was maintained through acts of cannibalistic terror. The Anasazi culture, so long admired, was one in which the bad and powerful controlled the weak and vulnerable. The great feats of art and astronomy, road building and city formation, were less sparks of human ingenuity and more the mark of organized social terrorism.

There was a distinct pattern to the spatial layout of the first cities. We can picture a model of the **preindustrial city**. At the city center were the political elite and sites of religious devotion, the temples and altars that forged a direct link between the sacred and profane (see Photo 16.1 and Figure 16.1). The city was like a map of the cosmos. Elites were in the center of the city, with the poor at the periphery. The homology between cosmos and city was a way to legitimize the social hierarchy. The city not only housed people: it was also a text that explained and justified the social world and embodied the wider cosmos.

Urban Vulnerability

The early cities were also vulnerable to ecological collapse. Over 4,000 years ago, the cities of Mesopotamia, including Ur, Uruk, and Umma, were brought under the unified control of the Akkadian Empire, centered on

PHOTO 16.1 The Mayan city of Tulum, now in Mexico, was a typical preindustrial city with a prominent central site. As with many early cities around the world, the architecture at the center of the city was monumental and symbolic, connecting the sacred and the profane, the religious and the political. It prospered from the thirteenth to the fifteenth centuries, but the arrival of the Spanish brought disease and death.

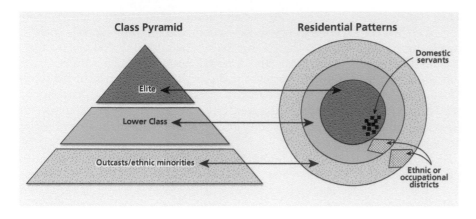

FIGURE 16.1 Model of the preindustrial city.

the city of Akkad. At least four generations of kings ruled over the sophisticated empire. And then, suddenly, the empire collapsed. One text, written a century after the collapse, noted,

For the first time since the cities were built and founded
The great agricultural tracts produced no grain.
The inundated tracts produced no fish,
The irrigated orchards produced neither syrup nor wine,
The gathered clouds did not rain.

Archaeological research has revealed that the collapse was the result of a prolonged drought. "The gathered clouds did not rain" is not a poetical conceit, but a record of severe drought.

Around the world, there is evidence of urban collapses caused by ecological change. Examples include the Mayan Empire, profiled in Box 16.1, and Angkor Thom in Cambodia.

Angkor Thom was the capital of the Khmer Empire, which controlled a large area of what is today Cambodia, Thailand, Laos, and Vietnam from the ninth to the fourteenth centuries. The city area encompassed more than 385 square miles. At its peak, it housed up to 1 million people, with more than 1,000 temples. Water collected from the hills was stored and distributed for a wide variety of purposes, including flood control, agriculture, and ritual bathing. A system of overflows and bypasses carried surplus water to the lake to the south. The network of reservoirs was extensive and supplied tremendous amounts of water to the city. The fall of the Khmer Empire in the fourteenth and fifteenth centuries was the result of the combination of a dramatic decrease in rainfall in the region and the deleterious impact of the built environments on the surrounding environment. The city's growth led to deforestation, which increased flooding.

The earliest cities, like the cities of the early twenty-first century, had an environmental vulnerability. The constant demand for food and water and the risk of climate change and ecological disruption mean that most urban empires, sooner or later, collapse (Photo 16.2).

PHOTO 16.2 Ruins of Carthage in North Africa. The city was destroyed in the Punic Wars between Rome and Carthage in 146 BCE. It became a Roman city and a center of early Christianity; it was then destroyed by a Muslim army in 698 CE. The ruins are silent testimony to the rise and fall of cities and urban empires.

Cities and Social Change

Urbanization is often described as the geographical redistribution of population from rural areas to urban areas. However, it is not just the spatial reorganization of society but also the social reorganization of space. Cities incubate social and political change.

The **merchant city**, for example, began to develop in association with the creation of a money economy, the extension of trade, and the emergence of a merchant class. In Europe, early merchant cities developed, such as Amsterdam, Bruges, Florence, and Venice (Photo 16.3). The merchant city was a vital cog not only in the development of international trade and commerce, but also in the creation of a commercial society where private interest was regulated, collective rules were established,

16.1 Climate Change and the Collapse of Mayan Cities

The Maya lived in the area in Central America that now consists of Yucatan, Guatemala, Belize, and southern Mexico. Mayan cities include Tikal and Uaxactun, Chichen Itza, Mayapan, Copan, and Palenque. Tikal's population is estimated to have been around 60,000, giving it a population density several times greater than that of the average European city at the same time in history. The Maya were highly accomplished in astronomy, with an intimate knowledge of the calendar. The cosmology of the Maya permeated their lives and structured their cities. Cities were designed to coincide with astronomical rhythms. At Chichen Itza, on the days of the spring and autumn equinox, during sunset, a sun serpent rises up the side of the stairway of the pyramid called El Castillo. In Mayan sites in the central and southern lowlands many temples have doorways and other features that align to celestial events.

Beginning in the eighth century and continuing for some 150 years, the great Mayan cities were abandoned, as wars raged and people fled. By 930 CE, the Mayan heartland had lost 95 percent of its population. This prolonged event, known as the "Mayan collapse," is one of the enduring mysteries of pre-Columbian America. Since the 1990s, archeologists have theorized that climate change contributed to the decline.

Recently, geographer Billie L. Turner II has used new geospatial technologies to discover evidence of a "megadrought," a prolonged period of significantly less precipitation. To support both farms and cities with populations of 60,000 to 100,000, Turner explained that the Maya had to cut down forests and increasingly manipulate wetlands, drawing water off into reservoirs and expanding agriculture into lowland wetlands. These actions consumed water that could not be spared during periods of drought. The Maya also unintentionally made their own agriculture less productive with their extensive deforestation. Removing trees, Turner explained, stopped the cycle by which the tree canopy would capture and return the naturally occurring nutrient phosphorus to the soil. Turner concludes, "The Maya had cut down so much of that vegetation and changed it in so many ways, they were amplifying the aridity that was already present."

REFERENCES

Coe, M. 2005. *The Maya*. 7th ed. New York: Thames & Hudson.

Faust, B., E. N. Anderson, and J. Frazier, eds. 2004. *Rights, Resources, Culture and Conservation in the Land of the Maya*. Westport, CT: Praeger.

Haug, G., D. Gunther, L. Peterson, D. Sigman, K. Hughen, and B. Aeschlimann. 2003. "Climate Change and the Collapse of the Maya Civilization." *Science* 5613:1731–1735.

Mckillop, H. 2006. *The Ancient Maya: New Perspectives*. New York: Norton.

Turner, B. L. 2020. "The Ancient Maya Response to Climate Change: A Cautionary Tale." *Peabody Museum*. https://peabody.harvard.edu/video-ancient-maya-response-climate-change-cautionary-tale.

PHOTO 16.3 Venice was an important merchant city with trade links throughout Europe and with the Ottoman and Chinese Empires.

and civic communities were forged. The people of the merchant cities created the notion of the public realm and civic society.

16.2 The Second Urban Revolution

The **second urban revolution** began in the late eighteenth century with the creation of the industrial city and unparalleled rates of urban growth. The **industrial city** was the crucible of early industrial capitalism, giving us factory life and class struggles. Factory production replaced household production, and industry replaced agriculture. The factories of Manchester and other cities in Britain in the early nineteenth century were wonders of the modern world—indeed, it was the world made modern; centuries-old agricultural dominance crumbled in a sudden seismic economic shift. In 1801, almost 70 percent of the British population lived in places with populations less than 2,500, but by 1851 over 40 percent were living in cities with a population greater than 100,000. New cities were built on greenfield sites, and major cities developed from tiny hamlets.

Although the cities of Britain were at the forefront of the Industrial Revolution, the rest of Europe and North America were not far behind. Cities like Essen, Cologne, Toronto, and Melbourne began to grow rapidly by the late nineteenth century. Toronto's population increased from 30,000 in 1851 to 181,000 in 1891. Swift economic change and population growth also occurred in cities such as Boston, Pittsburgh, Cleveland, Milwaukee, Cincinnati, and Philadelphia. Industrialization generated unprecedented levels of urbanization. In 1830, Cincinnati's population was 24,800, but by 1850 it had quadrupled to 115,400 and by 1870 it was 216,000. The population of Paterson, New Jersey, increased from 11,334 in 1850 to 125,600 by 1910.

Cities and Class Struggle

The industrial city created the arena for intense struggle between capital and labor. There were social commentators like Karl Marx who saw the industrial city as the harbinger of social revolution. There were also social reformers who used the appalling conditions of the industrial city to demand more environmental regulations, better public health, and a more livable urban environment. The city, in its various emblematic forms, creates the opportunities, contexts, and laboratories for new social developments, innovative political developments, and far-reaching economic changes. Cities are the accelerants of social, economic, and political change. Modernity emerges in the rise of the city.

16.2 Environment and the Industrial City: Reforming Victorian London

During the Industrial Revolution, few people attached importance to the quality of the urban environment. City authorities and citizens appeared to tolerate a great deal of filth in their cities and to ignore questions of environmental quality. Many accepted environmental degradation as the price for economic growth.

But there were consequences to unlimited and unregulated growth. Short-term consequences of environmental pollution in the industrial city were numerous: malformation of the bones caused by a lack of sunshine and a poor diet; skin disease from dirty water; smallpox, typhoid, and scarlet fever spread through dirt and human excrement in land and water; tuberculosis from bad diet and overcrowded conditions in tenements. Environmental conditions worsened with coal-filled air, polluted waterways, and garbage-strewn streets. Disease, frequently deadly, was rampant as housing squalor, poor nutrition, expensive healthcare, and the absence of decent public infrastructures combined to make death a frequent visitor to most households.

Long-term consequences were not understood at the time, but we now know that they include health problems associated with long-term exposure to pollutants and chemicals.

In late-nineteenth-century London, the Thames was a river so full of human, animal, and industrial waste of unprecedented volume that it was called the Great Stink. Engineers, scientists, and public officials set to clean up the river and the refuse in the streets. As a result of water sanitation efforts and improvements in water supply and water removal, the death rate in England declined from 20.5 per thousand in 1861 to 16.9 in 1901. London was a much cleaner place to live at the end of the nineteenth century than at the beginning.

Sanitary reforms in Victorian London aimed at improving public health, but as historian Michelle Allen has pointed out, such reforms were also connected to broader social concerns. At their most ambitious, sanitary reforms sought to lift up the poor and working classes, to moralize the population, and to create a more harmonious social order. Clean water, applied inwardly and outwardly, was both an instrument and a symbol of Victorian morality. Cleanliness was an outward sign of inward purity.

Reforms such as slum clearance, road building, and the building of sewer infrastructure altered not only the physical space of the city, but also its the social and symbolic meanings. A new paradigm of "cleanliness" accompanied modernization. Large-scale slum clearances became associated with "purification" and the removal of the urban poor, displacing thousands of working residents.

Allen's critical approach to sanitary reforms also uncovers the social, spatial, and textual discourses that inform the social order and the built environment and allows us to see the connection between ideas about filth and dirt, environmental issues, and the process of modernization in the city.

REFERENCE

Allen, M. 2008. *Cleansing the City: Sanitary Geographies in Victorian London.* Athens, OH: Ohio University Press.

16.3 The Third Urban Revolution

In the early twenty-first century we are in the midst of the **third urban revolution**. The first was concentrated in the fertile plains of only a few river basins. The second was restricted to cities in countries undergoing rapid industrialization. The third is a truly global phenomenon that picked up pace in the last half of the twentieth century. It is marked by five distinct characteristics.

Rapid Change

The first characteristic is the sheer scale and pace of urbanization. A majority of the world's population is now urban, and in many countries this new urban majority appeared in less than a generation. In 1900, only 10 percent of the world's population lived in cities. By 2010, it was more than 50 percent, and by 2050 almost 70 percent—more than two of every three people on the planet—will live in cities. In the early twenty-first century, that figure is almost 56 percent, or 4.2 billion people. Table 16.1 shows the urban percentages in 1950 and 2010, with estimates for 2050. The faster rates of urban growth occur in the Global South, particularly in the lowest-income countries. Urbanization is occurring more rapidly in the Global South. In 1900, only 10 percent of Mexicans lived in cities; by 2020, that figure increased to 79 percent. In the Sudan, just 6.8 percent of the population lived in urban areas in 1950; by 2020, that figure grew to 35.5 percent. In 2010, 3.5 billion people lived in cities, of whom 2.6 billion lived in the developing world. By 2050, these figures will be 6.2 billion and 5.1 billion, respectively. The future of the world is urban.

Big Cities

The second characteristic is the increasing size of individual cities. Throughout the world, cities have continued to grow larger. In 1800, there were only two cities that had more than 1 million inhabitants—London and Beijing; by 1900, there were thirteen. In the early twenty-first century,

TABLE 16.1 Percentage Urban, 1950–2050			
	1950	**2010**	**2050**
World	29.4	51.6	67.2
Global North	54.5	77.5	85.9
Global South	17.6	46.0	64.1

Source: Population Division of the Department of Economic and Social Affairs of the United Nations Secretariat, 2011.

TABLE 16.2 Ten Most Populous Megacities in 2023		
RANK	**CITY**	**POPULATION IN MILLIONS**
1	Tokyo	37.1
2	Delhi	32.9
3	Shanghai	29.2
4	Dhaka	23.2
5	São Paulo	22.6
6	Mexico City	22.2
7	Cairo	22.1
8	Beijing	21.7
9	Mumbai	21.1
10	Osaka	19.0

Source: https://worldpopulationreview.com/world-cities.

370 cities exceed 1 million in population, and over 90 have more than 5 million inhabitants. By 2030, it is estimated that there will be around 662 cities with at least 1 million inhabitants. The average size of cities has grown dramatically. One of the more visible aspects of contemporary urbanization is the rise of megacities, large urban agglomerations with more than 10 million inhabitants (see Table 16.2). They are a recent addition to the urban scene. In 1900, no metro region in the world had a population greater than 10 million. In 1950, only New York and Tokyo had populations of more than 10 million. Today, there are 39 cities with at least 10 million people. China is already home to six megacities, while India has five. Megacities continue to grow, and there are nearly ten cities with populations greater than 20 million, prompting the United Nations to introduce a new term, "metacity," to characterize them.

Rapid urbanization caused mainly by rural-to-urban migration is a cause of megacities. The underlying inequality between rural and urban conditions contributes to both push and pull factors, drawing people toward the cities. In 1950, the Nigerian city of Lagos had a population of only 320,000. By 1965, it had surpassed 1 million, and in 2002 it became the first sub-Saharan African **megacity** to top 10 million. With an annual population growth rate of 9 percent, it is one of the fastest-growing cities in the world. In 2022, the city had an estimated population of 21 million.

Megacities are not just big cities; they are a new, distinctive, spatial form of social organization that radically transforms the city–nature relationship. Their sheer size exerts a large and heavy environmental

16.3 The Metropolitan United States

The US Census Bureau employs the term **metropolitan statistical area (MSA)** to refer to urban areas with a core area of 50,000 or more people and economic links to surrounding counties. Using this statistical, rather than political, division of municipal boundaries, it is possible to measure the metropolitanization of the US population. In 1950, the metropolitan population constituted 56.1 percent of the total US population. By 2020, the figure was 86 percent. Growth was especially rapid in cities of the Sunbelt. Phoenix, Arizona, is one of the fastest growing cities since the beginning of the twenty-first century (see Photo 16.4)

and the Phoenix–Mesa–Scottsdale metro population is now the tenth largest in the country, with a population of 4.8 million.

The US population is increasingly and overwhelmingly concentrated in metropolitan areas. The geography of MSAs sometimes differs from the traditional political boundaries, especially in highly urbanized regions of the country. The counties of the state of New Jersey, for example, are part of seven different MSAs. Twelve of the counties are part of MSAs that include counties in New York State; five are part of MSAs that include counties in Pennsylvania. Only four (Atlantic, Cape May, Cumberland, and Mercer) are considered specifically New Jersey MSAs. Metropolitan realities are sometimes different from traditional political boundaries.

A further 8 percent of the US population lives in **micropolitan statistical areas**, which contain an urban core of at least 10,000 and, in total, have populations less than 50,000. Only 6 percent live outside these two types of urban areas. And despite some shift to selected rural areas in response to COVID-19, the United States is still overwhelmingly a metropolitan society.

PHOTO 16.4 Phoenix, one of the fastest growing metropolitan statistical areas in the United States.

REFERENCE

Short, J. R. 2012. "Metropolitan USA: Evidence from the 2010 Census." *International Journal of Population Research*, Article ID 207532. https://doi.org/10.1155/2012/207532.

footprint. Megacities can impose a heavy environmental toll. Continual city growth generates tremendous rural-to-urban land use changes and associated ecosystem transformations. The increasing population also puts extra pressure on the biophysical systems that provide land, air, and water.

Metropolitanization

Third, this revolution exhibits a marked **metropolitanization**. Improvements in transport have allowed dispersal of people and activities away from the tight urban cores of preindustrial and industrial cities. Large

metropolitan regions rather than individual cities are the new building blocks of both national and global economies. Three giant urban regions in the Asia Pacific region, Bangkok (15 million population), Seoul (25 million), and Jakarta (38 million), have between 35 and 75 percent of all foreign direct investment into their respective countries. In China, the three city regions of Beijing, Shanghai, and Hong Kong constitute less than 8 percent of the national population, yet they attract 73 percent of foreign investment and produce 75 percent of all exports. China is less a national economy than the aggregate of three large metropolitan economies. In the United States, ten **megapolitan regions**, defined as

16.4 Shanghai

After the 1842 Treaty of Nanjing, when the small fishing village of Shanghai was opened up as a port under colonial control, it became one of the largest shipping ports in the world. It exported food and raw material from the Chinese interior to world markets. Tea, silk, and raw materials were shipped through the city, which became China's major industrial center, with mills, factories, chemical plants, and shipyards. In the 1930s, there were 150,000 workers in textile factories, and the city's population approached 3 million, including 100,000 foreigners. At its pre-Communist peak it was the fifth largest city in the world, a cosmopolitan city with at least sixty different nationalities.

Shanghai became a Communist city on May 25, 1949, when troops of the People's Liberation Army marched in and took it over. The earliest economic plans of the party aimed to build up heavy industry. The city population quickly grew from 4 million in 1950 to almost 6.5 million in 1960, a staggering 50 percent increase.

During the Cultural Revolution, from 1966 to 1976, more than 30 million urban dwellers were forced to move to and work in the countryside. Shanghai was especially targeted. As a place of foreign influences and conspicuous consumption, it was treated as a dangerous hybrid place—Chinese, yet contaminated with anti-Communist tendencies and bourgeois sensibilities. From a peak of 6.4 million in 1961, the city's population declined to 5.4 million in 1978.

In 1984, fourteen coastal "open" cities, including Shanghai, were declared. As with other open cities, the reglobalization of Shanghai involved the creation of special economic and technological zones to foster concentrations in export-led manufacturing, high-tech industries, and financial services. From 1978 to 2005, the annual growth rate of GDP was close to 10 percent, and average wages grew sixfold. In 1990, a new open economic development zone called Pudong was created (Photo 16.5). An area of former rice paddies is now home to over 3 million people.

PHOTO 16.5 Shanghai. A new skyscraper in the Pudong neighborhood.

There is a frenzy to the remaking of the city: an expanding network of inner ring roads and outer motorways, new subway lines, tunnels and bridges, airport upgrades, and new high-speed trains. There are the new spaces of consumption: the malls, shopping centers, and gated communities. At the extremes are exclusive gated communities for the wealthy and enclaves of marginalized migrants. The post-Communist city now exhibits marked social and spatial segregation.

clustered networks of metropolitan regions that have a population of more than 10 million, constitute only 19.8 percent of the nation's land surface, yet comprise 67.4 percent of the population.

Population and economic activities have spread beyond the municipal boundaries. Since the 1960s, small towns across the world have grown into cities and big cities have sprawled into giant metropolitan regions.

16.5 Megalopolis

One of the largest contiguous areas of metropolitan counties is what is sometimes called **Megalopolis**, a region spanning 600 miles from north of Richmond in Virginia to just north of Portland in Maine and from the shores of the Northern Atlantic to the Appalachians (see Map 16.1). The region includes the consolidated metropolitan areas of Washington–Baltimore, Philadelphia, New York, and Boston and covers 52,000 square miles, with a 2020 population of 51.4 million. It contains just over 15 percent of the entire US population.

Megalopolis is overwhelmingly suburban, with three of every four persons living in the suburbs. Like other large city regions across the country, it is a place of increasing racial diversity, one of the most diverse in America. In some metro areas, there are more minorities living in the suburbs than in the central city. In the MSA of Washington, DC, for example, there were almost a million Hispanics living in the suburbs, compared to only 65,000 living in the city. The Hispanic experience in the Washington MSA is predominantly a suburban one. Across the entire region, the suburbs are becoming more racially diverse.

Megalopolis is the destination of significant amounts of immigration from overseas. Migrants are found both in central cities and in suburban areas where particular concentrations can be identified as immigrant gateways. In 2020 in Queens, New York, a traditional inner-city location, almost one in every two persons were foreign born. Tyson's Corner, Virginia, an archetypal edge city located off the Washington Beltway, has 22 percent foreign-born. Immigrant suburbs are now an important part of the metropolitan United States.

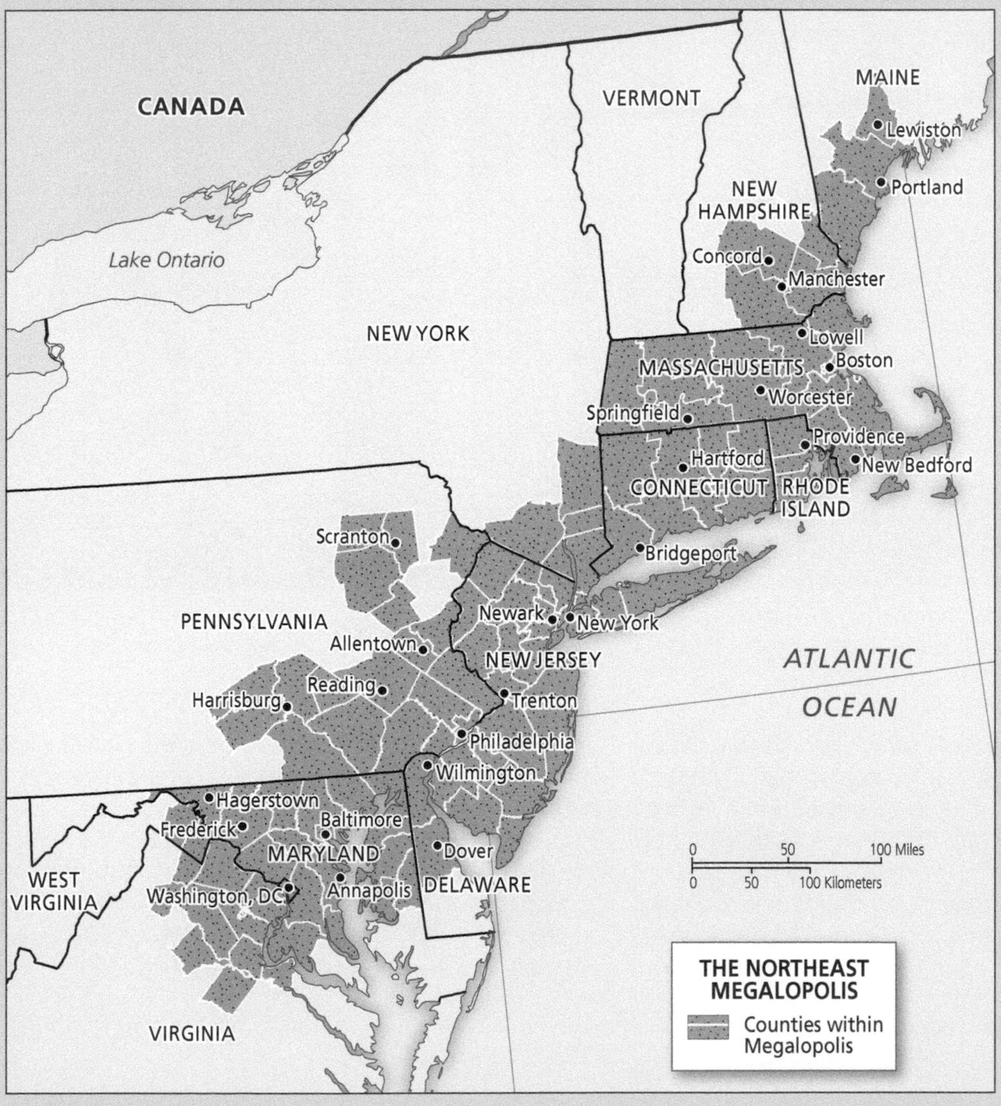

REFERENCES

Cartographic Lab, University of Maryland, Baltimore County. 2005. *Interactive Digital Atlas of Megalopolis.* http://www.umbc.edu/ges/student_projects/digital_atlas/instructions.htm.

Gottmann, J. 1961. *Megalopolis: The Urbanized Northeastern Seaboard of the United States.* New York: Twentieth Century Fund.

Short, J. R. 2007. *Liquid City: Megalopolis Revisited.* Washington, DC: Resources for the Future Press/Johns Hopkins University Press.

MAP 16.1 Megalopolis, United States.

Urban Sprawl

Fourth, a distinctive global urban trend is **urban sprawl**. Big-city regions are now characterized by more dispersed forms of urban development. The steady suburbanization of jobs and residences has extended the urban region farther out from central cities. In the United States in 1950, only 23 percent of the US population was living in suburbs. By 2020, this figure had increased to around 60 percent. More people live in metropolitan areas, and more of these people are living in the suburbs. The process, while particularly evident in the developed world, is not limited to it. Consider the case of Shanghai. Since 1991, new residential and industrial complexes have developed along the rural–urban fringe, aided and encouraged by local governments, which are key stakeholders in the land development system. The end result is a more widely dispersed urban system with land development often leapfrogging across the landscape. Shanghai is now a huge polycentric city.

Dispersed urban expansion triggers land cover changes, especially the disappearance or fragmentation of croplands and woodlands, the destruction of both low- and high-quality agricultural lands, and the emergence of a hybrid rural–nonfarm landscape (see Photo 16.6).

Sprawl is now associated with a host of negative environmental impacts, including:

- loss of agricultural land in the wake of low-density sprawling development;
- increased impermeable surfaces, which leads to flooding and large discharges of polluted and contaminated water that overwhelm drainage systems and damage ecosystems; and
- heavier vehicular traffic, which leads to increased air pollution and global warming.

Sprawl is a form of development that is very often too diffuse to support public transport or easy walking. The heavy, and in some cases total, reliance on private auto transport in the United States, for example, imposes a great environmental price in terms of air pollution and the increasing dedication of space for roads and parking. The reliance of a built form precariously balanced on one fossil fuel with large and fluctuating costs raises issues of long-term sustainability. There is also an emerging body of literature that points to the negative public health effects of suburban sprawl, including a link between the promotion of a driving lifestyle and both less physical activity and higher obesity.

Further, there is the fundamental issue of the long-term sustainability of sprawl. Low-density suburban sprawl is possible only with relatively cheap fuel and lack of accountability for its environmental impact. It is unlikely

PHOTO 16.6 Urban sprawl outside Baltimore, Maryland.

that the cheap gasoline that literally lubricated suburbanization will ever return. Where does that leave low-density suburban sprawl, which is so reliant on large-scale private car usage? The general answer: in a very precarious position. The long-term sustainability of low-density, energy-profligate sprawl with its heavy ecological footprint is now a matter for serious consideration.

Slums

Fifth, there is a distinct feature of urban growth in the rapidly growing cities of the developing world (also referred to as the Global South), namely, **slums** (Map 16.2). The term "slum" (also called "shantytown," "informal housing," and "squatter housing") refers to unplanned, often illegal, informal housing. The United Nations has a working definition of slums as marked by the following:

1. Inadequate access to safe water
2. Inadequate access to sanitation and infrastructure

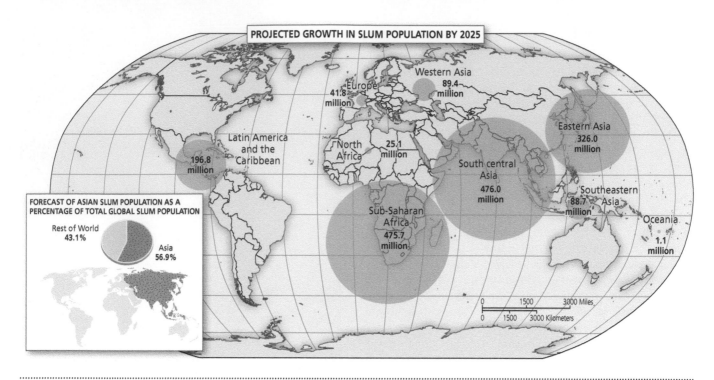

MAP 16.2 Projected growth in slum population by 2025.

3. Poor structural quality of housing
4. Overcrowding
5. Insecure residential status

The most recent global survey of slums was published in 2018 by the United Nations. The survey estimated that around 880 million people lived in slums, up from 790 million in 2000. The good news is that the overall percentage of people in the world who live in slums has been declining (although there are regions where few people live in slums and regions where significant numbers of people live in slums). In many Asian and Latin American countries, between 10 and 30 percent of urban populations live in slum households. In sub-Saharan Africa, more than half the urban populations of most cities live in slum households, and in some countries, such as Sudan, South Sudan, and the Central African Republic, that figure is over 90 percent. The projected number of people living in slums will continue to grow significantly. Nearly half a billion people in sub-Saharan Africa and nearly a half of a billion combined in South Asia and Central Asia are expected to live in slums.

Slums arise because of the inability of formal markets and public authorities to provide enough affordable and accessible housing. They grow because of continuing rural-to-urban migration.

There are numerous names for these slum settlements. They are called "ranchos" in Venezuela, "pueblos jóvenes" in Peru, "favelas" in Brazil, and "barong-barongs" in the Philippines. A UN survey estimated that around 1 billion people live in slums. The figure is estimated to rise to 2 billion by 2030 as people migrate to cities and urban populations grow. Regions of the world with the largest numbers living in slums are sub-Saharan Africa (196 million), South Asia (191 million), East Asia (190 million), and Latin America (110 million).

Some slums go through a development from temporary accommodation built on the most marginal sites to fully formed urban neighborhoods. In the 1950s, some rural migrants moved to Ciudad Nezahualcóyotl, an area on the outskirts of Mexico City, where they built their own homes on appropriated land using whatever materials they could find. Over time, more permanent buildings were constructed. The area was designated a municipality in 1963, allowing the formal provision of public services such as potable water, pavements, sewerage, and electric lighting. By the 1980s, public buildings such as hospitals and schools were being built. In 1995, the area exceeded 1 million residents. In this process, temporary shelter for recent migrants became a slum and then an integral part of the city. Peter Ward first studied 300 self-builders in neighborhoods in Bogotá and Mexico City in the 1970s. His follow-up study in 2007 allowed him to identify changes. He found that more than four of every five households stayed in the houses that they built, but densities rose along with the sizes of the extended families, and the values of the properties increased substantially. "Temporary" self-built housing provided stable family accommodation for over forty years as well as an important family asset.

There are different types of slums. In Abidjan, Côte d'Ivoire, where slum dwellers represent one-fifth of the city's population, there are three types of slums. Zoé Bruno, for example, contains buildings of permanent material and basic infrastructure and is different from the formal areas of the city only by the fact of the illegal land occupation. In Bilingue, the buildings are made of non-permanent materials and the area has low levels of infrastructure. In the worst areas, such as Alliodan, makeshift buildings have no infrastructure.

The urban explosion of the past fifty years has forced expansion onto new spaces on more vulnerable sites, such as steep hillsides and flood plains, areas with unstable soil conditions that are subject to landslides, and places with other environmental and social hazards. Many slum structures are erected on such marginal lands because standard, legal housing long ago claimed the best, most secure land in the city (see Photo 16.7). The poor are often forced

PHOTO 16.7 Informal housing on a steep slope on the Caribbean island of St. Vincent.

to settle on land subject to higher risks. Slum homes perched precariously on hillsides are often swept away by heavy rains, which also inundate slums located on floodplains. Heavy rain contributes to the landslides and the flooding, but the real causes are the fact that low-income groups could find no land site that was safe and the government failed to ensure a safer site or to take measures to make existing sites safer. The worst slums often occupy the most hazardous sites.

A major challenge is insecure residential status, which refers to the uncertain legal tenure of many slum dwellers, many of whom lack legal title to the land that they occupy. The lack of land tenure discourages residents from investing in infrastructure that could improve their lives, because their hold on their homes is tenuous and the risk of losing those investments is high should the government decide to demolish the slum.

Because many slum residents lack land tenure, they are therefore subject to government removal, especially if the land becomes valuable and sought by developers and politicians. Slum dwellers often have a precarious hold on their accommodations; the lack of legal ownership makes slum dwellers particularly vulnerable to government clearances. As one particularly brutal example of a process repeated throughout the world, on May 19, 2005, the government of Zimbabwe, under the dictatorial control of Robert Mugabe, launched a new urban policy. It was called Operation Murambatsvina, literally "Drive out the rubbish." As part of a nationwide campaign to beautify the cities, destroy the black market, reduce crime, and undermine the support base of its political opponents, the government authorized the bulldozing of squatter settlements that fringed the city's larger cities. The squatter settlements had long been the main source of opposition to his rule. The campaign destroyed the homes of 700,000 people and wiped out an informal economy that provided a livelihood for 40 percent of the population. Almost 2.4 million poor people were faced with increased economic hardship. As another example, in July 2012 the Lagos state government in Nigeria moved to evict residents of Makoko, a slum neighborhood of around 200,000 people living in dwellings built on stilts in a lagoon. The residents were given seventy-two hours to vacate before men started chopping down the dwellings. City officials argued that the slums were an environmental nuisance, impeded waterfront development, and undermined the megacity status of Lagos. They felt that a 2010 BBC television documentary program, *Welcome to Lagos*, which told the story of Makoko's residents, presented the country and the city in a negative light. In fact, the documentary was a celebration of the ingenuity and vibrancy of slum life in the city. While it noted problems of flooding, irregular power supply, and poor environmental conditions, it was lyrical in its invocation of the resiliency and vitality of marginal groups living in slums on the economic margins. Rather than depicting gloom

16.6 Inequality and Vulnerability in Slums

In many slums, the poorest urban dwellers are exposed to a myriad of environmental and social problems. The World Health Organization estimates that currently more than 860 million urban dwellers in developing cities have no access to clean water, sanitation, or drainage. In areas with limited access to clean water and sanitation, the child mortality rate is many times higher than in areas with adequate water and sanitation services. The primary barriers to accessing water in slums are not solely monetary or technical, but also legal, institutional, and political. Those who live in slums face insecure land tenure and exposure to hazards, and they often lack political voice.

Historically, governments tended to respond to slums by ignoring them, eradicating them, or relocating their residents. However, eradication, eviction, and involuntary relocation are insensitive measures. Slum dwellers have made both financial and social investments in their homes and communities, something that would be lost in attempts to clear slum neighborhoods.

One way some cities are addressing slums is through upgrade programs. Slum upgrading has gained prominence as a valid and cost-effective way to improve the living conditions of cities. This is reflected in SDG Target 11.1, which seeks to upgrade slums in situ, so they are safe and have basic services. Slum upgrading in situ refers to improvements to housing and/or basic infrastructure. It covers a wide range of possible interventions that can include the installation of basic infrastructure such as water, sanitation, waste, roads, storm drainage, and electricity, as well as housing improvements, improved access to healthcare, education and other social support services, and a regularization or recognition of land tenure.

Physical upgrades of slums have proven to make positive social and economic changes in many cities. Socially, upgraded slums improve physical living conditions, improve the general well-being of communities, strengthen local social and cultural capital networks, generate livelihood opportunities, improve quality of life, and increase access to services and opportunities.

Economically, upgraded slums can trigger local economic development, improve urban mobility and connectivity, and integrate an economically productive sphere into the physical and socioeconomic fabric of the wider city.

However, slum upgrade programs still face considerable challenges. First, mayors and local authorities will need to acknowledge the presence of slums and have the political will to improve them. Too often, political promises for upgrading are made just before an election and never materialize. Second, cities must mobilize financial resources to design and implement upgrades; this will usually entail coordinating with national governments for resources. Third, some studies have suggested that slum upgrades depend on the participation of slum residents, and this is most effective when initiated at the neighborhood level through individual or community projects. The participation of slum dwellers and community organizations is critical to slum upgrade programs—this must be a bottom-up process.

More recently, urban planners have begun to use the term "slum integration." Whereas upgrading is about improving access to basic urban services, slum integration is a social goal that embodies the idea of participation, inclusion, equality, agency, and sustainability. Slum upgrade programs by themselves do not necessarily address wider issues of housing affordability and land tenure. The lack of affordable housing is a main driver of high rates of illegal construction and the creation of slums in the first place. Cities will need to shift housing policy, urban planning, and building practices to become more sustainable.

REFERENCES

Short, J. 2018. *The Unequal City: Urban Resurgence, Displacement and the Making of Inequality in Global Cities*. New York: Routledge.

United Nations Human Settlements Programme. 2014. *A Practical Guide to Designing, Planning and Implementing Citywide Slum Upgrading Programmes*. https://unhabitat.org/a-practical-guide-to-designing-planning-and-executing-citywide-slum-upgrading-programmes.

and darkness, it highlighted the slum dwellers' incredible dynamism and energetic entrepreneurship.

The process of slum removal continued in the buildup to the 2014 World Cup and 2016 Olympic Games: many of the favelas in Rio de Janeiro came under assault from clearance, forced eviction, and destruction that removed 13,000 families from 123 neighborhoods in the city.

Across the urban world, then, we can identify a continuum from **slums of hope** on one end to **slums of despair** on the other. Slums of hope provide a platform for rural urban migrants to make their way in the city. In slums of despair, residents are imprisoned in webs of multiple deprivation. Both type of slums are vulnerable to assault and destruction from urban development projects.

Cited References

BBC. 2010. "Welcome to Lagos." https://vimeo.com/11206466.

Population Division of the Department of Economic and Social Affairs of the United Nations Secretariat. 2011. *World Population Prospects: The 2010 Revision and World Urbanization Prospects: The 2011 Revision*. http://esa.un.org/unpd/wup/index.html.

Ward, P. M. 2012. "A Patrimony for the Children": Low-Income Homeownership and Housing (Im)Mobility in Latin American Cities. *Annals of the Association of American Geographers* 102:1489–1510.

World Population Review. 2022. "World Cities." https://worldpopulationreview.com/world-cities.

Select Guide to Further Reading

Benton-Short, L., and J. R. Short. 2013. *Cities and Nature*. London: Routledge.

Bryan, G., E. Glaeser, and N. Tsivanidis. 2020. "Cities in the Developing World." *Annual Review of Economics* 12:273–297.

Davis, M. 2006. *Planet of Slums*. London: Verso.

Duncan, J. S. 1990. *The City as Text: The Politics of Landscape Interpretation in the Kandyan Kingdom*. Cambridge: Cambridge University Press.

Gruebner, O., A. Khan, S. Lautenbach, D. Muller, A. Kramer, T. Lakes, and P. Hostert. 2012. "Mental Health in the Slums of Dhaka: A Geoepidemiological Study." *BMC Public Health* 12:177. https://doi.org/10.1186/1471-2458-12-177.

Kaniewski, D., E. V. Campo, and H. Weiss. 2012. "Drought Is a Recurring Challenge in the Middle East." *Proceedings of the National Academy of Science* 109:3862–3867.

Nelson, A., and R. Lang. 2011. *Megapolitan America: A New Vision for Understanding America's Metropolitan Geography*. Chicago: APA Planners.

Sadik-Khan, J., and S. Solomonow. 2017. *Streetfight: Handbook for an Urban Revolution*. New York: Penguin.

Schneider, A., and C. E. Woodcock. 2008. "Compact, Dispersed, Fragmented, Extensive? A Comparison of Urban Growth in Twenty-Five Global Cities Using Remotely Sensed Data, Pattern Metrics and Census Information." *Urban Studies* 45:659–692.

Short, J. R. 2012. *Globalization, Modernity and the City*. London: Routledge.

Short, J. R., ed. 2017. *A Research Agenda for Cities*. Cheltenham, UK: Elgar.

Short, J. R. 2018. *The Unequal City: Urban Resurgence, Displacement and the Making of Inequality in Global Cities*. London: Routledge.

The Networks of Cities

Cities exist not only as singular places but also as nodes in a variety of networks. In this chapter we will explore regional, national, and global networks.

17.1 Regional Networks

One of the earliest forms of urban network is the **periodic market system**. Periodic markets occur in villages and towns when demand density is low and vendors are relatively mobile. At the simplest level, they may consist of a seller by the roadside with seasonal wares or produce (see Photo 17.1). There are also the larger weekly, monthly, and annual markets with larger numbers of vendors and customers. There is a geometry to this periodicity, as markets seek to either minimize travel for vendors by bringing them closer together (Figure 17.1A) or maximize demand by spreading markets across space (see Figure 17.1B). A study in northern Nigeria found that markets held on the same day are generally located 10.6 miles apart, whereas markets with meetings two days apart have an average spacing of 3.3 miles.

As demand grows, some traders stay in place, and eventually periodic markets are replaced by **fixed market systems**. Photo 17.2 shows a fixed market in Curaçao that used to be held weekly. There are still examples of periodic markets throughout much of the developing world. And even in the rich, developed world, farmers' markets continue to attract customers.

Central Place Theory

With rising demand, fixed markets tend to replace periodic markets. In 1915, C. J. Galpin, who was studying rural communities in Wisconsin, suggested that towns with the same number of services tend to locate at regular intervals. People in the surrounding rural areas tend go to the nearest town to shop for goods and services. The ideal pattern of a flat agricultural plain is highlighted in Figure 17.2, where a regular distribution of towns is surrounded by circular complementary regions. If we assume distance

PHOTO 17.1 Roadside sale in Jamaica.

PHOTO 17.2 Fixed market in Curaçao.

minimization by customers, the resultant complementary region of each town takes the form of a hexagon, because if we draw a straight line through shared areas it forms that shape. Walter Christaller (1893–1969) looked for a similar pattern in southern Germany in the 1930s. He identified

what he termed the **range** and the **threshold** of a good or service. The range of a good or service is the distance consumers are willing to travel. While a buyer of a car may travel long distances to seek out good bargains, the buyer of a carton of milk will only travel much shorter distances. The threshold of a good is the necessary minimum population to support the continued supply of that good. Car dealers need a larger population to support their business than do milk sellers. People buy cars more infrequently than milk. Higher-order goods and services have larger thresholds and larger ranges than lower-order goods and

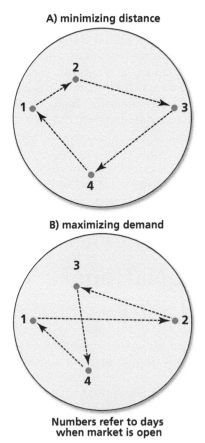

FIGURE 17.1 Periodic markets.

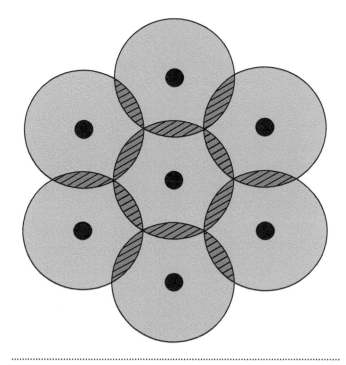

FIGURE 17.2 The Galpin model.

services. Market towns are distributed in a hierarchy, with higher-order goods and services located in the larger towns. Christaller visualized a range of hexagonal structures depending on whether the system was geared to optimizing markets, minimizing travel, or optimizing administrative boundaries (Figure 17.3).

Christaller's work, first published in 1933 and resuscitated in the 1960s, was seen as a contribution to economic and urban geography. More recently, he has been criticized as the context of his work has been uncovered. Christaller joined the German Nazi Party in 1940. He joined the Reichskommissariat für die Festigung deutschen Volkstums (Reichskommissariat for the Strengthening of German Nationhood), where he worked on plans to occupy the lands of Czechoslovakia, Poland, and the eastern USSR. His central place theory was employed to organize and settle the occupied lands. The seemingly innocent geometry of central place theory was devised to organize the Nazi occupation of an Eastern Europe emptied of its original inhabitants. He was "Hitler's geographer." The story of Christaller reminds us of the tangled connections between academics and politics, theory and practice, and the contested politics of space.

Central place theory was extended in a range of case studies. Brian Berry looked at the urban system in rural areas of the US Midwest, while Rosemary Bromley and R. J. Bromley uncovered the hierarchy in a region of Ecuador by looking at bus services. However, central place theory has fallen from favor as a major topic of study. The theory is more suitable for stable agricultural regions of the world. It gives little insight into the massive urbanization or rapid metropolitanization of the early twenty-first century. In addition, changes in communication technology and the rise of online shopping have undermined the idea of higher-order places, since luxury goods can be purchase in virtual space.

The basics of central place theory, and in particular the notions of complementary regions and of the range and threshold of goods and services, are still important tools for uncovering some of the socio-spatial relations in urban networks. The measurement and definition of complementary regions allows us to identify the effective influence of a city across space. The US Census Bureau, for example, uses commuting patterns to identify city regions.

The changing range and threshold of goods also allow us to understand some of the changes in local urban systems. In many rural areas, small towns are shriveling up and dying. Urban networks are thinning as goods and service provision move up the urban hierarchy to larger towns and cities. Small towns shrink in economic importance when the range and threshold of goods and services both increase. Urban decline is in part a function of changes in these variables.

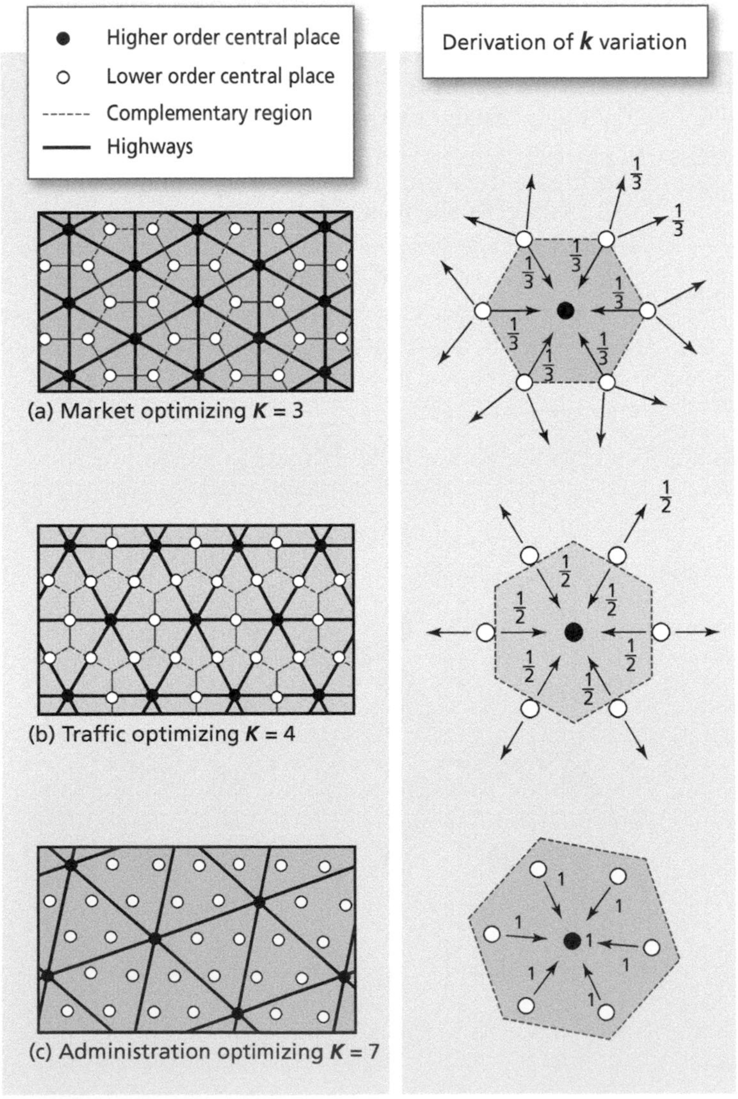

FIGURE 17.3 Market areas in central place systems.

17.2 National Networks

Cities are also part of national networks. There are numerous ways to measure a national urban network. One way is to consider the relationship between a city's population and its rank. In a **rank-size distribution**, also referred to as **Zipf's law**, there is a simple and regular relationship between a city's rank and its population. This relationship is found mainly in large countries with a large number of big cities, such as China, India, and the United States.

17.1 Urban Footprints

Another way to imagine the connection of cities to their wider region is to analyze resource consumption patterns. Ecologists have developed a measure, the **ecological footprint**, to refer to the total area of productive land required to support an ecosystem. The ecological footprint measures how much land and water area a human population requires to produce the resources it consumes and to absorb its wastes. It is measured in global hectares (gha) per capita. The ecological footprint of an urban region includes all the land necessary to support the resource demands and waste products of a city. The global average is around 2.6 gha.

London's ecological footprint was measured at 4.54 gha, slightly lower than the UK average of 4.64. More people in London use public transport than in almost any other city in the United Kingdom, reducing the relative size of the footprint. On the opposite end is Calgary, Canada, with a calculated ecological footprint of 9.8 gha. If everyone on Earth had the same ecological footprint as the average Calgary resident, we would need five Earths to maintain that level of resource consumption. Cold winters and a large city sprawl mean that many people need to heat their homes and use cars to get around Calgary. A city that is more reliant on private autos than public transport has greater energy needs and thus has a larger ecological footprint. Calgary's ecological footprint analysis demonstrates a city's need for a multimodal transportation system and improved jobs-to-housing balance to reduce its ecological footprint.

Comparing city and national footprints and biocapacity can shed more light on potential leverage points for improving sustainability. They can be useful in:

- helping governments track a city or region's demand on natural capital and compare this demand with natural capital available;

- informing a broad set of policies, ranging from transportation to building codes to residential development;

- highlighting the significance of long-term infrastructure decisions, amplifying future opportunities or risks;

- adding value to existing data sets on production, trade, and environmental performance by providing a comprehensive framework to interpret them;

- helping understand the link between local consumption and global environmental impact; and

- raising sustainability awareness and engagement among citizens.

Another measure is a city's carbon footprint, which is the total amount of greenhouse gases it produces. The basic unit is the kilogram or metric ton of CO_2. The global average is 1.19 metric tons per person. One study measured the carbon footprint of major cities. Seoul, Guangzhou, New York, Hong Kong, and Los Angeles had the largest footprints; London, Beijing, and Jakarta had smaller footprints. The study reports that 20 percent of global emissions come from just one hundred cities. These footprints are imperfect measures, but they constitute a start, and they have produced some interesting findings.

To learn more, visit https://www.footprintnetwork.org/our-work/cities/.

Urban Primacy

In contrast, a **primate distribution** occurs when one city dominates the national urban system. Primate cities are a different order of magnitude and significance from all the other cities in a national urban hierarchy. Geographer Mark Jefferson defined a primate city as "at least twice as large as the next largest city and more than twice as significant" (Galpin 1939, 227). John Rennie Short and Pinet-Peralta calculated a **primacy ratio** using the ratio of the population of the largest city in a country divided by the combined population of the next two largest cities. A value of 3.0 tells us that the primate city's population is three times the combined population of the next two largest cities. This is a substantial concentration of people in just one city. Table 17.1 notes all the countries with a primacy value above 5.0. Note that the highest levels of urban primacy are located in countries in the Global

South, although some countries in the Global North also have primate cities. Let us consider the most primate urban system to tease out some of the possible causal connections.

The population of Thailand exhibits the most pronounced degree of urban primacy. Of a population of almost 65 million, just over 15 million live in the urban region of Bangkok. Since 1782, the city has been the capital of the country. The court was based in the city, and there were few other towns of any size. Bangkok was and remains Thailand's unchallenged center of political, intellectual, and religious life. This embedded centrality of the city to the wider life of the country shaped subsequent growth, especially recent export-led growth. The city became the main transmission hub of economic globalization for the country and the wider region. The centralizing forces of globalization reinforced the national primacy. By 2005, the city was responsible for almost half

TABLE 17.1 High Urban Primacy

COUNTRY	PRIMACY
Thailand	9.48
Suriname	8.24
Togo	7.92
Uruguay	7.37
Nigeria	5.94
Uganda	5.94
Ethiopia	5.82
Mongolia	5.67
Peru	5.43
Guinea	5.27

of all gross domestic product. Thailand is one extended urban region centered in Bangkok. The city is an important global city not only for the country but also for the wider region of Southeast Asia.

Primate cities are home to the elites, the population center of gravity, and the economic hub of the national space economy. Primacy was reinforced as economies shifted from primary to secondary and then tertiary economic sectors, rural-to-urban migration increased, and foreign investment connected the local to the global. People, jobs, and investment moved to the major city.

A number of Global North countries, including Austria (3.2), Hungary (4.4), and the United Kingdom (4.4), also exhibit primacy. Vienna, Budapest, and London were all centers of far-flung empires rather than just capitals of individual countries. Nineteenth-century Vienna was one of the centers of political power in Europe, the capital of the vast Austro-Hungarian Empire. Vienna flourished as an imperial capital and the most important city in central

17.2 Estimating City Population

An obvious source of urban population data is the United Nations at https://unstats.un.org/unsd/Demographic/sconcerns/densurb/default.htm. This source has the benefits of ready accessibility, constant updates, and ease of use. The downside is that, as with all international urban data, it is deeply flawed. Different countries take censuses at different times, use different measures (e.g., some use formal urban political boundaries, while others give the data for functional urban agglomerations), and have wildly varying degrees of accuracy. Take the case of Nigeria and its major city of Lagos. At the UN source, accessed in February 2007, the city's population is available only for 1975 and is listed at just over 1 million. By 2006, official figures in Nigeria listed the population of Lagos at almost 9 million, but local officials accused the federal government of deliberately undercounting this southern Nigerian city and its encompassing region to favor the northern part of the country. Today, the population of Lagos is around 15 million.

A more user-friendly source, available at https://www.citypopulation.de, lists the population of all principal cities for most countries. The data are drawn primarily from official censuses and estimates, but even the webmaster reminds the user that the data "are all of varying, and some of suspect accuracy."

Another major problem with urban population data is the distinction between population estimates for the city and for the wider metropolitan region. London's population is calculated with reference to the city boundary. Almost 7.5 million

people live within this boundary. Yet its commuting range stretches much wider. The wider city region stretches from the Wash to the Isle of Wight. Within this urban agglomeration live 21 million people, or almost 35 percent of the United Kingdom's total population.

The "dirty little secret" of global urban research is the poor quality of comparative data, even basic data such as population. Population censuses are of varying degrees of accuracy, are taken at different times, and have differing definitions of city and metropolitan regions. The population figures for most cities around the world are more estimates than precise figures, averages across a wide band of error, and more appropriately used as rough comparisons rather than precise absolute values.

Hannah Ritchie and Max Roser note that while disagreement on the numbers can seem irrelevant, understanding cities, urbanization rates, the distribution, and the density of people matters because knowing the distribution of people in a given country is essential to ensure the appropriate resources and services are available where they are needed.

REFERENCES

Ritchie, H., and M. Roser. 2019. "Urbanization." Our World in Data. https://ourworldindata.org/urbanization.

Short, J. R., Y.-H. Kim, M. Kuus, and H. Wells. 1996. "The Dirty Little Secret of World Cities Research." *International Journal of Urban and Regional Research* 20:697–717.

Europe, but the loss of empire after the First World War (1914–1918) reduced Austria to a small German-speaking state, and Vienna found itself a large city in a small country. Budapest was also the center of a much larger country and empire. After 1920, Hungary forfeited two-thirds of its area. The end result was a large city in a much-reduced country. Nearly one-fifth of Hungary's population lives in Budapest, which is now more than nine times larger than the nation's second-largest city. London was not just the capital of the United Kingdom but also the central node in a worldwide formal and informal empire. The retreat from empire and decline of global economic dominance has left a huge city in a relatively small country. The imperial primacy has persisted in a postimperial society. The wealthy, the influential, and the movers and shakers live in the city; it is home to royalty, the political elites, those who manufacture the dominant forms of representation, and those who control much of the making and moving of money.

At the other end of the continuum of primacy values are the nonprimate distributions. Table 17.2 lists the countries with a recorded primacy value of less than 0.9. The case of Bolivia disproves the easy assumption of Latin American countries always having primate urban distributions. No simple generalizations can be made, because low primacy is found in the Netherlands as well as India, the United States as well as Venezuela, Canada as well as Benin.

The Case of Australia

Using aggregate data sets provides only a first glance at urban national networks. If we look at one example in more detail, more complex patterns are revealed. Consider

Australia, which by 2022 had a national population of 25.7 million. At first glance, it appears to have a nonprimate distribution, because the largest city, Sydney, has a population of 5.2 million, followed by Melbourne and Brisbane (see Photo 17.3) at 5.1 million and 2.5 million, respectively. The primacy value is 0.76. One could deduce that the urban system is evenly distributed. But on closer examination, as shown in Table 17.3, Australia is best depicted as a series of primate states in a federal system.

The case of Australia reminds us of the importance of scale, geography, and history. The primacy value for Australia is suggestive of nonprimacy, as if activities were evenly distributed throughout the urban system. Yet when we look at the level of individual states, we see there is a marked concentration of activity, enterprise, and population in large cities. The Australian Federation of 1901 brought together a series of **hyperprimate** economies that the passing of time has done little to change. The major cities continue to dominate their respective states. The geography and (white) history of Australia play an important part. An antipodean gulag for the British soon developed into a colonial enterprise organized and structured through the port cities of the different states. A classic case

TABLE 17.2 Low Urban Primacy	
COUNTRY	**PRIMACY**
Benin	0.58
South Africa	0.59
Venezuela	0.65
Netherlands	0.70
Egypt	0.72
Australia	0.74
China	0.78
United States	0.84
Bolivia	0.84
India	0.86
Canada	0.89

TABLE 17.3 Urban Primacy in Australia	
COUNTRY/STATE	**PRIMACY**
Australia	0.7
New South Wales	5.4
Queensland	5.3
South Australia	33.4
Tasmania	1.1
Victoria	14.6
Western Australia	11.9

of colonial spatial organization of a vast country was soon embedded into a series of primate states. The case of Australia shows that reliance solely on national data abstracted from issues of scale, geography, and history has limited explanatory value.

17.3 Global Networks

Cities are also part of global networks through which capital, goods, ideas, and people flow. Some of the flows have been identified and measured.

The Globalization and World Cities Network

A substantial body of material has emerged from the work of Peter Taylor and his colleagues at the **Globalization and World Cities (GAWC) research network**. The website (https://www.lboro.ac.uk/microsites/geography/gawc/) lists data sets as well as more than 470 research papers and is an indispensable guide to the metageography of urban networks. In 2000, GAWC collected data on the distribution of 100 global advanced producer service firms, which includes accountancy, advertising, banking/finance, insurance, law, and management consultancy, across 315 cities. They analyzed the resultant data matrix to identify a global urban hierarchy. In 2008, 2010, 2012, and 2020 they extended the analysis to 175 firms in 526 cities. The result was a fivefold hierarchy that identified cities as alpha, beta, gamma, high sufficiency, and sufficiency. Map 17.1 is a cartogram of alpha cities in the network of advanced producer services. Note how New York (NY) and London (LON) dominate. NYLON is an important pivot in the global networks (see Photo 17.5). However, these hierarchies are constantly changing. For example, in 2000, Shanghai was three steps below, in the alpha minus category, while Beijing was only a beta city. By 2020, both Shanghai and Beijing were classified as alpha plus, only one step below NYLON. Both cities are moving into the top tier as China's economic growth, both absolute and relative to the rest of the world, continues apace. The diagram highlights the centers of new metropolitan modernity, the rapidly growing cities of the Far East. Today, New York and London remain at the top of the alpha hierarchy; one step down in the alpha hierarchy are Hong Kong, Singapore, Shanghai, Beijing, Dubai, Paris, and Tokyo. Further down the hierarchy are beta

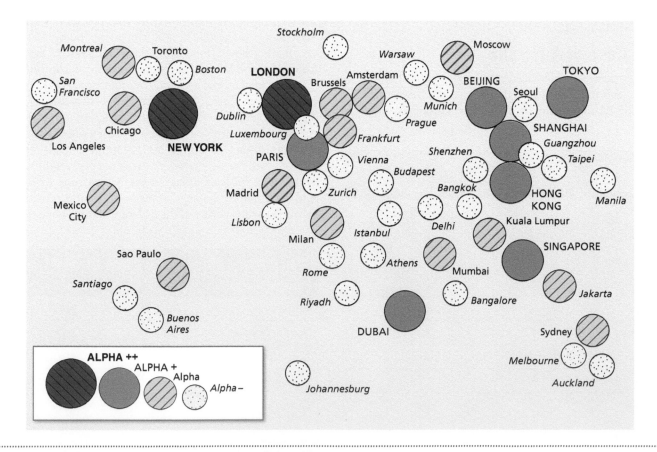

MAP 17.1 The global urban network according to Globalization and World Cities.

17.3 Inequality in World Cities

High-ranking world cities, such as New York and London, are known for their global power in finance, knowledge production, and social influence. Yet there is a darker side to their high-level status. London and New York have fallen into the hands of the super-rich often at the expense of everyone else.

According to Rowland Atkinson, while London has always been home to the wealthy, by the early twenty-first century it had become *the place* in which to invest capital. The result is a property boom economy and the expulsion of the urban poor.

Atkinson notes many of the newly constructed luxury apartments and condominiums function as a safe and secure place to absorb global investment capital. Foreign buyers such as Russian oligarchs and Saudi investors own property in many of the city's largest and most expensive redevelopments. Atkinson notes that at the St. George Wharf Tower, two-thirds of the owners are foreign buyers, and many of those units have been purchased through offshore companies or shell companies (Photo 17.4). Nearby, the Embassy Gardens offers luxury amenities such as a "sky pool," meeting rooms, spas, dog-washing stations, cinemas, and rooftop gardens. Atkinson calls these concentrated super-rich developments "alphahoods." Many flats sit vacant for months at a time. A single-bedroom apartment at Embassy Gardens sells for about $840,000. Atkinson argues that the impact of the super-rich on London's housing extends beyond the homes where they live. Their injections of cash drive up prices in residential markets and create an overheated and overpriced housing market that few native Britons can afford. Such a market means apartments sell in a day, above asking price, and often as all-cash offers.

With the collapse of the Soviet Union in the 1990s, post-Soviet kleptocrats saw London as a safe place for financial services. It has since has also attracted wealth from corrupt dictators and some terrorist organizations. London has become a preeminent destination to "launder" the money of the super-rich.

Consider the recent history of Witanhurst, a large mansion located on five acres in Highgate in North London. The house alone covers 90,000 square feet and contains sixty-five rooms. It was bought by a rich Kuwaiti, who sold it to the ruling family of Syria, the Assads. In 2008, it was bought by the Russian billionaire, Andrey Guryev (net worth $4.8 billion), through an offshore company located in the tax haven of the British Virgin Islands. It is the second largest private residence after Buckingham Palace and is currently valued at $450 million. Guryev also owns the five-story penthouse at

PHOTO 17.4 St. George Wharf Tower, London.

the top of St. George Wharf Tower, an exclusive residential block where no one in the 184 apartments is eligible to vote in the United Kingdom.

Atkinson estimates that around $146 billion dollars of property in England is owned by criminal capital. According to government figures, offshore companies now anonymously own about 44,000 properties in London. When these buyers sell their London properties, they essentially "launder" clean their investment.

Cities and towns have always struggled with inequality and the social and spatial realities of unequal access to power and resources. However, in the major world cities such as London, the past three decades have seen rising inequality.

REFERENCE

Atkinson, R. 2020. *Alpha City: How London Was Captured by the Super-rich*. New York: Verso.

PHOTO 17.5 City of London.

cities such as Washington, DC, Miami, Berlin, and Cairo. Gamma cities include Rotterdam, Belfast, Glasgow, St. Louis, and Baltimore. World cities are more defined by economic power than by population size.

Financial services continue to be concentrated especially in the global cities. At the apex are the supranational centers of London and New York, followed by a second tier of international centers that include Tokyo and Zurich, and host centers that attract foreign financial institutions, including such places as Sydney, Toronto, and Vienna.

The globalizing economy creates lots of information, narrative uncertainty, and economic risk that all have to be produced, managed, narrated, explained, and acted on. Global cities are centers of global epistemic communities of surveillance, knowledge production, and storytelling. Trust, contact networks, and social relations play pivotal roles in the smooth functioning of global business. Spatial propinquity allows these relations to be easily maintained, lubricated, and sustained in an efficient means of communication that helps solve incentive problems, facilitates socialization and learning, and provides psychological motivation. There are positive benefits of face-to-face contact. Global cities are sites of dense networks of interpersonal contact and centers of the important business social-capital trust vital to the successful operation of international finance.

Black Holes and Loose Connections

Networks also have holes. Table 17.4 lists eleven cities that met three criteria: they had a population of over 3 million, were not identified by GAWC as a world city, and did not share their national territory with a world city. They ranged from Tehran, Iran, with a population of 10.7 million, to Chittagong, Bangladesh, with a population of 3.1 million.

There are a number of reasons behind these very large cities' non–world city status. They are the **black holes** of advanced global capitalism, with many people but not enough affluent consumers or complex industries to support sophisticated producer services. There are also cases of catastrophic decline where there has been an almost complete collapse of civil society. In recent years, Khartoum and Kinshasa, for example, have witnessed social anarchy and the decline of the rule of law. War and social unrest have been the norm rather than the exception. These two cities typify those that have internally collapsed for all intents and purposes and have been abandoned or bypassed by global capitalism. Sustained social disruption reinforces the global disconnect.

Poverty and social anarchy do not explain all the cases. Tehran and Pyongyang, North Korea, for example, are cities where national ideologies have not encouraged global economic connections to the advanced capitalist economies.

We can identify three ideal types of large non–world cities: the poor city, the collapsed city, and the excluded city. There are clearly connections between these types, and in reality most of the cities listed in Table 17.4 have elements of all three in differing proportions. It is just as important to identify the black holes and loose connections as to point out the important nodes and connected cities of global urban networks.

Flows of People

Lisa Benton-Short and colleagues also sought to identify a global network, but their work was based on flows of people. They looked at immigration into cities around the world and established an index based on the percentage of

TABLE 17.4	Black Holes
CITY	**COUNTRY**
Tehran	Iran
Dhaka	Bangladesh
Khartoum	Sudan
Kinshasa	Congo
Lahore	Pakistan
Baghdad	Iraq
Rangoon	Myanmar
Algiers	Algeria
Abidjan	Ivory Coast
Pyongyang	North Korea
Chittagong	Bangladesh

17.4 Urban Sustainability Networks

Many cities are embedded in financial and trade networks. There are also networks and organizations to coordinate sustainability management among cities and local governments. For example, Local Governments for Sustainability is an international network of some 1,500 cities and regions that provides capacity building and collective learning for sustainability. The C40 Cities Climate Leadership Group provides access to tools, resources, and information that enables cities to take action to address climate change. The C40 now counts 1,500 cities as members and has encouraged or supported over 14,000 climate action programs in cities since 2017. The Rockefeller Foundation launched a global effort to increase urban resilience through its global "100 Resilient Cities Network," which provides funding for cities to hire a chief resilience officer, technical support, access to planning services, and networking with other member cities.

The number of cities now taking part in multiple networks is substantial and expanding. As cities plan for sustainability, companies and organizations have created a wide variety of networks and benchmarking systems that collect data; measure and rank sustainability and apply practical strategies, tools, and methodologies; and share best practices and lessons learned around sustainability. A few of these include the following:

ORGANIZATION	PURPOSE
Arcadis	This global design and consultancy firm ranks one hundred global cities on three dimensions of sustainability: people, planet, and profit.
C40 Cities on Climate	Facilitates networking and provides technical support
The Carbon Disclosure Project	This nonprofit organization operates a global disclosure system for companies, cities, and regions to report and track greenhouse gas emissions.
European Foundation for the Living and Working Conditions	Their indicators document describes a broad range of sustainability indicator measures; developed to aid indexing of urban sustainability performance.
Global Cities Institute at the University of Toronto	Free online platform with standardized indicators that allows for global comparison of city performance and knowledge sharing
Local Governments for Sustainability	Global network of over 1,500 cities, towns, and regions providing networking and technical assistance
Rockefeller Foundation	"100 Resilient Cities Network" supports planning and action to increase resilience from shocks and challenges for cities within the global network
U.S. Green Building Council	Features an online platform that helps cities set goals, implement strategies, track progress, and share performance data; recognizes city achievements with certification

REFERENCES

Acuto, M. 2013. *Global Cities, Governance and Diplomacy: The Urban Link*. London: Routledge.

Davidson, K., L. Coenen, M. Acuto, and B. Gleeson. 2019. "Reconfiguring Urban Governance in an Age of Rising City Networks: A Research Agenda." *Urban Studies* 56:3540–3555. https://doi.org/10.1177/0042098018816010.

Rockefeller Foundation. 2022. "100 Resilient Cities." https://www.rockefellerfoundation.org/100-resilient-cities/.

foreign-born, the total number of foreign-born, the percentage of foreign-born not from a neighboring country, and the diversity of immigrant home countries. The result was a threefold division into alpha, beta, and gamma cities. There are similarities and differences between the GAWC and the results of Benton-Short and colleagues. Among the similarities are that some cities appear in the same category in both analyses. New York and London are alpha cities and sit atop the apex of both hierarchies. They are centers of economic, political, and social power, and they attract a disproportionate number of immigrants. Other shared alpha cities include Toronto, Los Angeles, Sydney, and Amsterdam. Among the differences, while Miami, Melbourne, Vancouver, and Dubai are considered pivotal points of global migration reaching alpha status, in the advanced producer service category they only make it to beta status.

Cities as Spaces of Flow

Some of the differences relate to issues of data and data availability. But they also refer to the nature of the global urban network. Networks vary according to the flow. Some flows "pool" in some cities rather than others. Advanced producer services, for example, which can be considered command functions of the globalizing economy, are still concentrated in just a few cities because of the need for social interaction in global financial business deals. The flow of people, while matching the connections in command functions, also has slight differences. National regulations concerning immigration, the demand for labor, and the relative openness of societies to foreign migrants all play a part. The sheer need for labor, whether the global talent pool of specialized knowledge experts or cheap unskilled labor, varies throughout the urban network. The result is a similar but not identical match of command

centers with immigration hubs. There is no one fixed global urban hierarchy; it is more a global urban network with different configurations of hierarchy depending on the flows.

Money, people, ideas, goods, and practices do not just flow through the urban network; they are also transformed in the process. To be more accurate, we should use the term "space of transformative flows." Consider flows of people. The movement through the urban network is not a simple geographical movement; it involves cultural exchanges. These can refer to the new work habits and job practices of the temporary worker as well as the complex cultural transformations of long-term migrants as they adapt to a new milieu and in turn transform their surroundings. Rather than mere transfers, flows along the urban networks are transformative experiences, even in the flows of inanimate things such as money, goods, and services. A small amount of money in London or New York, when remitted back to Ghana or El Salvador, can become the source for a land and house purchase, enable a new business, or pay for better schooling. Flows through the urban network change the medium and the networks; as people adapt, ideas are tweaked, money is reimagined, and commodities are reappropriated.

Cited References

Benton-Short, L., M. Price, and S. Friedman. 2005. "Globalization from Below: The Ranking of Global Immigrant Cities." *International Journal of Urban and Regional Research* 29:945–959.

Berry, B. J. L. 1976. *Geography of Market Centers and Retail Distribution*. Englewood Cliffs, NJ: Prentice Hall.

Bromley, R., and R. D. F. Bromley. 1979. "Defining Central Place Systems through the Analysis of Bus Services: The Case of Ecuador." *Geographical Journal* 145:416–436.

Galpin, C. J. 1915. "The Social Anatomy of an Agricultural Community." University of Wisconsin, Agricultural Experimental Station, Research Bulletin 34.

Jefferson, M. 1939. "Why Geography? The Law of the Primate City." *Geographical Review* 29:226–232. (Quote is from p. 227.)

Short, J. R., and L. M. Pinet-Peralta. 2010. "Urban Primacy: Reopening the Debate." *Geography Compass* 3:1245–1266.

Select Guide to Further Reading

Alderson, A. S., J. Beckfield, and J. Sprague-Jones. 2010. "Intercity Relations and Globalisation: The Evolution of the Global Urban Hierarchy, 1981–2007." *Urban Studies* 47:1899–1923.

Brenner, N., and R. Keil, eds. 2006. *The Global Cities Reader*. New York: Routledge.

Castells, M. 1996. *The Rise of Network Society*. Oxford: Blackwell.

Derudder, B. 2008. "Mapping Global Urban Networks: A Decade of Empirical World Cities Research." *Geography Compass* 2:559–574.

Derudder, B., M. Hoyler, P. J. Taylor, and F. Witlox, eds. 2012. *International Handbook of Globalization and World Cities*. Cheltenham, UK: Edward Elgar.

Dobrinskaya, D. E., and I. A. Vershinina. 2018. "New "Connectography": Networks of Cities in the Global World." *Espacio* 39:7–7.

Jefferson, M. 1939. "The Law of the Primate City." *Geographical Review* 29:226–232.

Khanna, P. 2016. *Connectography: Mapping the Future of Global Civilization*. New York: Random House.

Khanna, P. 2021. *Move: The Forces Uprooting Us*. New York: Scribner.

Knox, P., ed. 2014. *Atlas of Cities*. Princeton, NJ: Princeton University Press.

Moomaw, R., and M. Alwosabi. 2004. "An Empirical Analysis of Competing Explanations of Urban Primacy Evidence from Asia and the Americas." *Annals of Regional Science* 38:149–171.

Neal, Z. P., and C. Rozenblat, eds. 2021. *Handbook of Cities and Networks*. Cheltenham, UK: Elgar.

Ren, X., and R. Keil, eds. 2018. *The Globalizing Cities Reader*. 2nd ed. New York: Routledge.

Sassen, S. 2016. *Global Networks, Linked Cities*. London: Routledge.

Taylor, P. J. 2004. *Global City Network*. London: Routledge.

Taylor, P., and B. Derudder. 2015. *World City Network: A Global Urban Analysis*. London: Routledge.

Van Meeteren, M., and A. Poorthuis. 2018. "Christaller and 'Big Data': Recalibrating Central Place Theory via the Geoweb." *Urban Geography* 39:122–148.

The Internal Structure of the City

In this chapter, we will focus on three specific themes of the internal structure of the city: the city as investment, as residence, and as political arena. The chapter ends with a discussion of the major contemporary trends of gentrification, suburban decline, city marketing, and the hosting of global events.

18.1 The City as Investment

The city is a site of investment. This investment is cyclical, because capital flows into the building of a city when relative rates of return are high. **Building cycles** occur, on average, every eighteen to twenty years. There is distinct periodicity to the building cycle. In the *trough*, there is limited investment and little building. During the *upswing*, the lack of supply just as demand picks up leads to higher levels of construction. At the *peak*, building activity is more feverish, land prices increase, and there is more speculative development (see Photo 18.1). But as the supply expands just as the demand falters, there is a *downswing*, and in more extreme cases a *crash*, as prices fall and supply outstrips demand. The cycle then begins again.

Different sectors of the property market may experience slightly differently timed cycles in varying places. The commercial office property market shows the largest peak and troughs. What was distinctive about the US housing market crash in 2008 was its national character. Previous US housing cycles had a more regional character.

Building cycles often align with transport improvements and changes. A building boom in the 1950s was associated with mass car ownership, giving a more dispersed and road-connected character to urban development compared to the booms in higher-density development spurred by the spread of trams and railways. It is more accurate, then, to speak of a building–transport cycle.

The built form of many cities is a visible legacy of the boom of the past. Each major building cycle is also associated with changes in architectural

PHOTO 18.1 Office development in Reading, UK. A pension company, eager to place large sums of money in secure, high-return investments, funded the project.

style. Capital is locked into place, with the architecture of the day acting as the equivalent of a date stamp. The very largest and oldest cities have examples of all the major building cycles. A transect through a major city is the equivalent of flipping through the pages of an architectural history text. Yet some cities, in contrast, experience massive growth only at specific times. The old part of Tallinn in Estonia is a reminder of the city's growth during the Hanseatic League (1300–1500s), Venice is filled with structures built during the Renaissance boom (Photo 18.2), and the Georgian architecture of Edinburgh is a sign of the city's vitality in the eighteenth century. Housing built in the 1920s boom was influenced by art deco and art moderne sensibilities, while housing and office buildings constructed in the immediate postwar era were more modernist inspired. More recent postmodern buildings indicate the latest rounds of investment, such as the frenzied growth that produced Shanghai's signature postmodern buildings.

18.1   Cities and Climate Change

Climate change will increase environmental shocks to cities. Many cities are already confronting climate change impacts, especially on their infrastructure systems. Climate change is causing damage to roads, buildings, and industrial facilities and poses an increasing risk to ports.

Globally, some 600 million people reside in coastal cities. These include cities such as Dhaka, Shanghai, São Paulo, and London. Coasts are affected by climate change in a variety of interconnected ways. Sea level rise is caused by the thermal expansion of sea water resulting from warmer temperatures and a net increase of water as a result of melting ice. Sea level rise will increase saltwater intrusion in proximate aquifers, groundwater and estuaries, affect coastal wetlands vegetation, and increase coastal erosion. The Intergovernmental Panel on Climate Change projects a global average rise in sea level of about two feet by 2100. The combination of sea level rise and changing precipitation patterns, which include larger, more intense storms, is forcing many cities to prepare for more frequent and severe flood events.

In his 2017 book *Extreme Cities: The Peril and Promise of Urban Life in the Age of Climate Change*, Ashley Dawson examines the impacts of climate change on the world's megacities, including Jakarta, Delhi, São Paulo, and New York. Using a political ecology perspective, he argues that neoliberalism and racial capitalism have already made cities places of stark economic inequality and, as a result, vulnerable populations such as people of color and those experiencing poverty will be even more vulnerable to floods, sea level rise, and extreme heat events. For example, in Indonesia the major brunt of climate change will be faced by the 26.5 million who live below the poverty line and have limited resources and capacity for resilience. In cities such as Jakarta, climate shocks and stresses will also force the near-poor population hovering marginally above the national poverty line to fall into poverty. Climate actions must be carefully designed so that they explicitly benefit the poor and near-poor populations and do not inadvertently increase vulnerability and inequality.

REFERENCES

Dawson, A. 2017. *Extreme Cities: The Peril and Promise of Urban Life in the Age of Climate Change*. New York: Verso.

Intergovernmental Panel on Climate Change. 2022. *Intergovernmental Panel on Climate Change Sixth Assessment Report: Impacts, Adaptation and Vulnerability*. https://www.ipcc.ch/report/ar6/wg2/.

Short, J. R., and A. Farmer. 2021. "Cities and Climate Change." *Earth* 2:1038–1045.

PHOTO 18.2 Renaissance architecture in Venice, Italy.

purchase homes. Because the system is based on long credit lines, any change in the financial structure reverberates in this sector of the housing market. Wider economic changes, such as major interest rate shifts, are reflected in the booms and slumps of the speculative housing sector.

Regarding housing in much of the world, incomes are too low, credit facilities for purchasing homes are lacking, and cash-strapped governments are unable and unwilling to provide accommodation. In many cities in the developing world, **self-built housing**, in which residents simply build their own residences, is an important form of housing production. Areas of self-built housing have a variety

18.2 The City as Residence

We can distinguish between production, supply, and demand for accommodation.

The Production of Housing

Housing is produced in a variety of ways. There is the individual contract, whereby a household hires builders to construct a custom-designed dwelling. This is normally expensive and limited to the very rich. The tiny size of this sector, however, is dwarfed by its architectural importance, as many new residential styles and designs first saw the light of day as individual contracts. Many of the early individual house projects of such famous architects as Frank Gehry, Frank Lloyd Wright, and Le Corbusier influenced changes in overall architectural style. More common is the **institutional contract**, whereby an institution or a government entity builds dwellings and then rents them out or sells them. Institutional contract production was particularly important during the boom in public housing from the 1950s to the 1970s, when local governments hired builders to construct large projects.

The two most important forms of production in the early twenty-first century are speculative and self-build. **Speculative building** involves builders acquiring land and constructing dwellings for a general demand rather than specific customers (Photo 18.3). This form of production is dependent on macroeconomic conditions, especially the interest rate, because builders use borrowed money and sophisticated credit facilities that allow households to

PHOTO 18.3 Speculative house building in Maryland.

of names, including slum, shantytown, informal housing, and squatter housing. Much of this self-built housing is considered illegal, because the occupiers hold no title to the land, do not pay taxes, and have constructed some form of shelter that does not meet building codes. Because they are illegal structures, they often lack public services such as clean water, sewage connections, and sanitation infrastructure. Around 2 billion people are estimated to live in such slums by 2030.

The Housing Stock

The rise and fall of housing stock depends on many factors. There is the natural aging of the housing stock, although the aging process can give older housing the allure of the established, much the same way that antiques have value. Building cycles involve not only the building of the new but also often the destruction of the past as old buildings are demolished. Urban renewal projects in the United States from 1949 to 1973 demolished 600,000 units, covering 1,000 densely packed urban square miles and ultimately displacing 2 million people. A program begun with the best of intentions, getting rid of substandard housing, ended up demolishing basically sound housing, destroying functioning neighborhoods, laying the seeds of subsequent inner-city decline, and concentrating poor people—especially poor Black people—in increasingly segregated urban neighborhoods. A similar process is occurring in the rapidly growing cities of the Chinese coast. Michael Meyer describes life in the vanishing backstreets of Beijing as they were demolished to make way for malls and new high-rise buildings. From 1990 to 2003, more than half a million residents were evicted from central Beijing, a process that accelerated just before the hosting of the 2008 Summer Olympics.

There are also environmental shocks to cities. Just before five o'clock in the afternoon, on January 12, 2010, an earthquake measuring a devastating 7 on the Richter scale occurred in Haiti. Over 300,000 people were killed, a similar number injured, and almost a million made homeless; much of the capital city of Port-au-Prince was reduced to rubble. Despite international relief efforts, in 2012 half a million people still lived in temporary shelter, often in appalling conditions. In October and November 2012, Hurricane Sandy etched a destructive path through the Caribbean and along the northeastern Atlantic seaboard. Fifteen thousand homes were destroyed in Cuba, and hundreds of homes were flooded along the New Jersey coast.

The Supply of Housing

Housing supply can be broken down into three main tenure types: public housing, private renting, and home-ownership. **Public housing** is housing provided by a government agency for its citizens. In much of postwar Europe, public housing was an important part of the housing stock. For example, by 1972 more than 80 percent of housing stock in Scotland was public (Photo 18.4). Public housing was, and in many countries still is, an important part of the social contract between the people and the government. In the United States, by contrast, public housing was always a much smaller percentage of the housing stock and often marginalized to house the very poorest. Every major US city had "projects," which tended to be large modernist units of public housing providing accommodation for the very poor. Many of them have since been demolished.

One public housing trend is privatization. In Britain after 1979, much of the public housing stock was essentially sold off to private residents. A total of 1.6 million units were sold to sitting tenants. Public housing as a proportion of total housing stock declined from around 30 percent in 1975, concentrated in the larger urban centers, to around 5 percent by 2020. In the early twenty-first century, more than 17 percent of households in the United Kingdom still live in social housing.

Across western Europe, public housing, sometimes known as **social housing**, is still an important part of the housing stock, at the highest levels ranging from 40 percent in the Netherlands to 20 percent in Sweden. In Sweden, 850,000 social housing units are provided by 300 local housing authorities. There are three allocation models: "universal," in which everyone has access to social housing,

PHOTO 18.4 Public housing in Glasgow, Scotland.

as in the Netherlands and Sweden, where rents are fixed at market rates with subsidies for the poorest; "restricted," where access is means-tested, as in France and Germany; and "residual," as in the United States, the United Kingdom, and Ireland, where it is reserved for the poorest.

A second form of tenure is **private renting**, whereby landlords rent out property to tenants. Through most of the nineteenth and early twentieth centuries, this was the dominant tenure form in cities of North America and Europe. Landlords tend to want to raise rents and keep maintenance costs low, while tenants want the opposite. History is studded with landlord–tenant conflicts. Generally, the conflict remains at the level of the individual landlord and tenant, but in a few rare cases the conflict can widen. In Glasgow, Scotland, in 1914–1915, a rent strike broadened into industrial stoppages. Because the country was at war, the government quickly intervened with legislation that limited rent increases and laid the basis for tenant protection and controls on rent that have lasted to the present day. In the United States, rent controls are in operation in around 140 cities. In New York City, rent control has been in operation since 1943 and covers around 185,000 residences, while rent stabilization, for apartments constructed after 1945, covers around 1 million residences.

Third, and in most affluent countries the largest form of housing supply, is **owner occupation**, also known as **homeownership**, whereby households purchase property. Housing is very expensive in relation to income, so immediate house purchase is beyond the reach of all but the wealthiest. Credit in the form of long-term mortgages is a vital prerequisite for mass homeownership. In the United States, for example, there is a steady encouragement of owner occupation through various inducements to lenders to provide mortgages. Government-backed agencies such as Fannie Mae and Freddie Mac provide liquidity to the system by purchasing mortgages from banks. Tax policies also encourage homeownership. In the United States, homeowners get exemption from the capital gains tax on house sales and tax relief on mortgage interest payments. These are massive government subsidies that are mostly regressive, in that the richest groups tend to benefit most.

Financial institutions play a key role in owner occupation. Their willingness to lend and the spatial patterns of their lending influence urban housing markets. In the 1970s, for example, many institutions in the United States issued fewer mortgages in lower-income minority neighborhoods. The process was known as **redlining**, and it resulted in a lack of housing finance to poorer neighborhoods in the inner city. It became both a cause and an effect of inner-city decline, because residential areas denied mortgages tended to deteriorate more rapidly. In 1975, Congress passed the Home Mortgage Disclosure Act in response to evidence that banks were not lending in inner-city minority neighborhoods. Two years later, the 1977 Community Reinvestment Act prodded banks to lend in inner-city communities. Promoting homeownership became a national fixation.

18.2 The Legacy of Redlining and Urban Segregation

In the United States, starting in the 1930s, low-income and minority communities were intentionally cut off from lending and investment through a system known as redlining. Federal lending agencies identified areas on maps with red lines where they would not allocate mortgages. The areas were predominantly minority communities. Later, private lending agencies such as banks pursued similar policies. Nearly ninety years later, those same neighborhoods suffer not only from reduced wealth and greater poverty, but also from lower life expectancy and higher incidence of chronic diseases.

Recent reports by researchers with the National Community Reinvestment Coalition compared areas that were redlined decades ago with 2020 Census data on levels of segregation and economic inequality. Their study found that 74 percent of the neighborhoods that were graded as high-risk or "hazardous" more than ninety years ago are low-to-moderate income today. Additionally, 64 percent of the areas graded "hazardous" are minority neighborhoods now. These same neighborhoods suffer from reduced wealth, higher rates of poverty, lower life expectancy, and higher incidence of diseases such as diabetes, asthma, and stroke.

While overt redlining is illegal today, having been prohibited under the Fair Housing Act of 1968, the effect has been a persistent inequality. Segregation—maintained through social attitudes and practices, enforced by legal measures, and implemented through the racialized evaluation of financial risk—has shaped residential patterns still visible today.

REFERENCES

Meier, H., and B. Mitchell. 2022. "The Method for Tracking the Origins of Housing Segregation." National Community Reinvestment Coalition. https://ncrc.org/redlining-score/.

Mitchell, B., and J. Franco. 2018. "HOLC 'Redlining' Maps: The Persistent Structure of Segregation and Economic Inequality." National Community Reinvestment Coalition. https://ncrc.org/wp-content/uploads/dlm_uploads/2018/02/NCRC-Research-HOLC-10.pdf.

18.3 Home Sweet Home

The home is of huge social significance. We spend much of our lives in the home, and our primary emotional connections are shaped in its domestic arena; where we live and how we live are important determinants of our social position, physical health, and individual well-being. Home is a central element in our socialization into the world. Despite its huge significance, there has been comparatively little work on the meaning of the home. However, recent work by geographers focuses on the home, where space becomes place and where family relations and gendered and class identities are negotiated, contested, and transformed. The home is an active moment in the creation of individual identity, social relations, and collective meaning. The home is a nodal point in a whole series of polarities: journey–arrival, rest–motion, sanctuary–outside, family–community, space–place, inside–outside, private–public, domestic–social, spare time–work time, feminine–masculine, heart–mind, being–becoming. These are not stable categories; they are both solidified and undermined as they play out their meaning and practice in and through the home. The home is a space riven by ambiguities, a place of paradoxes.

REFERENCES

Blunt, A., and R. Dowling. 2006. *Home.* New York: Routledge.

Cieraad, I., ed. 1999. *At Home: An Anthropology of Domestic Space.* Syracuse, NY: Syracuse University Press.

From the 1990s on, lending requirements and banking regulations were relaxed, leading to two developments. The first was a relaxation of strict requirements on borrowers. Previously, most mortgages required a 10 to 20 percent down payment and clear evidence of ability to pay. These were known as **prime mortgages**. By the mid-2000s, more mortgages were **subprime**, given to borrowers making no down payment and with fewer verifiable means of income. Inner-city neighborhoods became a prime site for predatory lending. These subprime mortgages grew in size and importance. By 2006, there were almost 7 million subprime loans.

The second development was the securitization of mortgages. Previously, mortgages had been kept on the books of the lending institutions. With securitization, mortgages were bundled up and sold as a commodity. The emphasis was on generating turnover rather than assessing risk, because the risk was passed on to the next buyer on the chain. It was in the interests of the originating company to allocate as many mortgages as possible and then sell them on. Credit agencies were supposed to assess the risk, but a government inquiry revealed massive abrogation of due diligence by the three main agencies. The result was that toxic mortgages infected the financial system. When the housing bubble burst in 2006–2008, 7 million mortgages went into foreclosure, and banks and other institutions had toxic assets on their books. The net effect was the financial crisis of 2008. The sorry tale reminds us of the importance of the housing finance sector in the overall economy.

The Demand for Housing

Demand for housing varies along a number of dimensions. Perhaps the most important is income and wealth. The private housing market covers a variety of income levels. At the top are luxurious apartments and palatial houses on sprawling estates. At the bottom are the cheapest apartments and smallest houses. In between are a variety of middle-income areas. The income hierarchy is reflected in the variegated housing market. The more income you have, the more housing you can consume, whether in terms of more space or more prestigious space. Most large cities, for example, have their exclusive neighborhoods where residency is a sign of wealth and achievement. At the other extreme, every city has its "bad" neighborhoods, with reputations sometimes undeserved but nevertheless an important part of the social geography of the city. The city's housing market can be considered a space-packing problem. The richest get to choose the best housing in the best areas. Groups with successively lower income get less choice, and as income declines, the choices narrow (see Photos 18.5 and 18.6).

PHOTO 18.5 Luxury home in exclusive Palo Alto, California.

Demand for housing also varies over the course of the life cycle. Young, single adults, for example, need less space than families of older adults with children. At the later stages of life, even families with children may consider moving to a smaller residence as their children mature and leave to start their own lives. This simple stage in the life cycle model of housing demand is undermined by two trends. First, in the highest-income countries, average family size is decreasing, so the demand for family-style housing is declining while the demand for smaller accommodation for single-person and childless households is increasing. The vast suburbanization of the US population from 1950 to 1990 was driven in part by the shift of families with children moving to single-family homes in the suburbs, the move in part a decision to get better educational opportunities for their children. Since the 1990s, there has been a move into the city, in part driven by more single-person and childless households for which a house in the suburbs is not as good a choice as a more central location. Second, throughout much of the world, the extended family, rather than the nuclear family, is the norm. Many adults tend to remain with their existing family rather than establishing new, independent households. Interestingly, the trend is also reappearing in the richer countries. In the United States, for example, more adult children are remaining longer with their parents than ever before and sometimes returning home after college; the shift is a function of a difficult job market and expensive housing.

Beyond income and life cycle stage, cultural factors also affect housing demand. Households vary in their ethnic, racial, and other forms of identity. In some cases, certain ethnic and racial groups are discriminated against. To take a prominent example, throughout the first two-thirds of the twentieth century, Blacks in the United States faced severe discrimination that limited housing choice in cities. Black concentrations in central cities, initially a sign of discrimination, also emerged through time as a platform for political power as the Black population found a voice in majority-minority cities such as Baltimore, where Blacks make up 62 percent of the population. Even after the end of formal discrimination, marked patterns of segregation persist. Cities around the world still bear the historical imprint and/or contemporary operation of racial and ethnic segregation. Although apartheid is no longer the formal policy of South African cities, there is still marked residential segregation by race in the social geographies of major South African cities.

Certain groups cluster together in distinct residential areas through choice sometimes as much as constraint. There are many reasons for the clustering, including feelings of community and security. A neighborhood of similar people can provide a safe platform for entry into a large city. Part choice, part constraint, racial and ethnic clustering is an important part of the social geography of the city. The geography shifts as new migrant streams enter; old neighborhoods are gentrified or commodified, as in the case of Chinatowns becoming places of Chinese restaurants and festivals more than places where Chinese people live. Some clusters became centers of entertainment as well as sources of identity, such as the well-known Little Italy in New York City and Little Havana in Miami. In Washington, DC, there is Little Ethiopia, home to the largest population of Ethiopians outside Addis Ababa, around 250,000. It is not just the formal place of residence of all American Ethiopians, but also the setting for restaurants and stores that sell Ethiopian food and goods. Little Odessa in Brighton Beach, New York City, is home to more than 350,000 people from Russia, Ukraine, and Georgia.

There is also clustering and segregation along religious lines. The geographer Frederick Boal has long studied the segregation between Catholics and Protestants in Belfast. The city was at the frontier zone between Irish and British realms. The segregation not only gave a sense of safety during the period of conflict; it was also part of the conflict, because it reinforced group stereotyping and the spatial basis and embodiment of community conflicts. Severe ethnic national and religious divisions can result in highly segregated cities with borders that scar and divide. In the most severe cases, the lines of segregation are patrolled and policed.

In recent years, sexual identity has also become a basis for residential clustering. In the Paddington district of Sydney, Australia, from the 1970s onward, an inner-city neighborhood of small family homes became a favored residence for openly gay households who were discriminated against in much of the city. The neighborhood became a gay-friendly district encouraging an open display of gay identity, sometimes called a gayborhood. As cultural attitudes shifted, the previously transgressive nature of the place was soon commodified and then incorporated the city's economy and national and international image. The annual Mardi Gras festival, initially supported only by the LGBTQ+ (lesbian, gay, bisexual, transgender, and queer/questioning) community as a celebration of sexual diversity, is now one of the city's biggest cultural events, drawing visitors from around the world. In other cities, such as San Francisco, sexual tolerance is now at the very heart of the city's identity. Elsewhere, in much of the world, where LGBTQ+ people face open discrimination, there are far fewer opportunities for neighborhoods of sexual diversity and tolerance to flourish. Homosexuality continues to be categorized as a crime in many societies.

Models of the Residential Mosaic

The housing market produces a residential mosaic of different types of neighborhoods. Urban models have been used to understand and predict urban land use.

Different models focus on different processes. Throughout the twentieth century, a number of different models were proposed. The **Burgess model**, named after the urban sociologist E. W. Burgess, was based on the experience of Chicago in the first third of the twentieth century, a time of massive immigration from overseas. Burgess saw a process of new migrants moving into the cheaper central city areas and more established residents moving farther out (see Figure 18.1). The result was a concentric ring pattern of rising socioeconomic status and increasing family size toward the city's edge. In the early twenty-first century, the built landscape around the central city still reflects the Burgess model; however, it is less accurate in describing suburban residential patterns that developed

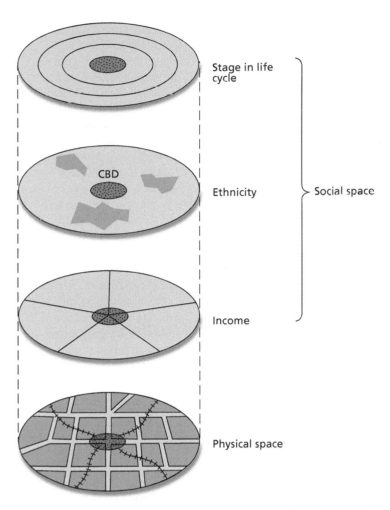

FIGURE 18.2 Murdie's composite model.

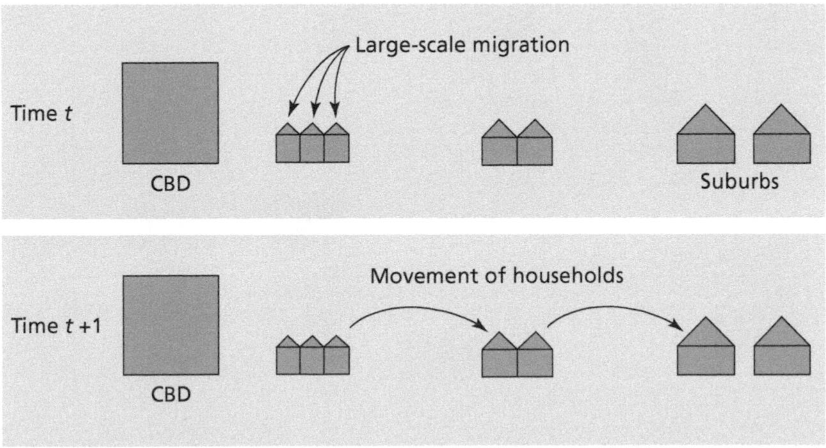

after the 1930s. Later, the land economist Homer Hoyt pictured socioeconomic groups arranged in different sectors of the city. Robert Murdie combined various observations into a layered model of concentric rings of households at different stages in the life cycle: socioeconomic groups in specific sectors and ethnic, racial, and other minorities in distinct clusters. The composite **Murdie model** is shown in Figure 18.2. These simple patterns have long been overlain by more complex patterns. An even more recent composite, the Hanlon–Short–Vicino model, is shown in Figure 18.3. Around the central area of this idealized US city, there are pockets of both concentrated poverty and gentrification, and in the inner ring there are suburbs in crisis, with declining prices and aging housing stock. Toward the city's edge, there are "boomburbs" and affluent enclaves. This model reflects the contemporary reality of

FIGURE 18.1 The process underlying the Burgess model (CBD = Central Business District).

the heterogeneous US metropolitan areas more accurately than the simplified picture of the traditional models.

These models apply only to the cities in the Global North, and particularly to cities in Australia, the United States, and Canada. Figure 18.4 presents the basic elements of a model for cities in the Global South. There, urban structure takes a different form because some cities may have a colonial legacy, often embodied in the Central Business District (CBD) and elite neighborhoods. Most cities have large informal settlements or slums that are found on the periphery of the city as well as distributed throughout it (Photo 18.7). The city center has the formal CBD as well as traditional markets and areas of slum housing. Some cities also have marked segregation between different ethnic and religious groups. Furthermore, some neighborhoods are undergoing gentrification. In the large cities, the route to the airport, often following the direction of elite residences, is also an area of new commercial and industrial building as well as new suburbs, many of them gated communities. Throughout cities in the developing world, there is a fuzzier demarcation between rich and poor and between commercial and noncommercial real estate than in the more developed world. Figure 18.4 presents the basic elements of a model for cities in the Global South.

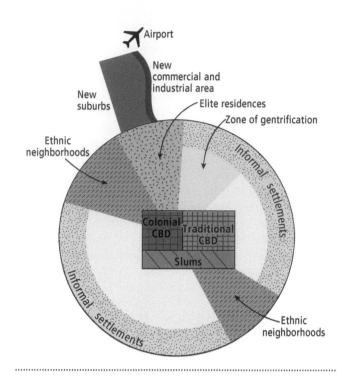

FIGURE 18.4 The Short model of city structure in the Global South.

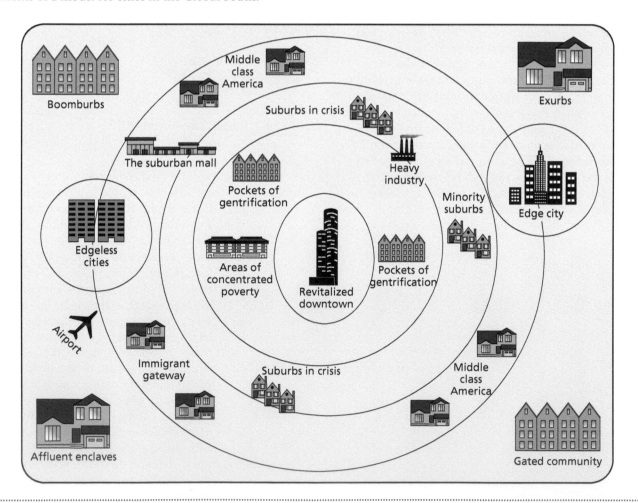

FIGURE 18.3 The Hanlon–Short–Vicino model.

Favela, or slum, in Rocinha, Brazil.

18.4 Measuring Segregation in Cities

The level of segregation in a city is measured in a variety of ways. The **index of dissimilarity** measures the unevenness in the distribution of two groups across spatial units. The index ranges from 0 to 100, with values closer to 100 indicative of more dissimilarity.

The **index of segregation** measures the distribution of one group compared to the total population. This index also ranges from 0 to 100, with values closer to 100 indicative of a more segregated distribution.

The **location quotient** measures the degree of concentration of one group compared to the total population; it is found by dividing the percentage of the subgroup by the total popu-lation of each and every spatial unit. Values greater than one suggest a higher concentration of the subgroup in that area.

Each of the different measures provides a different picture of residential segregation and concentration. All of them are very sensitive to the scale of the analysis. Larger observation units, such as the county level, may mask patterns of segrega-tion that are revealed only at the level of census tracts.

REFERENCE

Johnston, R. J., and K. Jones. 2010. "Measuring Segregation: A Caution-ary Tale." *Environment and Planning A* 42:1264–1270.

18.3 The City as Political Arena

The city is a site where different groups and stakeholders compete and cooperate, highlighting different scales of political economy. Let us consider the case of households.

Households

Households have diverse interests. As residents, they want a full range of public goods and services, includ-ing such **public goods** as clean water, efficient trans-port systems, and good schools. We can identify public bads as well as public goods. **Public bads** are all those unwanted facilities such as trash disposal sites, motorways, and halfway houses for convicted criminals. In many cases, the process is referred to as NIMBY: "not in my backyard." Households individually and collectively seek to attract public goods and deflect or relocate public bads. The process reflects the political power and economic strength of residents' organizations. The rich and powerful live in neighborhoods without toxic waste sites.

It is a consistent finding that toxic facilities are predominantly concentrated in lower-income and minority-dominated areas of the city, and major infrastructure projects with negative environmental impacts, such as urban motorways, are more commonly found in poor and minority neighborhoods. Study after study from around the world reveals a correlation between negative environmental impacts and the presence of racial/ethnic minorities. This is known as environmental injustice or environmental racism. Studies in the United States highlight that race is the most significant variable associated with the location of hazardous waste sites and that the greatest number of hazardous commercial facilities are located in areas with the highest proportion of racial and ethnic minorities. Three of every five Black and Hispanic Americans live in communities with one or more toxic waste sites.

In cities around the world, the poor and the marginal more often than not live in the areas with the worst environmental quality. Social inequalities are expressed and embodied in urban environmental conditions. In Europe, waste facilities are disproportionately located in lower-income areas and in places where ethnic minorities live. Urban environmental inequality is large and pervasive.

Poorer communities have less pleasant urban environments and often bear the brunt of **negative externalities**. It is through their neighborhoods that motorways are constructed, and it is in their neighborhoods that heavy vehicular traffic can cause elevated lead levels in the local soil and water. There is a direct correlation between socioeconomic status and the quality of the urban environment. Moving toward environmental justice would entail the removal of disproportionate risks for environmental bads, as well as the increase in environmental goods—such as solar energy for low-income public housing units or planting more trees in minority neighborhoods.

Households are also taxpayers, and as such they want to minimize their taxes while maximizing their benefits. Attitudes to taxation vary, with Scandinavian taxpayers more willing to bear a large load than those in the United States. In all cities in all countries, however, there are limits to the acceptable amount of taxation. Proposition 13, which passed in California in 1978, limited the tax based on property values. Prior to the legislation, increasing property values meant increasing taxes. In a period of rapid house

---

18.5 Energy and Equity

Too often, options for going sustainable are reserved for the wealthy. Electric cars or the installation of residential solar or wind power can have expensive start-up costs. Further, cost savings are often unavailable to renters because installing alternative energy systems is often a low priority for building owners, even if some subsidy is available.

Programs have been developed to increase access to clean energy for low-income and disadvantaged communities. California's Multifamily Affordable Solar Housing Program, introduced in 2008, incentivized installations of solar on multifamily low-income housing properties across California. The program, which was extended in 2015 to 2021, covered the costs of outfitting with solar. The program avoided 27,000 tons of CO_2 emissions. A follow-up program, the Solar on Multifamily Affordable Housing Program, funded by utility companies up to $100 million per year to the end of 2030, provides assistance and incentive to multifamily building owners and direct financial benefits to tenants.

Other cities are implementing low-income solar programs too. From 2013 to 2016, Washington, DC, spent $81 million on energy efficiency and renewable energy services for low-income residents in single-family homes and multifamily buildings. Through programs like Solar for All and Solar Works DC, low-income communities can also access no- or low-cost solar to reduce electricity bills and add more renewable energy to the city's grid. The programs also create job training and workforce development opportunities. In a strategic move that certainly increased program visibility among target communities, the city paid for Ben's Chili Bowl—an icon of the city's African American community—to install energy-efficient lighting. The long-lasting lighting will reduce energy use by approximately 50 percent and save the business about $1,200 in energy costs yearly. Washington, DC, has also installed 158 solar photovoltaic systems on the roofs of low-income residents. These efforts led to Washington, DC, being named the world's first LEED Platinum City in 2017.

REFERENCE

Keeley, M., and L. Benton-Short. 2018. *Urban Sustainability in the US: Cities Take Action*. New York: Palgrave.

price inflation, taxes skyrocketed. Households organized to place the proposition on the ballot. The effects were felt immediately in reduced public spending and public services. Although taxpayers enjoyed lower taxes on their homes, there was less spending on infrastructure such as roads and schools.

Households are also users of services. They want police and fire protection, libraries, and schools. And they want them to be of high quality. So residents have conflicting goals: they want low taxes and quality services. They will fight against increasing taxes but also resist a decline in services; they will work with local governments to improve services but still resist increasing taxes.

Exit and Voice

If there is a perception of declining services or increasing taxes, households have a variety of options. The most cited pair of options has been referred to as "exit" and "voice." Choosing the first, households can leave and move to another city or separate municipality. This is more common in the United States, where large metro areas are divided into separate municipalities, often with different tax regimes. People move, if they can, to higher-taxed areas if they believe that the result is better schools and other public services. If households stay, they can choose the second option, to voice their concerns, and if those concerns are not heeded, they can protest. There are three sources of protest: the call for goods and services, the expression of cultural identity (Photo 18.8), and the demand for political power. Collective action can range from the persuasive (petitions) and the collaborative (lobbying elected officials) to the confrontational (including allies, marches, and, in extreme cases, forms of civil disobedience).

When residents stage a protest, it represents their failure to influence the political process without direct action. The more powerful groups do not need to protest, because their interests are reflected in the political process. That is why we rarely see the very rich and the very powerful taking to the streets; they do not have to. Protest tends to come from the poorer and more marginalized populations.

Forms of civil disobedience were evident in the Arab Spring that swept through the cities of the Middle East in 2010–2011 and in the Occupy Wall Street movement that occurred in selected cities in Europe and North America the following year. Starting in 2020, US cities became the site of protests around racial inequality. The Black Lives Matter movement was initially founded in 2013 in response to the acquittal of Trayvon Martin's killer. It gained momentum after the killings of Eric Garner in New York City and Michael Brown in Ferguson, Missouri. In 2020, the police killings of George Floyd and Breonna Taylor reignited protests in numerous US cities. Since George Floyd's killing, more than 230 Black people have been killed by police. The *New York Times* estimated that 15 to 26 million people in the United States have participated in recent protests in hundreds of cities including Minneapolis, Los Angeles, Detroit, Dayton, and Washington, DC (Photo 18.9).

Black Lives Matter is a collection of organizations whose mission is to confront and eliminate white supremacy and build local power to prevent or intervene in violence inflicted on Black communities. The movement has opened the door for important conversations around police violence, mass incarceration, and systemic racism in the United States. These protests challenged all Americans to rethink how contemporary problems of inequality are rooted in historical failures to create a just society. These protests also remind us of the power of urban public space to function as a place for people to challenge the status quo and demand change.

Urban Regimes

Protests and uprisings in cities are rare. Much of the political life of the city consists of the more banal and everyday operation of political agreements, deal making, and compromise. Clarence Stone identifies what he calls **urban regimes**, the informal arrangements behind the formal workings of government. He developed the idea from his research on the

PHOTO 18.8 Political protest in Barcelona for Catalan autonomy.

informal partnership in Atlanta between City Hall and downtown business elites. More generally, urban regimes consist of informal governing coalitions that make decisions and get things done in a city. The combination of political and economic logic, with all the ensuing tensions, conflicts, and ambiguities, constitutes the local urban regime. The options and concerns of urban regimes vary over time and space—they may be inclusionary or exclusionary and will vary throughout the metropolitan regions. Suburban regimes, for example, are more concerned with preserving property values.

In the United States, three regime types have been identified: organic, instrumental, and symbolic. **Organic regimes** occur in small towns and suburban districts with a homogeneous population and a strong sense of place; their chief aim is to maintain the status quo. **Instrumental regimes** focus on specific targets identified in the political partnership between urban governments and business interests. An urban growth machine regime is concerned with securing urban business interests. **Symbolic regimes** occur in cities undergoing rapid changes, including large-scale revitalization, major political change, and image campaigns that try to shift the wider public perception of the city. We should see these three categories as ideal types, with any one city's regime exhibiting characteristics, albeit in different proportions, of each type.

18.4 Challenges in the Contemporary City

In the rapidly changing cities of the contemporary world, a number of transnational trends can be noted. Among the many, let us consider four ongoing challenges: gentrification, suburban decline, **city branding**, and the hosting of global events.

Gentrification

In 1964, urban scholar Ruth Glass coined the word **gentrification** to refer to a certain form of urban change. A strict definition of the term is the replacement of lower-income households by higher-income households (a demographic change) and a change in the physical landscape (such as renovations to existing buildings or the construction of new residential and/or commercial space). She employed the term to describe how middle- and upper-class residents moved into working-class areas in London. The term is now used to refer to the movement into city neighborhoods of higher-income households, whether through the replacement of lower-income households in the existing housing stock or in the form of movement into custom-designed buildings. The process is now found in cities across the world as the more affluent move into central and inner-city neighborhoods that were previously neglected or inhabited by the poor. It is promoted by developers and downtown elites eager to generate business and by city governments eager to lure higher-income households into the city to provide a stronger tax base and a population with higher disposable incomes. Gentrification is often trumpeted as revival, bringing life back into the city. Advocates of poor and lower-income people, in contrast, see gentrification as a form of class warfare over space. Gentrifiers tend to be richer, younger people, while those who are displaced tend to be older, poorer, and from racial minorities.

In Harlem, home to many low-income African Americans in New York City, major construction projects such as Towers on the Park, which opened in 1988, and individual households buying up brownstone dwellings have led to a distinct form of gentrification along the neighborhood's western corridor. Just across the Hudson River from lower Manhattan, a quick ferry ride from New York City's financial district, sits the small, formerly working-class city of Hoboken. The process of gentrification began there around 1980. The housing stock was effectively emptied of lower-income residents in favor of higher-income households, especially those working in the financial services sector in lower Manhattan, turning what was long a working-class town into a yuppie suburb. The increased employment opportunities in San Francisco from the early 1990s generated an extraordinarily tight housing market. The Mission District, close to downtown, was particularly vulnerable to gentrification, because it had relatively cheap housing stock. This area of working-class families and recent Latino immigrants quickly transformed as housing values, rents, and evictions all increased.

The Olde Towne East district, adjacent to downtown Columbus, Ohio, was like many late-nineteenth- and early-twentieth-century residential areas. Built close to the city, it contained both high-income and middle-income residents in a housing stock that contained some large dwellings. Post–Second World War suburbanization drew off the wealthier residents, and then a mixture of urban renewal and highway construction ripped the heart out of the neighborhood. Originally a stable residential area throughout most of the early twentieth century, by the 1960s the dwelling stock was in decline and most of the residents were low-income African Americans living in rented accommodations. However, being so close to the downtown and with some substantial housing remaining, the area soon attracted the attention of gay white men seeking to buy and improve housing. In many cities, gentrification has been associated with the coming out of a gay community. In Olde Towne East, the struggle, crystallized between the gay white newcomers and the longer-term residents, took a particular form as the new residents sought to have the area designated as historic. For the longer-term low-income residents, this was a financial penalty, because they were often unable to afford the costs of meeting historic preservation codes. The property cost increases forced out many of the older owners and low-income renters. The recent history of this area encapsulates a number of cleavages, including race, class, income, and sexual orientation, as two marginalized communities fought a turf war over the price and meaning of the residential space.

In the United States, most cities have failed to balance the need for revitalization with protecting and ensuring affordable housing remains in areas undergoing redevelopment. As a result, revitalization efforts usually displace low-income or minority residents.

Gentrification tends to occur when there is recentralization in central city areas. In the case of Sydney, Australia, the immediate postwar years saw a steady suburbanization. The city extended far into distant suburbs, increasing commuting times. Starting in the 1970s, more affluent younger and gay households began buying up the high-density terraced housing in inner-city neighborhoods such as Paddington. The central location reduced commuting times for those working in the city and provided very cheap housing, and the neighborhoods became safe for openly gay people. Paddington is now one of the more fashionable areas of the city (see Photos 18.10 and 18.11).

Gentrification can also be part of a more fundamental rewriting of urban space associated with more aggressive policing, in what is referred to as the punitive turn. This involves, according to proponents, making the streets safe; for critics, it is an aggressive form of policing directed specifically against the poor, minorities, and marginalized people in areas of the city undergoing gentrification. In selected neighborhoods, the gentrified city embodies all these active political moments in a socio-spatial urban transformation that marginalizes the poor and the homeless. Urban policies and market forces combine to promote

PHOTO 18.10 Gentrification in Paddington: before.

gentrification, marginalize the poor, and make the homeless disappear from public view.

Suburban Decline

An image persists, especially in the United States, of suburbs as bucolic idylls, different from the grit of the cities. Yet many older postwar suburbs struggle to survive, caught in the stagnant middle between the dynamic processes of urban decentralization, as more distant suburbs are developed, and recentralization, as investment shifts back into the city. The deindustrialization of older suburbs also leaves these places with a smaller economic base for opportunity and mobility. Once prized as the ideal location for families, many older inner suburbs now exhibit symptoms of aging. Those built during the postwar period of mass suburbanization are particularly outdated, because the housing stock now lacks the size and amenities to compete with newer housing on the outer fringe of the

PHOTO 18.11 Gentrification in Paddington: after.

metropolitan area or in the redeveloped heart of gentrifying cities. Caught between city gentrification and outer sprawl, many postwar suburbs are currently losing population and the battle for investment resources. Bernadette Hanlon and her colleagues write of the emerging feature of "suburban gothic" as numerous suburbs in crisis appear on the urban scene.

City Branding

The marketing of cities, also referred to as city branding, has a long history. The frontier town, the resort, the suburb, and the industrial city were all marketed with appropriate images to attract investment and people. In the contemporary world, three new forms can be identified.

The first is the marketing of the postindustrial city. In the wake of the deindustrialization of large swathes of urban North America, Australia, and Europe, formerly industrial cities were placed in the unenviable position of losing investment and jobs. The global shift of manufacturing creates new industrial cities in the developing world but leaves behind industrial cities in the developed world. These industrial cities, or, to be more accurate, these newly

postindustrial cities, became associated with the old, the polluted, the past, and the failed. A major theme of marketing these postindustrial cities is to distance them from their recent industrial past. In the case of Syracuse, New York, this involved a new logo for the city that replaced images of smoking factories with a postmodern building skyline and a transformation of a local lake from a dump site to a scene of regeneration. Formerly industrial cities are rebranded in more attractive packages that emphasize the new rather than the old, the fashionably postmodern rather than the merely modern, the postindustrial rather than the industrial, consumption rather than production, spectacle and fun rather than pollution and work.

Then there is Wollongong, an industrial city on the New South Wales coast of Australia. Massive steelworks dominated the urban economy. The steelworks shed 15,000 jobs from the early 1980s to the mid-1990s. The rate of job loss was only one major strand in negative imagery associated with the city in the national imagination and among foreign investors. City leaders decided to rebrand the city in the public imagination. An image campaign was built around the idea of "innovation, creativity, and excellence." An important part of the campaign involved planting half a million trees on the site of the steelworks while the council found funds to clean up the beaches and construct cycle ways. Such "greening" is now an integral part of a city's attempt to shed its hard industrial image for a softer postindustrial feel.

The second form is the marketing of a city as global. Because of the growing competition, there is constant need to upgrade and improve a city's image. At the top of the global hierarchy, as measured in terms of the concentration of advanced producer services, sit London and New York. Both cities actively position themselves as global cities. In 1964, New York hosted the World's Fair, and after a period of fiscal crisis and economic decline the city was rebranded in 1977 with the successful I ❤ NY campaign. In 2012, London hosted the Summer Olympic Games, presenting itself as a multicultural, cosmopolitan, cool city.

The rhetoric of global city branding is often associated with attempts to justify controversial and costly urban redevelopment projects, tax incentives, and business-friendly economic policies.

Darel Paul explores the politics of "global imagineering" in Montreal, Canada, where business interests triumphed, and Minneapolis–St. Paul, Minnesota, where similar interests were trumped by populist politics. In St. Petersburg, Russia, not only was the old Soviet name of Leningrad dropped, but also the city was actively promoted as an international hub of circulatory capital, host to corporate power, and stage for globalist megaprojects. St. Petersburg was rebranded as an entrepreneurial, economically competitive, globally connected city.

The third form of marketing is presentation as a business-friendly city. As capital becomes ever more mobile and more cities enter the competition, cities need to aggressively promote themselves as good places to do business. Here are some examples from advertisements of US cities in the business press:

Dallas	The city of choice for business
Milwaukee	The city that works for your business
New York	The business city that never sleeps
Phoenix	Moving business in the right direction

More specific themes around this general message include emphasizing a pro-business political climate, a skilled labor force, good infrastructure, and accessible location.

There is a growing body of critical work that highlights the silences of city marketing. The emphases of most marketing campaigns are on the city as a place of business, less on social inclusivity; more on the profitable city, less on the fair and just city. The dominant themes represent the power of the dominant groups. The more marginal are excluded. In an interesting twist on this finding, however, Jaime Hernandez and Celia Lopez suggest how slums can be used to positively brand a city, in this case Bogotá. They suggest that highlighting rather than ignoring the city's slums can invoke positive responses around ideas of genuine difference, cultural authenticity, and vernacular architecture. The poor and the marginal become less a political threat and more a marketing opportunity. In this case, the ethos of marketing is so pervasive that even the silences are uttered and the problems exposed, not as issues to be solved or addressed but as opportunities to sell and rebrand.

Signature Buildings and Global Events

Cities seeking a global reputation often employ big-name architects to design buildings and campaign to host global events.

The marketing of cities can involve the construction of buildings by prominent architects, so-called starchitects, showing the centrality of **starchitecture** to the symbolic capital of cities. To have a new building by a starchitect denotes cultural heft and global intentionality. Individual buildings and assemblages of certain built forms—museums, art galleries,

airports, and the like—built in the latest style by the hottest architects is an important way for a city to position itself. Perhaps the best-known and most-cited case study is Bilbao in northern Spain. It was a major banking and industrial center that from the 1970s experienced deindustrialization and massive job loss. The old industrial area along the river was a site of abandoned factories. The city authorities began an ambitious urban renewal policy. A centerpiece was a branch of the Guggenheim Museum designed by the famous American architect Frank Gehry. The museum, built along the river, is now an iconic image of the city and a classic case study of having a starchitect design a major cultural center to successfully rebrand a city (Photo 18.12).

Hosting events is now a significant part of marketing a city. So-called spotlight events highlight the city for a wider audience: the larger the event, the greater the potential marketing opportunity. The hosting of events such as film festivals, expos, and Olympic Games provides opportunities for a city's elites to stage a widely accessible and globally pervasive marketing campaign. In a detailed case study, Claire Colomb explores the marketing of Berlin between 1989 and 2011. She argues that the city was marketed through rebuilding and the staging of a new Berlin. In the inner city, sites were turned into tourist attractions. Place marketing and urban reinvention were intertwined in a symbolic politics of representation.

One of the most important mega-events is the Summer Olympic Games. From humble beginnings in Athens in 1896, when only 200 athletes from fourteen countries competed, with limited press coverage, the modern Olympic Games have now grown to a truly global spectacle involving most countries in the world.

Hosting the Games involves providing venues and improving urban infrastructure (Photo 18.13). Since Barcelona

PHOTO 18.12 Guggenheim Museum in Bilbao.

in 1992 and especially since Sydney in 2000, emphasis has also shifted to using the Games as way to undertake urban renewal of industrial and abandoned sites. For the 1992 Games, Barcelona built a new waterfront and upgraded a declining area of the city, as well as making numerous improvements throughout the metro area, including new roads, a new sewer system, and the creation or improvement of over 200 parks, plazas, and streets. For the 2000 Games, Sydney built a new road linking the airport to the downtown and constructed the main venue on a previously contaminated inner-city site. Beijing undertook a building frenzy with an estimated $40 billion of Olympic-related buildings and infrastructure, including a new expressway and ring roads, miles of rail and subway tracks, and a new $2.2 billion airport that is the largest in the world. For the London Games in 2012, much of the construction of the new sports venues was part of the regeneration of the lower Lea Valley, an area of postindustrial decline.

The Olympics also provide an opportunity to make a modern city. Athens spent close to $16 billion in large-scale public investments in water supply, mass transit, and airport connections to get ready for the 2004 Games. Hosting the Olympics provided the opportunity for citywide, coherent planning. But modernity can come at a cost. The changes often come with increased costs for city residents unless there is a specific commitment to redistributional outcomes. The Chinese government used the 2008 Games as an opportunity to modernize Beijing. One plan involved the destruction of the old, high-density neighborhoods of small alleyways in the central part of the city, seen by officials as a remnant of a premodern past. Almost 8 square

miles were destroyed and almost 580,000 people displaced as a result of this one program.

Hosting the Games provides an opportunity for massive urban renewal, major environmental remediation, dramatic infrastructural improvements, and the creation of a positive image across the globe. But there is also a darker side to using the Games as a driving force for revitalization and redevelopment. The 2016 Olympics held in Rio de Janeiro, Brazil, was branded as an opportunity to develop Rio's infrastructure and remake the city into a world city. But building some of that infrastructure involved forcibly removing some 700 families from the favela community of Vila Autodromo. The favela stood between the Olympic Park and the Athlete's Village and was slated to be demolished to become an access road. Rio's forced removal is not the only incident of injustice that has occurred while a city has redeveloped for the Games, but it has helped give momentum to anti-Olympics activists who criticize the Games for overspending, citizen displacement, greenwashing, and corruption. Activists and scholars are raising criticism of the costs of the Games, the impact on marginal groups, and the huge costs compared to the less-than-predicted benefits.

COVID and the City

COVID-19 is one of the many pandemics in recent history. Since the flu of 1918, at least eight pandemics with global social and economic effects have been documented. Pandemics are part of the modern world. About 14 million people die annually because of a pandemic.

There were many urban implications of this health crisis. The public space of cities is where people concentrate, mingle, and interact. Often prized as an important part of the convivial nature of cities, in a pandemic they become feared as places where contagion between strangers is more likely. The closure of urban public spaces, one of the more dramatic responses to COVID-19, especially impacts third spaces—those places that are neither work nor home—such as barbershops, bars, cafes, gyms, and restaurants. The longer-term impact may be in the form of less-crowded places with stricter controls over spacing distancing and limitations on direct human-to-human contact between strangers. The postpandemic city may be less convivial. The decline in the number and thickness of third places may lead to an increase in safety, but perhaps at the cost of social and psychological intimacy.

The postpandemic city will be faced with new design challenges to make cities safer. History has shown that pandemics can be a catalyst for building safer environments. The layout of the contemporary city emerged long after the impact of previous epidemics and pandemics was forgotten. Since the 1990s, there has been an emphasis on densely populated urban spaces, from crowded sports arenas to high-density pedestrianized areas. That was the old normal; postpandemic, the new normal will be more concerned with health impacts of density, hygiene, and the safety of urban public spaces. This is likely to become an embedded rather than a transitory condition as long as the risk of future pandemics remains a real possibility.

COVID-19 has involved marked changes in urban transport. First, there was an initial steep decline of all road traffic as fewer people traveled to work either by private car or by public transport. The decline of car traffic has created an opportunity to reinforce the shift away from car-dominant cities. Busy car thoroughfares were quieted with less traffic and less demand for street parking. This has offered a chance to reimagine urban public spaces away from the dominance of cars and car users toward greater emphasis on noncar spaces. Some city streets became "street eateries" as restaurants spilled out onto sidewalks to provide open-air, socially distanced places. The transformation of busy streets into pedestrianized spaces allowed social distancing and reimagining of what city streets could and should be. With COVID-19 lockdowns vastly reducing the use of roads and public transit systems, many city authorities took advantage by closing streets to cars, opening others to bicycles, and widening sidewalks to help residents maintain six-foot distancing (Photo 18.14). Cities across the world such as Boston, Vienna, Oakland, México, Milan, and Bogotá are adding and widening bike lines. The postpandemic city may be a more walkable and bicycle-friendly city than could have been imagined prepandemic. Although the shift away from cars was already taking place, the pandemic reinforced these trends.

One of the immediate impacts of the pandemic was a rapid and steep reduction in demand for public transport.

People worked from home and were wary of densely packed urban transportation. In general, public transport is perhaps cheaper than private transport, but in the pandemic city, it is seen as riskier. Health advice to avoid crowded, enclosed spaces and time spent in close contact with other citizens—both features of a normally functioning transit system—has made passengers more hesitant to re-enter the public transit system.

Both tourist and business travel were initially suspended and then reopened to varying degrees. Urban tourism may return to some extent, but business travel may fail to recover, with a consequent knock-on effect on convention hotels, convention cities, and the business services sector that deal with international business travel. After months and perhaps years of online operations, the business, government, and nongovernment communities may find it less necessary to fund business travel, especially as conference call technology becomes more sophisticated.

The pandemic has seriously affected urban economies and the lives of urban dwellers. Some have argued that it undermines the economic necessity for cities. However, the death of the city is regularly predicted. Like Mark Twain's premature obituary, it is greatly exaggerated. The city was thought to be redundant when the telephone was introduced and then the computer. What was the point of cities when people could communicate over the phone or by Internet? Hadn't they lost their rationale? The future was imagined as a global village of electronic cottages. In fact, the future was giant metro areas and dense cities.

After 9/11, some even thought that the threat of terrorism would lead to the suburbanization of financial services and drift away from the city. In the decades that followed, New York City continued to grow and prosper as a global financial center, as did London.

Cities, at least some of them, will survive the pandemic for the same reasons they survived the telephone, the Internet, and terrorist attacks. There are powerful economic forces at work. There are major economic benefits to agglomeration in cities. The pools of skilled labor allow the transfer of information, knowledge, and skill. The presence of subsidiary industries provides common goods and services; and the geographic proximity facilitates face-to-face contact that leads to the maintenance of trust and the exchange of information. These forces are even more powerful for the more dynamic sector of the economy, banking and financial services, advertising, and the vast range of cultural and creative industries that are in the business of ideas and the narration of information into knowledge. This cognitive capitalism, as it has been called, is built around face-to-face contact as an efficient means of communication that helps solve incentive problems, facilitates socialization and learning, and provides psychological

PHOTO 18.14 Streetery in Bethesda, Maryland, 2020.

motivation. The cities of cognitive capitalism will bounce back. Those that have more routine industries may not. Cties such as San Francisco, New York, and San Jose will continue to grow while cities like Detroit, Buffalo, and Schenectady may continue to falter. Manufacturing can be offshored; basic information processing can be done almost anywhere with an Internet connection. However, turning information into knowledge and ideas into products is best done in dense cities.

Cities are the perfect mechanism for finding out about new job opportunities and generating social ties. Even in the dating market, cities widen and deepen the dating pool and extend the range of choice of partners, for both one-night and long-term commitments, because social interaction needs geographic proximity. Despite the long tradition of antiurbanism that always sees the urban demise just around the corner, cities will also survive because they are our greatest invention. They continue to be the best platform to deal with a range of problems from climate change to ensuring economic growth.

The city will change. As online working becomes more common, more routinized, and more normal, the old office building filled with workers who could do much of their work at home may be a casualty of the COVID-19 pandemic. Some parts of the commercial real estate market may never recover. Commercial real estate, especially in expensive markets, will experience a major collapse in value. Office space will become less expensive. The downsizing of retail and commercial space will end the inflation in land values, perhaps making them more affordable. A revalorization of urban land markets may produce more affordable cities.

Cities are also sites of stark and growing inequalities, and these are reinforced as the pandemic cuts a more destructive swath through the less wealthy and less powerful. All the indications suggest a deepening of inequality as the wealthy face minor inconvenience, but the poorer face economic collapse. The progress in the urban Global South experienced in the past decade, in which cities advanced toward sustainability goals, inclusion, better governance, and poverty reduction, will most likely cease or even reverse given the significant economic and political impact of the global crisis. Cities in lower-income countries are less prepared to face the health crisis and economic aftermath of pandemics. Cities will experience the exacerbation of preexisting problems.

The crisis of COVID-19 will have a pervasive effect on urban poverty and may create a new precarity in cities across the world and especially in the Global South. Before the pandemic unfolded, the continued reduction of poverty was reaching a plateau. Between 2010 and 2015, world poverty—vastly concentrated in urban areas—decreased from 15 percent to 10 percent, but the rate decelerated from 2015 to 2019, with a reduction of only 1.8 percent. The United Nations estimates of the increase in poverty because of COVID-19 amount to 71 million people pulled back into extreme poverty, given the economic stagnation, job losses, and the decrease of about 40 percent in remittances to low- and middle-income countries. In sub-Saharan Africa, estimates suggest that urban poverty will increase by 44 percent. The increase in poverty with the resultant negative impacts on health, education, and food security will be among the most pressing issues in the post-COVID era.

Cited References

Boal, F. 2002. "Belfast: Walls Within." *Political Geography* 21:687–694.

Colomb, C. 2011. *Staging the New Berlin*. London: Routledge.

Glass, R. 1964. *London: Aspects of Change*. London: McGibbon and Kee.

Hanlon, B., J. R. Short, and T. J. Vicino. 2010. *Cities and Suburbs*. New York: Routledge.

Hernandez, J., and C. Lopez. 2011. "Is There a Role for Informal Settlements in Branding Cities?" *Journal of Place Management and Development* 4:93–109.

Meyer, M. 2008. *The Last Days of Old Beijing*. New York: Walker.

Paul, D. E. 2004. "World Cities as Hegemonic Projects: The Politics of Global Imagineering in Montreal." *Political Geography* 23:571–596.

Stone, C. N. 1989. *Regime Politics: Governing Atlanta, 1946–1988*. Lawrence: University Press of Kansas.

Select Guide to Further Reading

Barras, R. 2009. *Building Cycles: Growth and Instability.* Oxford: Wiley–Blackwell.

Castells, M. 1989. *The City and Grassroots.* London: Edward Arnold.

Gottlieb, M. 1976. *Long Swings in Urban Development.* New York: National Bureau of Economic Research.

Harvey, D. 2012. *Rebel Cities: From the Right to the City to the Urban Revolution.* London: Verso.

Knox, P., ed. 2014. *Atlas of Cities.* Princeton, NJ: Princeton University Press.

Short, J. R., ed. 2017. *A Research Agenda for Cities.* Cheltenham, UK: Elgar.

Short, J. R. 2018. *The Unequal City.* London: Routledge.

Short, J. R. 2018. *Hosting the Olympic Games: The Real Costs for Cities.* London: Routledge.

Short, J. R. 2023. *The Urban Now: Living in an Age of Urban Globalism.* Cheltenham, UK: Elgar.

Stone, C. N. 2008. "Urban Regimes and the Capacity to Govern: A Political Economy Approach." *Journal of Urban Affairs* 15:1–28.

ON GENTRIFICATION:

Duman, A. 2012. "Dispatches from 'The Frontline of Gentrification.'" *City* 16:672–685.

Lees, L., T. Slater, and E. Wyly. 2010. *The Gentrification Reader.* New York: Routledge.

Short, J. R. 2006. *Alabaster Cities.* Syracuse, NY: Syracuse University Press.

Wyly, E., K. Newman, A. Schafran, and E. Lee. 2010. "Displacing New York." *Environment and Planning A* 42: 2602–2623.

ON SUBURBAN DECLINE:

Hanlon, B. 2009. *Once the American Dream: Inner Ring Suburbs of the Metropolitan United States.* Philadelphia: Temple University Press.

Short, J. R. 2018. *The Unequal City.* London: Routledge.

Short, J. R., B. Hanlon, and T. Vicino. 2007. "The Decline of Inner Suburbs: The New Suburban Gothic in the United States." *Geography Compass* 1:641–656.

ON CITY BRANDING:

Dinnie, K., ed. 2011. *City Branding: Theory and Cases.* New York: Palgrave.

Short, J. R., L. M. Benton, W. B. Luce, and J. Walton. 1993. "Reconstructing the Image of an Industrial City." *Annals of Association of American Geographers* 83:207–224.

Glossary

Abrahamic religions Common designation for the religions of Judaism, Christianity, and Islam, which share Abraham as a central figure.

acid rain Rain that is contaminated by nitric and sulfuric acid. Caused by volcanic eruption and burning fossil fuels.

adopted ethnicity An expressed and chosen ethnic identity.

advanced producer services Includes business consultancies and accounting, advertising, financial, and legal services used by large firms and corporations.

affect In contemporary geographical discussions, refers to forces that make us feel, think, or act.

age of distraction The current period of time, marked by people becoming increasingly distracted by communication via mobile devices.

age of imperialism The period from 1880 to 1918 in which there was a renewed scramble for territory as old and new European powers struggled to incorporate more of the periphery. Africa was important in this race; hence, the name of one of the main processes, the scramble for Africa. Imperialism was evident both before and after this designated period. It was just more intense in this period as more countries were involved.

agglomerations Clustering of industries, companies and workers that can reduce costs, create spillover effects, improve efficiency and enhance innovation.

agrarian transition Occurs as agriculture shifts from subsistence to commercial and moves from meeting local market demands to provisioning national and global food supply chains.

agricultural productivity The amount of food that can be grown given a set of limited resources, such as arable land and labor.

anarcho-communist Theory of societal organization that favors eradicating government in favor of communal ownership of society's resources.

animism (animist religions) The belief that the Earth is a living entity filled with sacred space and spirits. It is prominent among hunter-gatherers.

Anthropocene The current geological period in which human activity is deemed responsible for the profound and fundamental restructuring of the environment.

Arab Spring The independent democratic uprisings across the Arab world in 2011.

arable agriculture The practice of growing crops in soils.

Asian Tigers The first batch of newly industrializing countries in East Asia, after Japan. Generally taken to include Hong Kong, Singapore, South Korea, and Taiwan. Also describes those countries that developed later, including Malaysia and Thailand.

assimilationist model Robert Park's description of how new immigrants are socialized to a new society and in turn provide the dynamic raw material for the creation of a new urban society.

asteroids Rocky objects that orbit the Sun and occasionally hit the Earth. An asteroid strike is probably responsible for a mass extinction during the Mesozoic era, as its impact triggered volcanoes and earthquakes.

atlas A collection of printed maps.

authoritarian regime A system of government in which power is concentrated among a few, but not so deeply entrenched as in a totalitarian regime.

axial age A period that lasted from 900 to 200 BCE, during which the great religious traditions of the world came into being across the globe—Confucianism and Daoism in China, Hinduism and Buddhism in India, monotheism in Israel, and philosophical rationalism in Greece.

Baumol's disease Named after economist William Baumol, who first noted that although innovation and efficiency occur rapidly in the manufacturing sector, leading to price reduction, this trend is less pronounced in the service sector.

Bennett's law The phenomenon of how consumption of starch staples declines as household income increases.

Bible Belt A region encompassing all or parts of the southeastern and south-central United States marked by greater churchgoing and formal religious belief than the rest of the country. Protestant and Evangelical religions are particularly strong in the region.

Big Bang theory Describes the origin of the universe as emanating from an explosion of energy that sent matter expanding outward.

bilateral oligopolies A food supply chain in which large-scale producers combine with large-scale retailers to supply food goods.

biodiversity An indicator that describes and measures the number of distinct plant and animal species in a given geographical area.

biofuels Fuels that are made from living matter, such as ethanol and biodiesel.

Black Death A recurrence of the bubonic plague in the fourteenth century that killed up to 60 percent of the population of Europe.

black holes Cities with a population of over 3 million that are not identified by Globalization and World Cities as a world city and do not share a national territory with a world city.

body mass index (BMI) A measurement defined by a person's weight in kilograms divided by the square of their height in meters. Useful in comparing levels of obesity, defined as having a BMI over 30, across the world and over time.

BRICS An acronym for the countries Brazil, Russia, India, China, and South Africa, which are grouped together because they are all large countries with fast-growing economies.

bubonic plague A pandemic that spread the disease in the sixth, seventh, and eighth centuries across modern-day Europe and the Middle East, killing more than twenty-five million people.

budding Migration that involves small subgroups breaking away to settle new places.

building cycle Cyclical explanation of how cities are developed and grow, as they go through troughs and peaks of investment.

Burgess model Depicts a concentric ring of the residential mosaic, with the poor at the center and the rich on the periphery. Named after Chicago sociologist E. W. Burgess.

buyer-driven chain A food supply chain comprising a small number of retailers who consolidate their supply around a few suppliers.

bystander effect The phenomenon of individuals not responding to help those in distress.

capital In standard economics and finance, describes real assets such as factories, property, and machinery, or financial assets such as bonds, securities, and stocks. However, human geography seeks to examine the social relationship behind capital.

carbon dioxide A gas produced in the burning of carbon. One of the main greenhouse gases.

carbon economy Since the nineteenth century, economic growth and much of our transportation have been fueled by the growing use of carbon-based energy resources, especially coal and oil.

carbon footprint The amount of carbon dioxide produced by human activities.

carrying capacity A theory that the world can support only a specific level of human population as well as sustain a limited amount of its growth.

caste system The social hierarchy in Hindu culture that divides the population into different social categories known as castes.

casual racism Often found in early geographical works, reflecting the belief of the time that certain races were viewed as inferior to those more dominant in the world.

central place theory A theory attempting to describe the organization of regional networks in which market towns are distributed in a hierarchy, with higher-order goods and services located in the larger towns.

centrifugal forces The influences that disrupt a state's power across space, including political and economic inequality.

centripetal forces The influences that unify a state's power across space, including through external aggression, which may stimulate national bonding against a common enemy; federal structures, which allow the safe expression of regional and other subnational differences; and national mass media and education, which create a shared culture and language.

chain migration Migration in which people tend to move along channels and to destinations, followed by friends and families.

Chicago school A body of work in the field of urban sociology created during the first half of the twentieth century and centered on the University of Chicago.

city branding The marketing of cities with distinct messages such as postindustrial cities, global cities, business-friendly cities, or other forms of branding in an attempt to shift a city's previous image.

climate change Significant long-term change in regional and/or global climates.

code-switching How people can change the language they speak according to where they are, who they are with, and the topic of conversation.

Columbian Encounter The arrival of Europeans to the Americas, resulting in vast demographic changes to Indigenous populations, environmental transformation, and an exchange of goods between the Old and New Worlds.

commodification Describes how material becomes a resource by being valued and traded, assigned a price equivalent, and entered into the arena of things bought and sold.

commodified agriculture The use of agriculture as a means of profitability rather than just a source of local sustenance.

communal economy Sector of the economy that involves the cashless exchange of goods and services.

comparative advantage The ability of a region or country to produce goods more efficiently than others.

conic projection map A map created by charting the Earth on a cone, then flattening it with lines of latitude as concentric circles. Ptolemy was the first to implement this map-making technique.

contact language A language that develops as a form of communication between speakers of two or more other languages.

core of believers Most faiths have a hard core of true believers and a wider periphery of occasional practitioners and more casual believers.

core–periphery model A dynamic economic model that classifies countries as either core, periphery, or semiperiphery by the economic interactions between them.

corporate industrialized food production system A system of industrial food production dominated by large corporations.

cosmography Study of the Earth through the understanding of the interconnected cosmos, incorporating other disciplines such as astronomy, astrology, and even magic.

creole A pidgin that has native speakers.

crime maps Maps of local areas used to depict data on crime statistics, including the spacing and timing of criminal activity.

crisis mapping The mapping of humanitarian crises such as environmental disasters and wars.

cultural geography Looks at the spatial expression of culture. It is, for example, concerned with landscape as a socially contested, materially produced cultural artifact.

cultural globalization The flow of ideas and cultural practices, such as speaking and writing English, around the world.

cultural industry Also known as the cultural products sector or cultural creative sector. It is made up of the artistic industries of music, dance, theater, literature, the visual arts, crafts, and many newer forms of practice such as video art, performance art, and computer and multimedia art.

cultural region A region sharing a similar cultural feature such as language, ideas, religion, or cultural practices. Often divided into a core and a periphery.

culture   The collective materials, beliefs, ideas, and practices represented by a particular group of people, or in a particular location, but can also be globally representative.

cumulative causation   Describes how growth in a region can spread out into other lagging regions through the diffusion of innovations and by providing markets, which leads growth to feed on itself.

dark energy   The mysterious energy force that accelerates the expansion of the universe, overcoming the effects of gravity.

decolonialization   The process by which former colonial territories achieve independence.

deconstructivism   A postmodernist style of architecture that involves the use of nonrectilinear surfaces.

deep ecology   A belief that humans should alter their lifestyles to be more biocentric and should question the morality of economic growth.

deep state   State organizations and interests that persist despite changes in governments.

deepwater wells   Offshore oil wells that extract oil from 5,000 feet or more below the surface.

deindustrialization   A marked decline in the size and importance of industry in an economy. Often caused by manufacturing job losses.

democracy   A system of government in which political power arises from the majority will of the people.

democracy index   An index produced by the Economist Intelligence Unit to gauge the level of democracy. It uses sixty indicators related to electoral process, civil liberties, functioning of government, political participation, and political culture.

demographic cumulative causation   A process whereby population movement from a declining to a growing region exacerbates these regional differences.

demographic dividend   A relative and absolute increase of younger, more productive workers. Occurs when birth rates fall to a point that requires less investment in the very young, but before more investment in the elderly is required.

demographic transition   A global change in mortality and fertility that occurred around 1800 and began the shift away from higher to lower birth rates and from higher to lower death rates.

dependency ratio   A measure of the number of individuals below the age of fourteen and over the age of sixty-five when compared to the total working-age population.

dependent population   The ratio of retired persons to working population, calculated by dividing the number of people over the age of sixty-five by those aged fifteen to sixty-five and multiplying the result by one hundred.

deterritorialized   When referring to culture, describes how cultures can be removed from pertaining only to specific locations.

development state   Describes countries that have successfully crafted their economic policies to move from the periphery of the global economy closer to the core.

dialect maps   Maps that show how the same language varies by dialect.

diaspora (diasporic communities)   Any group of people living outside their homeland.

digiphrenia   Created by social media, the phenomenon of being in multiple places and more than oneself all at once.

diglossia   Occurs when two varieties of the same language are used in different contexts.

disease mapping   The mapping of the incidence, causes, and consequences of disease and ill health.

distance-decay effect   The negative relationship between distance and similarity. Similarity between things tends to decline as distance increases. For example, linguistic similarity declines between peoples as distance between them increases, and early imperial expansion was marked by the decay of influence of effective military outreach and political control.

domestic economy   Sector of the economy that constitutes the amount, type, and division of labor within the home.

double exposure   Harmful impacts on a geographical area caused by both climate change and economic restructuring.

Dutch disease   When countries rely on nature's bounty to the point where it hampers and restricts the innovation necessary for long-term economic growth.

earthquake   A violent shaking of the ground caused by shifts in the Earth's crust or volcanic eruptions.

Easterlin hypothesis   Named after the economist Richard Easterlin, who suggested that the fortunes of a cohort depend on its size relative to the total population.

Easterlin paradox   A theory suggesting that once basic needs are met, there is little difference in reported happiness with increasing incomes.

easy oil   The rapid and cheap extraction of oil.

ecofeminism   Highlights the gendered nature of many environmental ideologies and practices and tries to engage feminist and ecological issues in a shared dialogue.

ecological footprint   The ecological footprint measures the amount of land and water area a human population requires to produce the resources it consumes and absorb its wastes.

ecology   A study of the relationships of organisms to each other and their environments.

economic development   The improvement in an economy leading to more and better jobs and higher standards of living.

economic union   An agreement between countries that attempts to extend the size of their economic market.

ecosystem   A group of interconnected organisms that function as a unit.

ecstatic worship   A type of religious activity that involves festive events or ceremonies filled with music and dance.

enclosure movement   The period from 1760 to 1820 when much of England's land was privatized, replacing a system of traditional rights and obligations with agrarian capitalism.

environmental change   Change in the type and quality of the environment. May result from natural processes such as volcanic eruptions or human behaviors such as the introduction of farming, burning fossil fuels, or greater car use.

environmental determinism   The idea that certain climatic conditions are especially favorable to human progress, first argued by Yale geography professor Ellsworth Huntington in the early twentieth century. Often used, incorrectly, as evidence for why certain races and cultures were less advanced than others.

environmental Kuznets curve A graphical representation depicting the relationship between economic growth and environmental quality.

environmental legislation Legislation enacted to protect the environment and safeguard environmental standards.

environmental movement A social movement concerned with protecting and conserving the environmental quality and diversity.

environmental racism The practices and processes whereby minorities live and work in the poorest environmental conditions, especially low-income minority groups.

environmental stressors Changes that put pressure on the livability and sustainability of the environment.

epistemological break Epistemology is the theory of knowledge; an epistemological break implies a fundamental reordering of what constitutes knowledge and how knowledge is understood and used.

eutrophication The enrichment of ecosystems with chemical nutrients, which creates algal blooms that result in dead zones.

evolution The process by which living organisms are thought to have developed and diversified from earlier forms of living organisms.

faith communities A group of people who share the same religious beliefs and principles.

famine Widespread extreme hunger caused by food shortage.

female participation rate Measures the percentage of women, usually those aged fifteen to sixty-four, employed in the labor force.

feminist critique An argument that drew attention to the bias in academia and the dominance of men in geography.

fertility The birth rate of a population.

first-past-the-post An electoral system in which the politician with the majority of votes in a district wins that district.

first urban revolution A period of time between 11,000 and 5,000 years ago that was marked by the development of the world's first cities.

first wave of globalization A distinct period of globalization that followed the Columbian Encounter of 1492 and inaugurated a truly global economy.

fiscal crisis Occurs when a state has more expenditures than revenue and lacks the means to reverse the trend.

fish farming The harvesting of fish in enclosures. The main species in this form of aquaculture include salmon, tuna, cod, and halibut.

fixed market system A system in which permanent marketplaces are established to replace period markets.

flat world A condition of relatively reduced economic distance and less friction between different national economies.

flight capital The movement of money to a location that is considered safer and less transparent.

food desert Parts of a city or region with very limited access to affordable, nutritious food.

food supply chain A system of connected suppliers that distribute food throughout the world.

forced migrations Movements of people forced to relocate, often caused by war, political conflict, or social unrest.

Fordism A system of production, generally controlled by a small number of very large companies operating under oligopolistic conditions, that employs mechanized and repetitive manufacturing processes.

formal economy Sector of the economy that involves the sale of labor in the marketplace and is recorded in government and official statistics.

fracking Short for hydraulic fracturing, a process for removing natural gas in which high-pressure liquid is injected into the ground to crack rocks and release gases and oil.

gendering of space and place The processes by which gender differences are created and reinforced in spatial arrangements and in the character of places.

genetic variation The diversity of the genetic materials found in DNA.

genetically modified organism (GMO) When the genes of plants raised for crops are engineered to make them more resistant to disease and predators to produce higher yields.

gentrification The replacement of lower-income households by higher-income households.

geographic profiling Using crime maps to determine the probable location of criminals and future criminal activity, based on the assumptions that criminal activity is place-specific and that criminals like to use known areas. The technique has been expanded to analyze invasive species and disease outbreaks and even to identify well-known but anonymous artists.

Geographic Information Systems (GIS data) A computer storage and retrieval system for displaying and analyzing geographically coded data.

geographical imagination A way of thinking about the world that focuses on space, place, location, and their interaction.

gerrymandering The manipulation of voting district boundaries to engineer specific political outcomes.

Gini index A measure of income inequality named after the Italian statistician Corrado Gini, who proposed the measure in a 1912 paper. It is generally expressed from 1 to 100, with 1 representing maximum equality and 100 maximum inequality.

global climate change Long-term, worldwide climate change.

global epistemic community The transnational network of knowledge-based workers.

global languages Languages used across the world.

global shift The movement of employment from the industrial regions and cities of the developed world to the cities and regions of the developing world.

global warming An increase in the Earth's surface or atmospheric temperatures, often associated with an increase in atmospheric greenhouse gases.

Globalization and World Cities (GAWC) A research network that compiles data on the distribution of global advanced producer service firms across 315 cities to identify a global urban hierarchy and network.

globalization The growing interdependence of national economies and societies through the international exchange of goods services and capital, cultural flows and forms of political integration.

glocalization A business strategy to create worldwide operations attuned to local markets and conditions.

Great Migration The movement of US African Americans between 1916 and 1970 from the rural South to the urban industrial North.

Green Revolution A series of innovations and transfers, especially marked from the 1960s to the 1970s, that increased agricultural production through new high-yielding varieties of crops (especially maize, rice, and wheat) and liberal helpings of pesticides and fertilizers.

greenhouse gas A gas emitted into the atmosphere that produces a greenhouse effect by trapping heat in the atmosphere.

gross national happiness A measure of national collective happiness.

ground truthing The identification of finer-grained processes to see the effects and impact on households and their vulnerabilities and coping strategies.

hard border A country's nonporous boundary.

hard power The ability to coerce others through the superiority of military forces.

heartland theory Sir Halford Mackinder's theory that whoever ruled the central part of the Eurasian landmass ruled the world.

hegemony Dominance or leadership of one or more state by another, either directly or indirectly, through cultural, economic, or political control.

highway to India Sea route that reduced travel times between western Europe and East Asia by linking the Mediterranean and Red Seas through the Suez Canal.

homeownership When households purchase residential property to own.

hominids Primates, including humans, which first appeared on Earth around 50 million years ago.

homogenization thesis A theory about consumer activity that assumes the same things are consumed in the same ways in very different parts of the world.

Hotelling's law Refers to the fact that is it often rational for producers to make goods similar to those of their competitors. When applied to location, it leads firms to cluster together.

Hubbert curve Named after a geophysicist who employed a theory to predict the decline of US oil production, suggesting that nonrenewable resource use rises quickly from zero to peak production levels and then tapers off, forming a bell-shaped curve.

human geography A branch of geography dealing with the interaction between human activity and the spaces around them.

humanist critique A critique that developed in the 1960s, arguing that there was a lack of concern with personal geographies that ignored feeling, perception, and intersubjectivity.

hunting-gathering societies Groups of people more dependent on hunting fauna and gathering seeds and berries than on settled agriculture.

hukou A Chinese government registration system that defines rural and urban residents.

hybrid identities A state of multiple identities. Often occurs in immigrant communities, as when Turkish immigrants to Germany have children, and those children are German citizens but are brought up in a Turkish home—their hybrid identity is German Turkish.

hyperprimate An urban system marked by extreme primacy (i.e., the dominance of one city).

illegal economy Sector of the informal economy that includes the shadowy world of the semilegal and illicit.

immigration regimes The set of rules that governments implement to govern the entry, settlement, and assimilation of foreign migrants.

imperial incorporation The annexation of areas and peoples into an empire.

imperial overstretch The tendency for empires to expand beyond their ability to maintain economic dominance and military power.

imperialism A doctrine or policy that, to improve a home country's economy, argued for protection of home industries and the possession of overseas colonies to ensure cheap raw materials and a secure market to which competitors' access would be denied.

import substitution An economic development strategy for bolstering local growth marked by protectionist policies.

imposed ethnicity When a group of people are named and treated as ethnically different.

internally displaced persons (IDP) People forced to leave their homes who remain inside the country's border.

index of dissimilarity Measures the unevenness in the distribution of two groups across spatial units. Ranges from 0 to 100; values closer to 100 indicate greater dissimilarity.

index of segregation Measures the distribution of one group compared to the total population. Ranges from 0 to 100; values closer to 100 indicate a more segregated distribution.

Indic region One of the main hearths of the axial age, where Jainism, Buddhism, and Sikhism were founded. Centered in northern India's Indus valley.

industrial city Cities developed from industrial capitalism and factory growth.

Industrial Revolution The development of new mechanized manufacturing processes, occurring first in Britain between 1760 and 1850 and involving the creation of industrial and urban societies.

industrialized agricultural system Agriculture that relies heavily on machinery, technology, and chemicals.

infant mortality rate A measurement of the number of deaths of children under one year of age in a specific geographic location.

informal economy Sector of the economy that is not recorded in government and official statistics, where few, if any, taxes are paid.

institutional contract When an institution or a government entity builds dwellings and then rents them out or sells them.

instrumental regimes Focus on specific targets identified in the political partnership between urban governments and business interests.

intergenerational inequities The differences in wealth, power, status, etc., that can exist between generations of a society.

intervening opportunities The opportunities between an origin and possible destination that can disrupt the flow of movement between them.

isothermal map A map composed of lines connecting points of similar temperature.

just-in-time production A system of producing goods in which they are made to meet demand.

Keeling curve A graphical representation of the changes in the concentration of atmospheric gases since 1958.

kleptocracies States geared toward the enrichment of a tiny political elite.

Kondratieff cycles A series of fifty-year cycles identified by Nikolai Kondratieff, each of which is associated with key innovations that structure society and space.

Kuznets curve A diagram that depicts rising inequality, with early economic growth turning into increased equality as growth continues. A modified Kuznets curve depicts an uptick in inequality. Named after economist Simon Kuznets.

labor Describes organizations of workers in their dealings with government and employers.

land bridge Dry land that allowed humans to pass between continents that rising sea levels have since separated.

land deficit Difference between the amount of land needed to feed the world's population and the amount of land available.

land grab When land is taken by force or by illegal or semi-legal methods.

language domain A context for language use defined by place, role, relationships, and topic.

language shift Change in the degree of use from one language to another.

latitude Lines of latitude that run parallel to the equator, depicting positions north or south of it. Expressed in degrees, minutes, and seconds.

legitimation crisis When the state loses its popular appeal and its ability to govern.

Lewis turning point When the supply of cheap labor runs out within a specific geographic location. Named after economist William Arthur Lewis.

life expectancy The length of time a person is likely to live.

light year Equivalent to 5.8 trillion miles, the distance light travels in an Earth year, calculated from the speed of light in a vacuum (186,000 miles/second).

limits to growth An idea that gained prominence following the publication of an influential book in the 1970s that suggested the Earth and its resources could support only a finite amount of continued growth.

linguistic landscape Refers to the display of language in the physical world around us. Examples include street signs, advertising slogans, and words on billboards.

Little Ice Age From 1640 to 1690, world temperatures cooled, leading to declining agricultural yields and increased hunger.

liturgical language A language used in religious worship and ceremonies or in their sacred texts.

location quotient Measures the degree of concentration of one group versus the total population.

longitude Also known as meridians, lines of longitude depict positions east and west on the Earth from a defined point. Expressed in degrees, minutes, and seconds.

lunar cycle The 27.3 days it takes for the Moon to revolve around the Earth. The basis for our division of time into months and the basis of the lunar calendar. Responsible for the movement of tides.

malapportionment The imbalance between the numbers of voters in similarly represented constituencies. An example: The five largest states in the United States hold more than a third of the entire US population but have only 10 percent of seats in the US Senate. The five least populous states have only 1 percent of the nation's population but hold 10 percent of Senate seats.

male gaze The way things are seen and understood from an overtly masculine perspective.

Malthusian checks Thomas Malthus suggested that when population increases beyond needed food supplies, checks such as misery, poverty, and famine bring the population back into alignment with available resources.

Marxist political economy An approach that draws on the work of Karl Marx to look at issues in capitalist economies such as the endemic crises of capitalism and the conflict between capital and labor.

Mason–Dixon Line A line drawn across America in the 1760s by two surveyors, Charles Mason and Jeremiah Dixon, to settle a dispute between Maryland, Pennsylvania, and Delaware. It took on a wider significance as a division between North-leaning and South-leaning states during the Civil War and as a marker of different racial attitudes and practices.

mass extinction Unusually high levels of species extinctions. For example, rising sea levels caused the extinction of many species of shallow-water fauna during the Cambrian period.

megacity Large urban agglomerations with more than 10 million inhabitants.

megafauna Large animals weighing more than 220 pounds, which were slower than other animals and therefore especially vulnerable to human hunters immediately at the end of the last Ice Age, leading to the eventual extinction of many.

Megalopolis Region spanning 600 miles from north of Richmond, Virginia, to just north of Portland, Maine, and from the shores of the northern Atlantic to the Appalachians, which includes the consolidated metropolitan areas of Washington–Baltimore, Philadelphia, New York, and Boston.

megapolitan regions The extended metro areas of very large cities.

melting pot The metaphor that portrayed different immigrant groups blending into a more homogenous society.

Mercator projection A map projection of the Earth depicting the surface area of degrees of latitude as equal in size.

merchant city Cities that developed based on a money economy in association with the extension of trade and the emergence of the merchant class.

metropolitan statistical area (MSA) Urban area with a core area of at least 50,000 residents and economic links to surrounding counties.

metropolitanization Dispersal of people and activities away from the tight urban cores of preindustrial and industrial cities.

Metz model Identifies four phases of mobility: human travel, settlement, mass mobility, and an era in which travel time, trip rate, and distance traveled hold steady.

micropolitan statistical area Urban area with a core area of at least 10,000 and fewer than 50,000 residents.

modernist architecture Architectural style dominant from the 1950s to the 1980s. Concerned with form following function, truth to material, and a revulsion against bourgeois decoration. Led to designs that incorporated clean-cut lines and soon became serialized into tall, flat-topped buildings worldwide.

modernity Of or relating to anything modern.

monotheism The belief that there is only one god.

moral statistics The tabulation of crime, pauperism, and a wide range of other social phenomena. First appeared in an

1833 essay by André-Michel Guerry; such statistics were an important part of nineteenth-century thematic mapping.

mortality rates   The measurement of the number of deaths.

mountaintop removal   A process for extracting coal by removing mountain peaks to gain access to their coal seams and then dumping into neighboring valleys.

Murdie model   Depicts a concentric ring pattern for stages in the life cycle, a sectoral pattern for socioeconomic states, and a clustered distribution of ethnicity. Named after geographer R. A. Murdie.

nation   Community of people with a common identity, shared cultural values, and a commitment and attachment to a particular area.

national imaginaries   A distinct form of geographical imaginary that refers to the spatial nature and representation of the nation-state.

negative externalities   The cost that a party not participating in an economic transaction must bear as a result of the transaction. For example, the cost borne by residents of a particular neighborhood resulting from air pollution emitted from a nearby factory.

neighborhood effect   A local effect found in voting outcomes in which neighborhood allegiances can trump sectional ones.

neoclassical architecture   A type of architecture that echoes the classical architecture of Rome and Greece.

neoliberalism   An economic ideology that promotes deregulation, minimal or small government, low taxation, and free trade.

nongovernment agencies   Organizations that are not sanctioned or affiliated with any particular government or state, but may be funded by particular countries. Examples include the International Monetary Fund, World Bank, and World Trade Organization.

nonporous boundaries   Partitions that do not allow for the easy movement of people and goods. The border between the former East and West Germany is one example.

obesity   Describes the state of individuals with a body mass index greater than 30.

obesogenic   Conditions that tend to increase obesity.

occidentalism   The writing and description of the world from the viewpoint of those from or who live in the West.

oil crunch   The large and growing gap between the supply of oil and its demand.

oil shock   An unexpected shift in the supply or demand of oil, which leads to a significant price change.

organic regimes   Occur in small towns and suburban districts with a homogeneous population and a strong sense of place with a chief aim of maintaining the status quo.

Organization of the Petroleum Exporting Countries (OPEC)   An oil cartel dominated by Middle East oil producers.

orientalism   Writing and describing the world from the viewpoint of those from or who live in the East.

ornamentalism   The trumping of class and status over ethnicity in worldviews. A wrinkle in the occidentalist thesis.

Out of Africa hypothesis   A theory that all humankind are descendants of early humans originally found in Africa.

The idea is substantiated by the finding that genetic variation decreases in humans found farther away from Africa.

overfishing   Exploitation of the supply of fish in a certain geographic area that depletes its fish stocks.

overkill   The extinction of megafauna, such as the mammoth and the mastodon, that occurred on every continent and was experienced by every community of large terrestrial vertebrates as a result of humans' increased technological sophistication in hunting.

overpopulation   A state in which there are more individuals than resources available to sustain them at healthy levels.

overweight   Describes the state of individuals with a body mass index greater than 25.

owner occupation   A type of housing supply that results when households purchase their residential property.

pandemic   An outbreak of an infectious disease more widespread than an epidemic.

paracme   Technically, the point beyond a peak; in diffusion studies, the abandonment of an innovation.

participation rate   The percent of those in a given demographic, such as women, who are working in formal employment.

pastoral farming   Raising livestock to produce meat, wool, and dairy products.

patriarchal society   A society in which males are dominant over females.

patriarchy   The belief system and practice that men are society's authority figures and women should be restricted to home, hearth, and the bearing and rearing of children.

peak oil   A hypothetical point in time at which global oil production reaches its maximum, or peak, declining thereafter.

Peasants' Revolt   A social upheaval in England in 1381 against a proposed reduction of wages.

periodic market system   System of nonpermanent marketplaces established in villages and towns by mobile vendors.

permafrost   A subsurface layer of soil that usually remains frozen year-round. Common in the higher latitudes of the Northern Hemisphere.

pidgin   A language created by contact between two or more other languages.

plate tectonics   A theory describing the origin and movement of the Earth's crust and its numerous plates, which float and slowly move atop a molten rock layer (the mantle).

Pleistocene overkill   The extinction of many large mammals around 11,000 years ago. The main reasons are thought to be human overhunting and climate change.

political ecology   The study of the struggle over resources, who controls them, and how the costs and benefits are apportioned.

polyculturalism   How various groups interact and influence each other, with both immigrant groups and host societies metamorphosing. Distinct from multiculturalism.

polytheism   The belief that there are many gods and deities.

porous boundaries   Partitions that allow for the relatively easy movement of people and goods. The US–Canada border is a good example.

postmodernism   The rejection of universal, stable, and objective ways of knowing.

potato blight   The Irish famine that began in 1845, leading to millions of deaths and substantial migration out of Ireland.

preindustrial city The form of cities before the onset of the Industrial Revolution. Sometimes also used with reference to cities before 1800 and especially before 1700.

primacy ratio The ratio of the population of the largest city in a country divided by the combined population of the next two largest cities.

primary sector The sector of the economy that includes agriculture, forestry, mining, and fishing.

primate distribution A national urban network in which one city dominates.

prime mortgages Residential mortgages that require a 10 to 20 percent down payment and clear evidence of an ability to pay according to the loan terms.

pristine myth The mistaken belief of a vast wilderness before the first European explorers and conquerors arrived. Instead, it is theorized that a new wilderness developed as Indigenous populations were depleted.

private renting A type of housing supply whereby landlords rent out property to tenants.

producer services The knowledge-based segment of the service industry made up of advertising, banking services, financial services, business consultancies, and information technology. An example is when a household employs an attorney or agency to complete their tax returns.

producer-driven chain A food supply chain in which producers tend to establish cartels.

product cycle The trajectory of a brand-new product.

profit cycle The process of how sales of new products impact how profits grow, mature, and decline over the life of the product.

proportional representation An electoral system in which representatives are elected in proportion to the number of votes cast.

protectionism Any form of economic policy that seeks to support local companies at the expense of foreign corporations, usually through quotas, tariffs, or some combination thereof.

Ptolemy One of the earliest geographers, he mapped both the universe and the terrestrial world, using a conic projection technique to represent the habitable world.

public bads Necessary but unwanted civic facilities and institutions such as trash disposal sites, motorways, and halfway houses for convicted criminals.

public goods Collectively provided goods and services.

public housing A tenure type of housing in which a government agency provides accommodation for citizens.

push and pull factors Determinants of migration. Pull factors are the positive aspects of a place, such as better opportunities. Push factors that make people move away from places include war, social conflict, and lack of opportunities. Pull factors are reasons to go to a place, push factors are reasons to leave.

quantitative geography An approach that emphasizes measurement and calibration.

quaternary sector The knowledge-based part of the economy that includes business advisory services, consultancies, information technology, and research and development.

quinary sector Those in the information- and knowledge-based economies who make the final executive decisions.

racial/ethnic clustering Occurs when different groups gather as a result of similar race or ethnicity.

racialized groups The social construction of groups of people as racial categories.

range When referring to a good or service, the distance consumers are willing to travel to make a purchase.

rank-size distribution The regular relationship between a city's population and its rank in the urban hierarchy. Also known as Zipf's law.

rationality crisis When the state makes enough poor decisions that other crises emerge.

redlining A process by which US institutions restricted their mortgage allocations in lower-income minority neighborhoods.

refugee A person who, because of fear of being persecuted for reasons of race, religion, nationality, membership of a particular social group, or political opinion, is forced to live outside the country of his or her nationality.

relative space The social dimension of space, such as the different income of different regions, often contrasted with the space of latitude and longitude.

religio-ecological The combination of environmental and religious beliefs that sees environmental protection as God's work. The human occupancy of the Earth has spiritual responsibilities and ethical obligations.

remittances The money that temporary and permanent migrants send back to their home countries.

remote sensing Gaining information about the Earth using aerial sensor technologies such as satellite recordings and monitoring.

renewable energy Energy from effectively limitless sources such as wind, tide, and the Sun.

replacement rate The number of births required to keep a given population stable.

resource curse Describes how valuable resources can create short-term booms but may depress long-term economic growth. Also known as the "paradox of plenty" or the Dutch disease.

reterritorialized In reference to culture, describes how cultures are being reformed in new places away from their traditional hearths.

rewilding A movement that aims to create landscapes and ecosystems similar to those that existed in Paleolithic times.

Ring of Fire The horseshoe-shaped distribution of intense seismic activity around the Pacific Ocean. This region of active plate-tectonic activity is responsible for 90 percent of the world's earthquakes and 75 percent of volcanic eruptions.

Romanticism Grew as a response to the commodification of land and views nature as a source of profound aesthetic experiences and a vehicle for a personal connection to the infinite.

rural-to-urban migration The movement from rural areas to cities.

Rust Belt A region in the upper Midwest and Northeast of the United States that used to be dominated by manufacturing and heavy industry. The term denotes the deindustrialization and shrinking of the industrial labor force in the region since the mid-1970s.

salinization An increase in the amount of salt in soils that reduces their fertility and, at high levels, makes them toxic. Affects 10 to 20 percent of world soils.

seasonal affective disorder The tendency for more negative moods among normally healthy people during distinct seasons.

second urban revolution A period that began in the late eighteenth century with the creation of the industrial city and unparalleled rates of urban growth.

second wave of globalization A distinct period of globalization that occurred from 1865 to 1970, facilitated by lower tariffs, an international labor market, relatively free capital mobility, expanded transportation, and the effects of two world wars.

secondary sector The sector of the economy that mainly consists of the manufacturing of goods.

self-built housing Housing that is built by residents themselves.

Semitic region One of the main hearths of the axial age. Centered in the Middle East, it is the location where Judaism, Christianity, and Islam were founded.

service sector A segment of the economy that is made up of workers who provide services as opposed to producing goods.

settled agriculture Describes when a group of people remain in a specific location and survive through means of growing food, rather than hunting and gathering.

shadow economy A reference to the informal economy in which individuals pay no taxes.

sharing economy A private market hybrid that relies on collaborative consumption. Owners rent assets such as cars and rooms to users, the transaction often facilitated by social media. Examples include Uber and Airbnb.

shipping container The creation and worldwide usage of a standardized container. Its usage resulted in dramatically reduced shipping times and transportation costs.

slave trade The forced migration of millions of Africans to the New World from the sixteenth to the nineteenth century.

slum Unplanned, informal, and often illegal housing in cities that arises because of the inability of formal markets and public authorities to provide enough affordable and accessible housing. Slums are also referred to as "shantytowns," "informal housing," and "squatter housing."

slums of despair Slum areas where people are on a downward spiral.

slums of hope Slum areas where people are improving their living conditions and those of their children.

social conflict An increase in social disorder, including violent conflict that may occur during a significant youth bulge.

social cost Costs imposed on society.

social Darwinism Theory that argues that people are also subject to the survival of the fittest. First proposed by Herbert Spencer in the late nineteenth century.

social economy Sector of the economy that consists of not-for-profit activities. Sometimes referred to as the "third sector."

social geography A branch of human geography concerned with determining the relationship between social phenomena and geography.

social housing Housing provided by a government agency.

social statistics The quantification of social problems. Used by mapmakers to track urban inequality and conduct important research on urban environments.

soft border A porous boundary.

soft power The ability to co-opt others to your point of view without the use of force.

space-time convergence The decrease in time taken to cover a specified distance, reducing the cost of transporting goods and people and seemingly pulling places closer together.

spatial diffusion The movement across space and through time of an innovation, idea, or practice. Swedish geographer Torsten Hägerstrand did early work on the topic.

speculative building Involves builders preparing land and constructing dwellings for general demand rather than specific customers.

speech community People who share a similar language.

starchitecture The construction of buildings by big-name architects. Often used to market or brand a city.

state A separate and distinct spatial unit of political authority.

statecraft Geopolitical strategies used by a nation-state. According to early-twentieth-century writer Sir Halford Mackinder, geography was used to create statecraft.

subaltern geographies Human geography by and about more marginal groups, often ignored in the history of human geography.

subprime mortgages Residential mortgages that require little or no down payment with few verifiable means of income.

suburbanization The movement of people from their permanent residences in the central city to the suburbs.

sulfur dioxide A gaseous oxide of sulfur (SO_2). In prehistory, thought to have poisoned the air following colossal volcanic eruptions, contributing to mass extinctions of animal life in the Mesozoic era. Its main modern source is burning fossil fuels. Affects the human respiratory system.

surveillance Has a wider meaning, but here used to refer to International Monetary Fund monitoring of a member's economic policies and evaluation of their economy.

sustainability Meeting the needs of the present generation without compromising the needs of future generations.

symbolic ethnicity Refers to an allegiance to cultural traditions such as food, religious celebrations, and rites of passage.

symbolic regimes Occur in cities undergoing rapid changes, including large-scale revitalization, major political change, and image campaigns that try to shift the wider public perception of the city.

technopoles Contemporary industrial complexes established around high-technology industry.

tectonic plates The thin plates that comprise the Earth's crust and float and slowly move on a molten layer (the mantle).

territorialization When referring to culture, signifies the connection between culture and specific places.

tertiary sector The sector of the economy that comprises the service industry.

thick market A location that contains a large pool of labor and specialized firms.

third urban revolution The global phenomenon in which we currently reside. Marked by five distinct characteristics: rapid urbanization, increases in the size of individual cities, increased metropolitanization, increased urban sprawl, and the proliferation of slums.

third wave of globalization A distinct period of globalization that emerged after 1970 and is associated with an even more marked space-time-cost compression, global production chains, and a deregulated global financial system, creating a "flatter" world with easier and hence quicker international economic links.

threshold Referring to a good or service, the necessary minimum population to support its continued supply.

tides The ceaseless and regular vertical motion of our oceans and seas created by the lunar cycle and the gravitational pull of the Moon.

time-cost compression As economies mature, the importance of the exchange of information leads to innovations that lower the costs of communications via telephone, Internet, and the like.

Tobler's law A geographical law stating that "everything is related to everything else, but near things are more related than distant things."

toponymic colonialism When colonial powers named features of acquired territory and erased preexisting local names.

total fertility rate (TFR) The total number of children a woman would have during her life if fertility rates remain unchanged from the year in question.

totalitarian regime A system of government in which the state has control over wide and deep swathes of social, political, and economic life.

tough oil When the extraction of oil becomes more expensive and risky, often producing an increased interest in alternative fuel sources.

tourist gaze The set of expectations that people have and the way they look when they visit places as tourists.

traditional market A type of food supply chain that is dominated by smallholders who cater to domestic and local markets.

transhumance The traditional movement of mountain peoples with their animals from summer to winter pastures.

transnational architecture A type of architecture that incorporates styles that transcend specific places, particularly those that cross national boundaries.

transnational corporations (TNCs) Describes different types of large companies that operate in more than one country and can be further classified as multinational, international, global, or integrated networks.

triangular trade A trading circuit of sailing ships that linked western Europe, West Africa, and the New World. Manufactured goods were shipped to Africa, slaves were transported to the Americas, and commodities were shipped back to Europe.

tropical zone Geographical location that covers the equator and extends to the lines of latitude known as the Tropics of Cancer and Capricorn.

tsunami A large sea wave resulting from a volcanic explosion or other movement of the Earth's plates.

unequal exchange The trade between economies producing primary commodities and manufacturing economies, which tends to provide more wealth to the latter.

urban regimes Informal arrangements that surround and complement the formal workings of governmental authority, which consist of informal governing coalitions that make decisions and get things done in a city.

urban sprawl The dispersed form of urban development toward suburban areas and farther from central cities.

urbanization The geographical redistribution of population from rural to urban areas; also marked by social reorganization.

Uruk expansion A broadening of the influence of Uruk, one of the major cities in early Mesopotamia.

value-added chain The position of goods and services on a chain of increasing price and value. For instance, making shoes has a lower value than making computer chips, so when a region shifts from shoe to chip manufacturing, it is considered a move up the value-added chain.

vernacular architecture A style of architecture that uses local resources and traditions in the design and construction of buildings.

volcanic hotspot A very hot area of the earth's mantle where molten magma can break through the surface to form a volcano. Often associated with island-building. Examples can be found in Galapagos, Hawaii, and Iceland.

Von Thünen model Describes land use around cities; suggests a concentric ring pattern with more expensive land and thus more intensive agriculture closer to the city. Proposed by Johann Heinrich von Thünen in the early nineteenth century.

wheat belt Countries located in climates amenable to the production of wheat, including Canada, Russia, Argentina, and Australia.

workfare The welfare-to-work initiatives that have developed since the 1990s.

world population The number of people inhabiting the world, currently at 8 billion.

yield gap The difference between the possible food supply productivity and actual productivity.

youth bulge A period of time marked by a rapid increase in the number of people between the ages of fifteen and twenty-four. It occurs in the middle stages of a demographic transition and is the result of a rapid reduction in child mortality before a rapid falloff in fertility.

Zelinsky model Describes three phases of mobility: premodern traditional society with limited mobility, advanced society of increased mobility and circular movement, and the super advanced society in which there is increased acceleration of movement, but a potential peak.

Zionism A movement formally established in 1896/97 as a political organization promoting the creation of a Jewish state in the Middle East. Now the term refers to the support of Israel as a Jewish state.

Zipf's law The regular relationship between a city's population and its rank in the urban hierarchy.

Credits

All photographs © John Rennie Short unless otherwise specified.

Chapter 1

Photo 1.3: Sheila Terry/Science Source. Photo 1.4: William Channing Woodbridge (Cartographer), Alexander von Humboldt (Author). Restoration by Jujutacular and Durova/ Wikipedia. Photo 1.5: Library of Congress Prints and Photographs Division Washington, D.C. 20540 USA. Photo 1.6: Courtesy of the Newberry Library, Chicago, Illinois. Photo 1.7: Astronaut Photograph/NASA/JSC. Map 1.1: UN HDI data– 2021. Map 1.2: http://www.worldmapper.org/. Map 2.1: M.Bitton/Wikipedia.

Chapter 2

Figure 2.2: Gerardo Ceballos et al., Accelerated modern human– induced species losses: Entering the sixth mass extinction. Sci. Adv.1,e1400253(2015). Photo 2.3: NASA Johnson Space Center/Wikipedia.

Chapter 3

Chapter 3 Opener: Photo by John Rennie Short. Photo 3.1: Ori/ Wikipedia. Figure 3.2: Andres Etter, Clive McAlpine & Hugh Possingham (2008) Historical Patterns and Drivers of Landscape Change in Colombia Since 1500: A Regionalized Spatial Approach, Annals of the Association of American Geographers, 98:1, 2–23. Map 3.1: Richard Aspinall (2010) Geographical Perspectives on Climate Change, Annals of the Association of American Geographers, 100:4, 715–718. Photo 3.6: https://www.un.org/sustainabledevelopment/. The content of this publication has not been approved by the United Nations and does not reflect the views of the United Nations or its officials or Member States.

Chapter 4

Figure 4.4: Inga Spence/Science Source.

Chapter 6

Figure 6.1: www.radicalcartography.net. Photo 6.1: Division of Political and Military History, National Museum of American History, Smithsonian Institution. Photo 6.2: © 2023 Artists Rights Society (ARS), New York/ADAGP, Paris. Photo 6.3: AP Photo/Czarek Sokolowski.

Chapter 8

Figure 8.4: www.geographyfieldwork.com.

Chapter 9

Photo 9.2: NASA/GSFC/METI/ERSDAC/JAROS. Photo 9.5: Benny Marty/Shutterstock. Photo 9.6: AP Photo/Paul Sakuma.

Chapter 10

Photo 10.2: Flik47/Shutterstock.

Chapter 11

Photo 11.8: Mr. Tickle/Wikipedia. Photo 11.9: Ovedc/Wikipedia. Map 11.3: Jon T. Kilpinen.

Chapter 13

Photo 13.1: SeanXu/iStockphoto.

Chapter 14

Photo 14.3: Handout/Alamy Stock Photo.

Chapter 15

Photo 15.1: Thomas koch/Shutterstock.

Chapter 17

Map 17.1: Courtesy of the GaWC Research Network, Loughborough University. Photo 17.5: IR Stone/Shutterstock. Photo 17.4: Gareth Jones/Wikipedia.

Chapter 18

Photo 18.7: Vadim Ozz/Shutterstock. Photo 18.9: Allison C Bailey/ Shutterstock.

Index

Page numbers followed by *t* indicate tables; *b* indicate boxes; and *italicized* page references indicate figures.